AF868393

Food and Agricultural Byproducts as Important Source of Valuable Nutraceuticals

Chukwuebuka Egbuna • Barbara Sawicka
Johra Khan
Editors

Food and Agricultural Byproducts as Important Source of Valuable Nutraceuticals

Springer

Editors
Chukwuebuka Egbuna
Chukwuemeka Odumegwu Ojukwu University
Awka, Nigeria

Barbara Sawicka
Department of Plant Production
Commodities Science
University of Life Sciences in Lublin
Akademicka, Lublin, Poland

Johra Khan
Department of Medical Laboratory Sciences
College of Applied Medical Sciences,
Majmaah University
Majmaah, Saudi Arabia

Health and Basic Sciences Research Center,
Majmaah University
Majmaah, Saudi Arabia

ISBN 978-3-030-98759-6 ISBN 978-3-030-98760-2 (eBook)
https://doi.org/10.1007/978-3-030-98760-2

© The Editor(s) (if applicable) and The Author(s), under exclusive license to Springer Nature Switzerland AG 2022
This work is subject to copyright. All rights are solely and exclusively licensed by the Publisher, whether the whole or part of the material is concerned, specifically the rights of translation, reprinting, reuse of illustrations, recitation, broadcasting, reproduction on microfilms or in any other physical way, and transmission or information storage and retrieval, electronic adaptation, computer software, or by similar or dissimilar methodology now known or hereafter developed.
The use of general descriptive names, registered names, trademarks, service marks, etc. in this publication does not imply, even in the absence of a specific statement, that such names are exempt from the relevant protective laws and regulations and therefore free for general use.
The publisher, the authors and the editors are safe to assume that the advice and information in this book are believed to be true and accurate at the date of publication. Neither the publisher nor the authors or the editors give a warranty, expressed or implied, with respect to the material contained herein or for any errors or omissions that may have been made. The publisher remains neutral with regard to jurisdictional claims in published maps and institutional affiliations.

This Springer imprint is published by the registered company Springer Nature Switzerland AG
The registered company address is: Gewerbestrasse 11, 6330 Cham, Switzerland

Book Description

Food and agricultural by-products are leftovers or wastes from parts of foods, fruits, vegetables, and animal sources which are obtained after processing. Agricultural by-products include peels and rinds from citrus fruits, pineapple, mango, and banana. Other notable ones are pomace from apple, olive, red beet, and those from winemaking. Also, whey from milk, straws, hulls, and brans from grains are among the top agricultural by-products. These by-products often impact the environment and the social-economic sectors when they are disposed of. But with the recent advances in biotechnology and scientific research, scientists have found usefulness in some of these by-products as sources of valuable nutraceuticals, a term used to refer to chemical entities present in foods that have the propensity to impact health for disease prevention and treatment. This book entitled "*Food and Agricultural By-products as Important Source of Valuable Nutraceuticals*" presents detailed information about major agricultural by-products that are rich in nutraceuticals. The nature and the type of nutraceuticals that they contain and their health-promoting benefits were presented. The editors and chapter contributors are renowned experts from key institutions around the globe. This book will be useful to students, teachers, food chemists, nutritionists, nutritional biochemists, food biotechnologists, among others.

Key Features

- Highlights the health promotion benefits of nutraceuticals
- Presents information on agrifood by-products as sources of nutraceuticals
- Discusses functional nutraceuticals from peels, rinds, pomace, hull, bran, etc.

Contents

About the Authors

Chukwuebuka Egbuna is a chartered chemist, chemical analyst, and academic researcher. He is a member of the Institute of Chartered Chemists of Nigeria (ICCON), the Nigerian Society of Biochemistry and Molecular Biology (NSBMB), the Society of Quality Assurance (SQA) (USA), and the Royal Society of Chemistry (RSC) (UK). Egbuna is also an editor of new Elsevier series on Drug Discovery Update. The series includes books, monographs, and edited collections from all areas of drug discovery, including emerging therapeutic claims for the treatment of diseases. He has also edited books with Springer Nature on functional foods and nutraceuticals and John Wiley on poisonous plants and phytochemicals. He has been engaged in a number of roles at New Divine Favor Pharmaceutical Industry Limited, Akuzor Nkpor, Anambra State, Nigeria, and Chukwuemeka Odumegwu Ojukwu University (COOU) in Nigeria. He has collaboratively worked and published quite a number of research articles in the area of phytochemistry, nutrition, and toxicology and related areas. He is a reviewer and editorial board member of various journals, including serving as a website administrator for the Tropical Journal of Applied Natural Sciences (TJANS), a journal of the faculty of Natural Sciences, COOU and as an advisory board member under toxicology for the book publisher, Cambridge Scholar Publishing. His primary research interests are in phytochemistry, nutrition and toxicology, food and medicinal chemistry, and analytical biochemistry. Africa Centre of Excellence in Public Health and Toxicological Research (ACE-PUTOR), University of Port-Harcourt, Port Harcourt, Rivers State, Nigeria

Barbara Sawicka obtained her PhD from the Agricultural University in Lublin, Poland. Since 1999 she has been a Professor at the University of Life Science in Lublin and was promoted to Head of the Department of Plant Production Technology and Commodities Sciences in 2012. She is an honorary member of the advisory board of Journal Archives of Agricultural and Environmental Science, a member of the advisory board of International Journal of Research and Methodology in Social Science, and also the board of STUDIA NATURAE, for the Annals Universitatis Pedagoogicae Cracoviensis. She has over 280 original research papers, 8 books, and

1 patent. Her research areas of interest are Agronomy, Agribusiness and Rural Development, Bioengineering, Environmental Protection, Food Sciences, Food Safety, and Commodity Products. Department of Plant Production and Commodities Science, University of Life Sciences in Lublin, Lublin, PolandAkademicka 15 street20-950 Lublin, professor

Johra Khan PhD, is an assistant professor and principal research investigator at the College of Applied Medical Laboratory Sciences, Majmaah University, Saudi Arabia. She is a biotechnologist and academic researcher with interest in the chemistry of natural products, functional foods and nutraceuticals, analytical chemistry, and studies related to both communicable and noncommunicable diseases. She is a recipient of many awards and a member of many scientific bodies such as the Middle East Molecular Biology Society (MEMBS), the International Society of Global Health (ISoGH), Science Advisory Board (SAB), and the Royal Society of Biology (UK). She has written many articles, books, and book chapters with different publishers such as Springer, CRC, IEEE, and Elsevier. She has supervised many undergraduate and postgraduate students in the last 10 years. She has collaborated on key research projects with scientists from different institutions including those outside of the Kingdom of Saudi Arabia. She is a reviewer and editorial board member of both local and international journals of repute. Department of Medical Laboratory Sciences, College of Applied Medical Sciences, Majmaah University, 11952, Majmaah, Saudi Arabia

Chapter 1
The Role of Nutraceuticals as Food and Medicine, Types and Sources

Abhay Prakash Mishra, Neeti Srivastav, Anita Singh, Manisha Nigam, Raffaele Pezzani, Chukwuebuka Egbuna, Chukwuemelie Zedech Uche, and Johra Khan

1.1 Introduction

The term "nutraceuticals" is not new, as it has been adopted many years back. An attempt was made in United States for creating a functional component through fortification by adding iodine to salt in order to prevent goitre. If we talk in the terms

A. P. Mishra
Adarsh Vijendra Institute of Pharmaceutical Sciences, Shobhit University, Saharanpur, UP, India

Department of Pharmacology, Faculty of Health Science, University of Free State, Bloemfontein, South Africa

N. Srivastav
ABESIT College of Pharmacy, Ghaziabad, UP, India

A. Singh
Sunder Deep Pharmacy College, Ghaziabad, UP, India

M. Nigam (✉)
Department of Biochemistry, Hemvati Nandan Bahuguna, Garhwal University, Srinagar, Uttarakhand, India

R. Pezzani
Phytotherapy Lab (PhT-Lab), Endocrinology Unit, Department of Medicine (DIMED), University of Padova, Padova, Italy
e-mail: raffaele.pezzani@unipd.it

C. Egbuna
Chukwuemeka Odumegwu Ojukwu University, Awka, Nigeria

C. Z. Uche
Department of Medical Biochemistry and Molecular Biology, Faculty of Basic Medical Sciences, University of Nigeria, Nsukka, Enugu State, Nigeria

© The Author(s), under exclusive license to Springer Nature Switzerland AG 2022

C. Egbuna et al. (eds.), *Food and Agricultural Byproducts as Important Source of Valuable Nutraceuticals*, https://doi.org/10.1007/978-3-030-98760-2_1

of marketing, "nutraceutical" has been developed as nutritional supplement and sold in the market with an intention to treat or prevent diseases. Therefore, we can say that a "nutraceutical" can be considered either a food or its part. They have a broader range, starting from isolated nutrients to dietary supplements and even genetically engineered "designer" foods, processed foods and herbal product. Nutraceuticals play a potential role in the improvement of health [1]. It can even delay aging processes, give protection from chronic diseases, and increase expectancy of life [1]. Nutraceuticals are the substances that are usually not patent protected but used to prevent diseases. Apart from nutrition aspect, these substances are used for health purposes. Currently, there are over 470 nutraceuticals/functional food products with reported health benefits [2].

Thus, as far as definition is concerned, we can say that a "nutraceutical" is entire food or its part that benefits our health [3]. Considering their advantages, they: (a) help in preventing diseases; (b) help consumers to fulfill the daily dose requirements of vitamins and minerals; (c) in comparison to conventional pharmaceuticals, are less toxic; (d) are cost effective and easily available, without necessity of medical prescription.

Of course, there are also limitations to the use of nutraceuticals. Food laws regulate the quality and manufacturing processes. The major limitations are: (a) potential presence of adulterations due to the lack of pharmaceutical quality control; (b) the purity and dosage of the nutraceuticals are sometimes ambiguous; (c) some products are expensive due to their production, elaboration, preparation.

According to European Medicines law, a nutraceutical can be included in the category of medicine for two reasons: (a) it can be used for the prevention, treatment and cure of a condition or disease; (b) it can be administered in order to restore, correct or modify the physiological functions in human beings [4].

1.2 Classification of Nutraceuticals

There are many ways to classify nutraceuticals. Nutraceuticals can be classified based on:

1. Availability of scientific data
2. The source of food
3. According to the mechanism of action
4. Based on chemical nature

J. Khan
Department of Medical Laboratory Sciences, College of Applied Medical Sciences, Majmaah University, Majmaah, Saudi Arabia

Health and Basic Sciences Research Center, Majmaah University, Majmaah, Saudi Arabia

Broadly, nutraceuticals can be classified as:

1. Potential nutraceuticals
2. Established nutraceuticals

A potential nutraceutical is one that gives good health or medical benefits. A potential nutraceuticals comes under the category of established one after having sufficient clinical data to demonstrate its benefits [5]. Most of the nutraceutical products come under the category of potential one due to lack of literature availability. The food products that fall under the category of nutraceutical are:

- Probiotics
- Prebiotics
- Dietary fiber
- Omega 3 fatty acids
- Antioxidants

Nutraceuticals can be studied under the three broad categories [6]:

- **Nutrients**: These are the substances that have established nutritional functions which include vitamins, minerals, amino acids and fatty acids.
- **Herbals**: These are the concentrates and extracts of herbs and other botanical products.
- **Dietary supplements**: These include reagents derived from other sources like steroid hormone precursors, pyruvate, chondroitin sulphate that serve specific functions such as nutrition for sports, weight loss supplements and replacements for meal.

1.2.1 Nutrients

This includes prebiotics, probiotics, dietary fiber, fatty acids that are polyunsaturated in nature, various natural antioxidants. Most of the nutraceutical products come under the category of potential ones due to lack of literature availability.

1.2.2 Herbals

Since ancient times, herbs have been used as natural remedies for curing many disorders. According to the traditional medicinal system, herbs have innate abilities to treat illness and serve mankind. Various research related to the biological activity and toxicity of phytochemicals present in herbs have been carried out

- Herbals/phytochemicals: include herbs and botanical products.
- *Aloe vera*: it is used as an anti-inflammatory, emollient and wound healing.
- Evening Primrose oil: it plays an important role in treating atopic eczema and also act as a supplement for linoleic acid.

- Garlic: It is used as antibacterial, antifungal, antithrombotic, anti-inflammatory like allicin.
- Ginger: It is used as a carminative, antiemetic and in the treatment of dizziness.
- Ginseng: It is an adaptogen.
- Green tea: It is important for boosting up both types of immunity—humoral as well as cell mediated. It is also a potent antioxidant.
- Others: Herbs that may include vegetables, fruits, peels, whole grain, nuts and various seeds may contain phenolic compounds, sulphur compounds, pigments other phytoconstituent like terpenoids etc.

1.2.3 Vitamins

These come under the category of dietary supplements that contain nutrients derived from food products. The ingredients present in these products include metabolites, vitamins, minerals, herbs, and amino acids.

1.2.4 Functional Foods

These are the enriched foods that are available in their original form to the consumer. The enrichment of food is called nutrification. They fulfill the need of various nutrients like vitamins, carbohydrate, fat, proteins to body. The features of functional food is that it should be:

1. In their naturally occurring form,
2. One of an essential part of our daily diet,
3. Regulate a biological process to prevent disease.

1.2.5 Dietary Fiber

Dietary fiber can be called roughage or bulk. Roughages are the plant parts that are difficult for the body to digest. Fats, proteins or carbohydrates can be broken down and absorbed, but fiber can't be digested by the body, rather passes as such through GIT and then finally gets eliminated. Dietary fiber is present in abundance in fruits, vegetables, legumes and whole grains. It is very essential to prevent or relieve constipation. Fiber can be differentiated into soluble and insoluble form [7].

1.2.6 Phytochemicals

Plant bioactive constituents are collectively referred to as phytochemicals. They are in much demand in pharmaceutical industry. Various forms of food like beans, grains, vegetables are rich source of phytochemical and thus serve as a potential nutraceutical.

The term phytochemical emphasizes the plant source of some of the health-protecting compounds. The terms "phytochemical" and "phytonutrient" can be interchanged for describing the active components of plants. They have antioxidant, immune-boosting and other health-promoting properties of active compounds. They may include terpenes, carotenoids, limonoids, and phytosterols.

1.3 Roles of Phytochemicals

A summary of the roles of phytochemicals are presented below:

1. They are a substrate for various biochemical reactions.
2. They act as a cofactor of enzymatic reactions.
3. They are the inhibitors of enzymatic reactions.
4. They are used as an absorbent thereby bind and eliminate the undesirable constituents in the intestine.
5. Helps to increase the absorption and stability of essential nutrients.
6. They act as a selective growth factor for beneficial bacteria.
7. They destroy harmful intestinal bacteria.
8. Act as scavengers.

1.4 Role of Nutraceuticals in Prevention of Diseases

This section describes the benefits of nutraceuticals in the prevention and treatment of diseases.

Nutraceuticals and Cardiovascular diseases Components like antioxidants, dietary fibers, Omega-3 poly-unsaturated fatty acids, minerals, vitamins, are used for the prevention and treatment of cardiovascular disease. In grapes, polyphenol is present that prevents and controls arterial diseases.

In onion, grapes, red wine, apples, and cherries, flavonoids are present that block the Angiotensin enzymes and give strength to those small capillaries that are responsible for the transport of oxygen and nutrients to all cells.

Rice bran lowers the level of serum cholesterol in the blood, thus helps in the reduction of the level of bad cholesterol (LDL) and increases the level of good cholesterol (HDL). Rice bran contains lutein and zeaxanthin that helps in improving vision and reduces the possibility of cataracts. Other components of rice bran important for eye health are omega-3, omega-6, omega-9 and folic acid.

There is a higher mortality risk associated with low consumption of fruits and vegetables among patients suffering from CVD and other disorders [8].

Heart attack and lung cancer Corn contains high quantity of fiber and folates. Corn maintains the homocysteine, whose high level is directly responsible for the damage of blood vessels, heart attack, stroke, or peripheral vascular disease. It has been observed that if folate be consumed 100% of daily value, then it may reduce chances of heart attacks by 10%. Cryptoxanthin which is a natural pigment, is one of the components obtained from corn reduces the occurrence of lung cancer.

Obesity Obesity is a problem across the globe and is defined as accumulation of unhealthy and extra amount of body fat. It gives rise to various risk factors like CHF, angina pectoris, rise in blood pressure, disorders of respiratory system, thrombosis of renal vein, osteoarthritis, formation of cancerous cells, hyperlipidemia, and reduced fertility.

Diabetes Dietary fibers from psyllium are used to control sugar in diabetic patients and also reduces lipid levels in hyperlipidemia cases.

Cancer Flavonoids prevent estrogen-induced cancers by blocking the enzymes producing estrogen. They may even prevent prostate/breast cancer. Soy-foods, curcumin from curry and soya isoflavones possess cancer preventive properties. Lycopene gets accumulated in the skin, testes, adrenal and prostate where it protects against cancer.

Anti-inflammatory Curcumin, which is a polyphenol of turmeric possesses anti-carcinogenic, antioxidative and anti-inflammatory properties. Anti-tumor activity is found in beet root, cucumber, spinach leaves, and turmeric rhizomes. For the treatment of auto immune diseases and for reducing inflammation, Gamma linolenic acid is used which is present in green leafy vegetables and also in the oils obtained from vegetables (evening primrose oil, blackcurrant and hemp seed oil). Glucosamine and chondroitin sulfate are used to prevent osteoarthritis.

Parkinson's disease Vitamin E present in the food is important in order to prevent Parkinson's disease. Even creatine may modify Parkinson's disease. Significant results were obtained in preliminary studies done for nutritional supplements. However, due to insufficient scientific data, it is still not recommended for Parkinson's disease.

Osteoarthritis Osteoarthritis (OA) is a major form of joint disorder. Individuals having pain in joints due to OA and other joint disorders may experience reduced physical activity resulting in weight gain. The weight gain put additional stress on joints. Osteoarthritis is prevented with the use of chondroitin sulfate (CS) and glucosamine (GLN) [9].

Table 1.1 Roles of nutraceuticals

Physiological property	Proposed effect	Health benefits
Soluble dietary fiber	Gastric emptying is delayed and prolongs the intestinal phase	Safety contribution
	Nutrients uptake in small intestine is delayed	Lower down the blood cholesterol level [11]
	Prevent the reabsorption of bile acid	Protects against breast cancer [12]
	Prevent the digestive enzymes from reaching lipid substrates, inhibits enzyme activity	Lowers glucose, insulin (type II diabetics) [13] and lipid level after meal
Interaction or binding	Binds with bile acids	Lower blood cholesterol level
	Interacts with digestive enzymes	Lowers glucose, insulin and lipid level after meal
Fermentation	Growth of health promoting bacteria	Act as anti-inflammatory and protects against colorectal cancer [14]
	Short chain fatty acids are produced	Lower down the blood cholesterol level [15] and protect against cancer
Insoluble dietary fiber	Increase stool weight [16]	Reduce the incidence of colorectal cancer and intestinal diseases
	Transit time is speed up	Reduces absorption time for nutrients and lowers the level of glucose, insulin and lipid

Adrenal Dysfunction Adaptogens are substances in plants that are nontoxic which help the body to resist various stressors like physical, chemical or biological. Their use in Chinese and Ayurvedic traditions were explained long years back. Adaptogens include herbs *Eleutherococcus senticosus*, *Ocimum sanctum*, *Ginkgo biloba*, *Panax ginseng* and, *Withania somnifera* and the mushroom *Cordyceps sinensis*. Following is a short description of each.

Chinese have been using *Ginkgo biloba* for last so many years in various indications like vertigo, short-term memory loss, attention or vigilance loss. They have antioxidant and neuroprotective properties, including slowing the progression of dementia.

Ocimum sanctum known as holy basil or tulsi is used in Ayurvedic medicine and has been shown to relieve from stress. The prominent roles of nutraceuticals are explained in a nutshell in the tables given below [10] (Tables 1.1, 1.2, 1.3, and 1.4).

1.5 Plant Food Residues/Peels as Source of Nutraceuticals

Plant-based food residues such as leaves, peels, stems, seed flours, waste-waters, etc., which are generally abandoned, discarded like garbage in the environment and often fed to livestock or used as combustion feedstock for biofuels, constitute the

Table 1.2 Role of nutrients in providing health benefits

Nutrients	Health benefits
Fat soluble vitamins	
Vitamin A	Effective antioxidant, required for the growth and development, keeps eyes, mucous membrane and the skin healthy, also helps in the prevention and treatment of certain cancers and skin disorders [17, 18]
Vitamin D	Required for the formation of bones and teeth and helps in the calcium absorption [18]
Vitamin E	Acts as an antioxidant and helps in the formation of blood cells, muscles, lung and nerve tissue [19], it is immune system enhancer
Vitamin K	It is essential for blood clotting and helps in wound healing [20]
Water soluble vitamins	
Vitamin C	Prevents common cold [18, 21]
Vitamin B1	Essential in neurologic functions [22]
Vitamin B2	Helps for producing energy to perform other chemical processes in the body, helps to maintain healthy eyes, skin and nerve function
Vitamin B3	Helps in the conversion of food into energy and thus maintain proper brain function [23]
Vitamin B6	Helps in the production of essential proteins [24] and its conversion into energy
Vitamin B12	Helps in the production of genetic material of cells, helps to form red blood cells [18, 25] it synthesizes amino acids and is aids in metabolism of fats [26], protein and carbohydrates [27]
Folic acid	Genetic materials of cells are formed, required during the first trimester of pregnancy to prevent birth defects [28], helps in RBC formation [18], protects from cardiac diseases [18]
Pantothenic acid	Required for the steroid, cholesterol and fatty acids synthesis [29] also important for intra-neuronal synthesis of acetylcholine
Minerals	
Calcium	Plays a vital role for building bones and teeth thus maintaining its strength, also important for nerve, muscle and glandular functions [30]
Iron	Important for the production of energy and helps in the transportation of oxygen to tissues [31]
Magnesium	Essential for healthy nerve and muscle function and bone formation, may help prevent premenstrual syndrome (PMS) [32]
Phosphorous	Important constituent for building healthy bones and teeth, helps information of genetic material [33], produces energy and even stores it
Trace elements	
Chromium	Helps in the conversion of carbohydrates and fats into energy [34]
Cobalt	One of an essential component of vitamin B12, cobalt when ingested is metabolized *in vivo* for the formation of B12 coenzymes
Copper	Important constituent for the production of hemoglobin and collagen [35], helps in proper functioning of the heart, produces energy, helps in the absorption of iron from digestive tract [36]
Iodine	Required for the proper functioning of thyroid [37]
Selenium	Essential for proper functioning of the heart muscle [38]

(continued)

Table 1.2 (continued)

Nutrients	Health benefits
Zinc	Plays a significant role in cell reproduction, provides normal growth and development in children, wound healing, aids in the production of sperm and testosterone [39]
Other important bioactives	
Biotin	Needed for various metabolic functions
L-carnitine	Required for the oxidation of fatty acids [40], promotes certain organic acid excretion and even the rate of oxidative phosphorylation is increased
Choline	It is a lipotropic agent which is used to treat fatty liver [41] and in the conditions of disturbed fat metabolism
Vitamin F	Required for the proper development of various membranes, responsible for the synthesis of prostaglandins [42], leukotrienes and different hydroxyl fatty acids
Inositol	Being a lipotropic agent, it is necessary for the transport of amino acid and movement of potassium and sodium [43]
Taurine	Helps in bile acid conjugation, photoreceptor activity of retina, white blood cell formation, act as CNS neuromodulator, for platelet aggregation, cardiac contractility, sperm motility [44] also act as an antioxidant [45]

Table 1.3 Phytochemicals and their uses

Chemical constituents	Uses
Carotenoids	
Lycopene	They reduce cholesterol levels, antioxidants, protects against cancer [46]
β-carotene	Antioxidants, protection of cornea against UV light [47]
Lutein	Protect eyes against age related muscular degenerations [48], cataracts, anticancer activity (colon)
Tocotrienol	Improves cardiovascular health, fight against breast cancer [49, 50]
Saponins	Very effective against colon cancer [51], reduces cholesterol level
Polyphenolic compounds	
Flavanones	Demonstrates both anti-oxidant and anticancer activity
Flavones	Possess antioxidant activity and effective against cancer
Flavanols	Antioxidant activity
Curcumin	Strongly anti-inflammatory and strongly antioxidant, effective anti-clotting agent [52]
Glucosinolates	Anticancer activity, protect against bladder cancer [53]
Phytoestrogens	
Isoflavones	It lowers LDL cholesterol, antioxidants, protects against prostate, breast, bowel and other cancers [54]
Lignans	Protect against development of cancer like colon and breast cancer [55]
Fatty acids	
Omega 3 fatty acids	They are the potent controllers of the inflammatory processes, help in maintenance of brain function [56] and reduce cholesterol disposition
Fatty acids	Lower down the chances of coronary heart disease [57]
Prebiotics/probiotics	They help to improve gastrointestinal health [58]
Minerals	They are the important constituents of balanced diet
Polyols sugar alcohols (xylitol, sorbitol)	They may reduce the risk of dental cavities [59]

Table 1.4 Commercially available nutraceuticals [60]

S/no	Product	Category	Benefit	Manufacturer
1	Glucon D	Glucose	Energy provider	Heinz
2	Proteinex	Protein and nutritional supplement	Both physical and mental growth	Pfizer
3	GRD	Nutritional supplement	Tissue growth and repair	Zydus
4	Tropicana	Energy drink	Full of nutrients	Tropicana products
5	B-protein	Nutritional supplement	Formation of haemoglobin	British biologicals
6	Cod liver oil	Omega 3 fatty acids	Immunity booster	Sanofi
7	Revital	Health supplement	Improve health and immunity	Ranbaxy
8	Pediasure	Nutritional supplement	Helps in child growth	Abott
9	Threptin diskette	Protein supplement	Ensures protein nutrition in body	Raptakos Brett & Co

bulk of agro-wastes. Fruits, vegetables and cereal processing accounts for over 80% of these waste residues, as they are commonly produced in considerably large quantities during industrial food processing and domestic consumption [60, 61]. Both fruits and vegetables have been long recognized as rich sources of beneficial nutraceuticals, which are replete in all parts including the stems, barks, peels, stalks, flowers, fruits, leaves, roots, latex, hulls, seeds, fruit rinds, and pods [62–64]. Notably, some fruits and subterranean vegetables have been reported to possess significant concentrations of biologically active phytochemicals such as polyphenols, flavonoids, tannins, saponins, phlobatannins, anthraquinones, phytosterol, alkaloids, terpenoids, chalcones, etc. [65–67]. Therefore, it is important to note that these supposed leftovers are indeed rich sources of phytochemicals and may serve as better sources of dietary fibres and antioxidants than their respective edible portions. Consequently, their recovery and proper utilization could be employed in the production of novel nutraceuticals and other functional foods (Table 1.5).

In addition, given the vast amount generated every year, utilizing these bioactive rich residues can provide efficient, alternative, low-cost, and environmentally-friendly nutritional supplements, particularly in resource-limited settings. As a result, screening, and extraction of these plant-based residues may result in the inadvertent isolation/discovery of nutritional and therapeutically beneficial compounds.

1.6 Regulatory Aspect

Nutraceutical market is governed by the Dietary Supplement Health and Education Act (DSHEA) passed in 1994. Food Safety and Standard Authority regulates the registration procedure of food industry as well as its licensing, packaging and storage.

Table 1.5 Selected plant-based waste residues and their nutraceutical constituents

S/n	Plant source	Classification	Waste-residue	Nutraceutical content	References
1	Orange, mandarin, lime, lemon etc. (*Citrus R.*)	Fruits	Peel (flavedo and albedo)	Dietary fibres, essential oils, carotenoids, flavanones and polyphenols: Hesperidin, naringin, lutein, zeaxanthin, eriocitrin, narirutin	[67–69]
2	Banana (*Musa acuminata* or *Musa balbisiana*)		Peel	Polyphenols: Gallocatechin, anthocyanins, delphindin, cyaniding, catecholamine. Flavonols: Rutin, Quercetin-7-rutinoside, Quercetin-3-rutinoside, Kaempferol-3-rutinosid, Kaempferol-7-rutinoside, Isorhamnetin-3-rutinoside, Myricetin-3- rutinoside	[70, 71]
3	Apple (*Malus pumila R.*)		Peel and pomace	Flavonols, Epicatechin, catechins, hydroxycinnamates, apigenin, phloretin glycosides, quercitin glycosides, procyanidins, chlorogenic acid, anthocyanins	[72–75]
4	Mango (*Mangifera indica*)		Kernel	Gallic acid, ellagic acid, gallates, gallotannins, condensed tannins	[76, 77]
5	Grape (*Vitis vinifera L.*)		Seed and skin	Cinnamic acid, coumaric acid, caffeic acid, ferulic acid, chlorogenic acid, neochlorogenic acid, *p*-hydroxybenzoic acid, protocatechuic acid, vanillic acid, gallic acid, proanthocyanidins, quercetin, resveratrol, pullulan, apigenin	[78–81]
6	Papaya (*Carica papaya L.*)		Peel and seed flour	Fibre protein, Chlorogenic acid, Caffeic acid Ferulic acid, *p*-Hydroxybenzoic acid, *p*-Coumaric acid, Myricetin, Quercetin, Kaempferol	[61, 82]
7	Litch *(Litchi chinensis* Sonn)		Pericarp, seeds and flower	Cyanidin-3-glucosides, cyanidin-3-rutonoside, malvidin-3-glucoside, gallic acid, epicatechin-3-gallate	[83–85]
8	Avocado (*Persea americana L.*)		Peel and seeds	Epicatechin, catechin, gallic acid and chlorogenic acid	[86]

(continued)

Table 1.5 (continued)

S/n	Plant source	Classification	Waste-residue	Nutraceutical content	References
9	Palm fruit (*Elaeis guineensis A.*)		Fruit bunches, kernel shells, trunks, fronds and pressed fibre	Tocopherols, Tocotrienols, sterols, squalence, phenolic antioxidants	[87–89]
10	Pomegranate (*Punica granatum L.*)		Peel, seeds, flowers, bark, buds and leaves	Ellagic acid, punicalagin, punicalin and gallagic acid, ellagitannins	[90–92]
11	Pineapple (*Ananas comosus*)		Peel and pomace	Dietary fibres, flavanones, bromelain	[77, 93]
12	Guava (*Psidium guajava L.*)		Skin and seeds	Catechi, Gallic acid nomogentisic acid cyanidin-3-rutomoside, kaempferol galangin, carotenoids	[77, 94, 95]
13	Tomato (*Solanum lycopersicum*)	Vegetable	Skin and pomace	Resveratrol, carotenoids, Cis-lycopene, Beta carotene, Trans-lycopene, Lutein, Ascorbic acid, Quercetin, Kaempferal	[96, 97]
14	Cucumber (*Cucumis sativus L.*)		Peel	Pheophytin, caryophyllene, phellandrene	[97]
15	Garlic (*Allium sativum L.*)		Husk	Ferulic acid, gallic acid, hydroxybenzoic acid, caffeic acid, *p*-coumaric acid, Di-Ferulic acid, chlorogenic acid, Caffeic acid-*O*-glucoside, Coumaroylquinic acid	[97, 98]
16	Onion (*Allium cepa*)		Skin	Quercetin 3,40-*O*-diglucoside and quercetin 40-*O*-monoglucoside	[97]
17	Carrot (*Daucus carota*)		Peel	Phenols, beta-carotene, chlorogenic acid, dicaffeoylquinic acids	[99–101]
18	Potato (*Solanum tuberosum L.*)		Peel	Gallic-acid, caffeic acid, vanillic acid Chlorogenic acid, protocatechic acids, *p*-Hydroxybenzoic	[97]
19	Barley (*Hordeum vulgare*)	Cereal crop	Bran	β-Glucan	[102]

(continued)

Table 1.5 (continued)

S/n	Plant source	Classification	Waste-residue	Nutraceutical content	References
20	Rice (*Oryza sativa*)		Bran	γ-Oryzanol, ferulic acid	[103, 104]
21	Wheat (*Triticum aestivum*)		Bran	Phenolic acid, antioxidant, ferulic acid	[105]

1.7 Conclusions

This chapter provides an insight on the concept of nutraceuticals for the better understanding of their potential benefits in improving health. Briefly, we can say that nutraceuticals serve as a support for the body by enhancing the body immunity to diseases. Nutraceuticals are available in agro byproducts such as peels of fruits and vegetables.

References

Egbuna C, Tupas GD (eds) (2020) Functional foods and nutraceuticals—bioactive components, formulations and innovations, 1st edn. Springer Nature, Cham, 646 pp

Brower V (1999) Nutraceuticals: poised for a healthy slice of the healthcare market. Nat Biotechnol 16:728–731. https://doi.org/10.1038/nbt0898-728

Dudeja P, Gupta RK (2017) Nutraceuticals: food safety in the 21st century. Academic, Cambridge

Om P, Gulati A, Ottaway PB (2006) Legislation relating to nutraceuticals in the European Union with a particular focus on botanical-sourced products. Toxicology 22:75–87. https://doi.org/10.1016/j.tox.2006.01.014. Epub 2006 Feb 17

Kumar G, Kalam N, Ansari SH (2013) Current concepts and prospects of herbal nutraceutical: a review. J Adv Pharm Technol Res 4(1):4–8. https://doi.org/10.4103/2231-4040.107494

Kalia AN (2005) Textbook of industrial pharmacognosy. CBS Publishers, New Delhi

Jain N, Ramawat KG (2013) Nutraceuticals and antioxidants in prevention of disease. Research Gate 2573

Egbuna C, Mishra AP, Goyal MR (eds) (2021) Preparation of phytopharmaceuticals for the management of disorders. Academic, New York. https://doi.org/10.1016/C2019-0-01420-3

Ganapathy M, Bhunia S (2016) Nutraceuticals: the new generation therapeutics. Adv Tech Biol Med 4:2. https://doi.org/10.4172/2379-1764.1000179

Verma G, Mishra M (2016) A review on nutraceuticals: classification and its role in various diseases. Int J Pharm Ther 7(4):152–160

Gunness P, Gidley MJ (2010) Mechanisms underlying the cholesterol-lowering properties of soluble dietary fibre polysaccharides. Food Funct 1(2):149–155. https://doi.org/10.1039/c0fo00080a

Yikyung P, Louise B, Amy FS, Albert H, Arthur S (2009) Dietary fiber intake and risk of breast cancer in postmenopausal women: the National Institutes of Health–AARP Diet and Health Study. Am J Clin Nutr 90(3):664–671. https://doi.org/10.3945/ajcn.2009.27758

Chunye C, Yuan Z, Jing X, Hongting Z, Jun L, Rong F, Wenyi Z, Lijia Y, Yu Q, Shihui C, Yong Z, Ying W, Jing W, Mantian M, Jian W (2016) Therapeutic effects of soluble dietary fiber

consumption on type 2 diabetes mellitus. Exp Ther Med 12(2):1232–1242. https://doi.org/10.3892/etm.2016.3377
Scheppach W, Bartram HP, Richter F (1995) Role of short-chain fatty acids in the prevention of colorectal cancer. Eur J Cancer 31A(7–8):1077–1080. https://doi.org/10.1016/0959-8049(95)00165-f
Hiroshi H, Satoko H, Yoritaka A, Shuhachi K (1999) Short-chain fatty acids suppress cholesterol synthesis in rat liver and intestine. J Nutr 129(5):942–948. https://doi.org/10.1093/jn/129.5.942
Dingra D, Michael M, Rajput H, Patil RT (2012) Dietary fibre in foods: a review. J Food Sci Technol 49(3):255–266. https://doi.org/10.1007/s13197-011-0365-5
Elena D, Gaetana C, Sara A, Chiara T, Amedeo F, Gaetano A, Daniela P, Maria GS, Augusto O (2015) Vitamin A, cancer treatment and prevention: the new role of cellular retinol binding proteins. BioMed Res Int 2015:624627. https://doi.org/10.1155/2015/624627
Akram M, Munir N, Daniyal M, Egbuna C, Găman M, Onyekere PF, Olatunde A (2020) Vitamins and minerals: types, sources and their functions. In: Egbuna C, Dable Tupas G (eds) Functional foods and nutraceuticals. Springer, Cham. https://doi.org/10.1007/978-3-030-42319-3_9
Saliha R, Syed TR, Ahmed F, Ahmad A, Abbas S, Mahdi F (2014) The role of vitamin E in human health and some diseases. Sultan Qaboos Univ Med J 14(2):157–165
Nader P, Gholamreza H, Reza Y, AliAsghar H (2019) Wound healing effects of topical vitamin K: a randomized controlled trial. Indian J Pharmacol 51(2):88. https://doi.org/10.4103/ijp.IJP_183_18
Amanda B, Nicole W (2016) Vitamin C in the prevention and treatment of the common cold. Am J Lifestyle Med 10(3):181–183. https://doi.org/10.1177/1559827616629092
Carlos AC, Ospina M, Orlando N (2020) Vitamins in the nervous system: current knowledge of the biochemical modes of action and synergies of thiamine, pyridoxine, and cobalamin. CNS Neurosci Ther 26(1):5–13. https://doi.org/10.1111/cns.13207
Valeria G, Matteo S, Isabella S, Maria VC (2019) Niacin in the central nervous system: an update of biological aspects and clinical applications. Int J Mol Sci 20(4):974. https://doi.org/10.3390/ijms20040974
Marcelina P, Seth S, Hanjo H (2018) Vitamin B6 and its role in cell metabolism and physiology. Cell 7(7):84. https://doi.org/10.3390/cells7070084
Mark JK, Prem P (2004) New insights into erythropoiesis: the roles of folate, vitamin B12, and iron. Annu Rev Nutr 24(1):105–131. https://doi.org/10.1146/annurev.nutr.24.012003.132306
Joseph B, Antonysunil A, Jinous S, Ponnusamy S (2020) Low vitamin B12 and lipid metabolism: evidence from pre-clinical and clinical studies. Nutrients 12(7):1925. https://doi.org/10.3390/nu12071925
Lubna M (2014) The metabolic processes of folic acid and vitamin B12 deficiency. J Health Res Rev 1(1):5–9. https://doi.org/10.4103/2394-2010.143318
Butterworth CE Jr (1996) A bendich, folic acid and the prevention of birth defects. Annu Rev Nutr 16:73–97. https://doi.org/10.1146/annurev.nu.16.070196.000445
Aleksandra C, Sylwia SJ, Areta H, Jacek T, Swierczynski J, Sledzinski T, Ewa S (2020) The pathophysiological role of CoA. Int J Mol Sci 21(23):9057. https://doi.org/10.3390/ijms21239057
Pravina P, Didwagh S, Mokashi A (2013) Calcium and its role in human body. Int J Res Pharmaceut Biomed Sci 4(2):659–667
Dunn J, Mythen MG (2016) Physiology of oxygen transport. B J A Educ 16(10):341–348. https://doi.org/10.1093/BJAED/MKW012
Nahid F, Elham E, Mahboube V, Naser T, Manizhe HY (2010) Evaluating the effect of magnesium and magnesium plus vitamin B6 supplement on the severity of premenstrual syndrome. Iran J Nurs Midwifery Res 15(1):401–405
John AR (2013) RNA function and phosphorus use by photosynthetic organisms. Front Plant Sci 4:536. https://doi.org/10.3389/fpls.2013.00536
Andrea T, Tatiana B (2010) Nutrient utilization in humans: metabolism pathways. Nat Educ 3(9):11

Harris ED, Rayton JK, Balthrop JE, DiSilvestro RA, Quevedo M (1980) Copper and the synthesis of elastin and collagen. Ciba Found Symp 79:163–182. https://doi.org/10.1002/9780470720622.ch9

Wai YC, Owen MR (1980) The role of copper in iron metabolism. Ann Clin Lab Sci 10(4): 338–344

Hye RC (2014) Iodine and thyroid function. Ann Pediatr Endocrinol Metab 19(1):8–12. https://doi.org/10.6065/apem.2014.19.1.8

Carina B, Andreas G, Sandra K, Sebastian B, William M, Gil H, Christian S (2015) Selenium and its supplementation in cardiovascular disease—what do we know? Nutrients 7(5):3094–3118. https://doi.org/10.3390/nu7053094

Deborah AF, Haim B (2020) The role of zinc in male fertility. Int J Mol Sci 21(20):7796. https://doi.org/10.3390/ijms21207796

Nicola L, Marta F (2016) Carnitine transport and fatty acid oxidation. Biochim Biophys Acta Mol Cell Res 1863(10):2422–2435. https://doi.org/10.1016/j.bbamcr.2016.01.023

Mehedint MG, Zeisel SH (2013) Choline's role in maintaining liver function: new evidence for epigenetic mechanisms. Curr Opin Clin Nutr Metab Care 16(3):339–345. https://doi.org/10.1097/MCO.0b013e3283600d46

Mohsen M, Simin N, Meydan JB, Macauley J, Blumberg J (1985) Influence of dietary vitamin F and selenium on the ex-vivo synthesis of prostaglandin E2 in brain regions of young and old rats. Prostaglandins Leukot Med 18(3):337–346. https://doi.org/10.1016/0262-1746(85)90066-6

Simmons DA (1989) Mechanism of glucose-induced (Na+, K+)-ATPase inhibition in aortic wall of rabbits. Diabetologia 32:402–408. https://doi.org/10.1007/BF00271258

Lorenzo B, Roy MK, Hay AM, Morrison EJ, Stefanoni D, Xiaoyun F, Tamir K, Steve K, Larry JD (2020) Mars stone, impact of taurine on red blood cell metabolism and implications for blood storage. Transfusion 60(6):1212–1226. https://doi.org/10.1111/trf.15810

Jang HY, Kong HS, Park CK (2006) Effects of taurine on sperm characteristics during *in vitro* storage of boar semen. Asian Australas J Anim Sci 19(11):1561–1565. https://doi.org/10.5713/ajas.2006.1561

Adetunji CO, Akram M, Mtewa AG, Jeevanandam J, Egbuna C, Ogodo AC et al (2021) Biochemical and pharmacotherapeutic potentials of lycopene in drug discovery. In: Egbuna C, Mishra AP, Goyal MR (eds) Preparation of phytopharmaceuticals for the management of disorders. Academic, New York, pp 307–360. ISBN 9780128202845. https://doi.org/10.1016/B978-0-12-820284-5.00015-0

Wilhelm S, Helmut S (2012) β-carotene and other carotenoids in protection from sunlight. Am J Clin Nutr 96(5):1179S–1184S. https://doi.org/10.3945/ajcn.112.034819

Liwen F, Kailai N, Hui J, Wei F (2019) Effects of lutein supplementation in age-related macular degeneration. PLoS One 14(12):e0227048. https://doi.org/10.1371/journal.pone.0227048

Nesaretnam K, Selvaduray KR, Razak A (2010) Effectiveness of tocotrienol-rich fraction combined with tamoxifen in the management of women with early breast cancer: a pilot clinical trial. Breast Cancer Res 12:81. https://doi.org/10.1186/bcr2726

Ramaswamy K, Gupta SC, Kim JH, Aggarwal BB (2012) Tocotrienols fight cancer by targeting multiple cell signaling pathways. Genes Nutr 7(1):43–52. https://doi.org/10.1007/s12263-011-0220-3

Cheng YT, Chen YH, Chien YW, Huang YH, Lins SH (2010) Effect of soy saponin on the growth of human colon cancer cells. World J Gastroenterol 16(27):3371–3376. https://doi.org/10.3748/wjg.v16.i27.3371

Kim DC, Ku SK, Bae JS (2012) Anticoagulant activities of curcumin and its derivative. BMB Rep 45(4):221–226. https://doi.org/10.5483/bmbrep.2012.45.4.221

Akram M, Jabeen F, Riaz M, Khan FS, Okushanova E, Imran M, Shariati MA, Riaz T, Egbuna C, Ezeofor NJ (2021) Health benefits of glucosinolate isolated from cruciferous and other vegetables. In: Egbuna C, Mishra AP, Goyal MR (eds) Preparation of phytopharmaceuticals for the management of disorders. Academic, New York, pp 361–371

Ahmad A, Bernhard B, Yiwei L, Bin B, Dejuan K, Ali S, Banerjee S, Sarkar FH (2013) Perspectives on the role of isoflavones in prostate cancer. Am Assoc Pharm Sci 15(4) 991–1000. https://doi.org/10.1208/s12248-013-9507-1

Calado A, Neves PM, Teresa S, Ravasco P (2018) The effect of flaxseed in breast cancer: a literature review. Front Nutr 5:4. https://doi.org/10.3389/fnut.2018.00004

Erika F, Milène V, Jennifer TM, Sébastien T, Blachère JC, Bégin ME, Brenna JT, Windust A, Cunnane SC (2006) Omega-3 fatty acids, energy substrates, and brain function during aging. Prostaglandins Leukot Essent Fatty Acids 75(3):213–220. https://doi.org/10.1016/j.plefa.2006.05.011

Siri-Tarino PW, Sun Q, Hu FB, Krauss RM (2010) Saturated fatty acids and risk of coronary heart disease: modulation by replacement nutrients. Curr Atheroscler Rep 12(6):384–390. https://doi.org/10.1007/s11883-010-0131-6

Verna EC (2010) Use of probiotics in gastrointestinal disorders: what to recommend? Ther Adv Gastroenterol 3(5):307–319. https://doi.org/10.1177/1756283X10373814

Loveren CV (2004) Sugar alcohols: what is the evidence for caries-preventive and caries-therapeutic effects? Caries Res 38:286–293. https://doi.org/10.1159/000077768

Gupta S, Parvez N, Sharma PK (2015) Nutraceuticals as functional foods. J Nutr Ther 4:64–72. https://doi.org/10.6000/1929-5634.2015.04.02

Santos CM, Abreu CMP, Freire JM, Queiroz ER, Mendonça MM (2014) Chemical characterization of the flour of peel and seed from two papaya cultivars. Food Sci Technol (Campinas) 34(2):353–357. https://doi.org/10.1590/fst.2014.0048

Aref HL, Salah KBH, Chaumont JP, Fekih AW, Aouni M, Said K (2010) *In vitro* antimicrobial activity of four *Ficus carica* latex fractions against resistant human pathogens. Pak J Pharm Sci 23:53–58

Rajaei A, Barzegar M, Mobarez AM, Sahari MA, Esfahani ZH (2010) Antioxidant, antimicrobial and antimutagenicity activities of pistachio (*Pistachia vera*) green hull extract. Food Chem Toxicol 48:107–112. https://doi.org/10.1016/j.fct.2009.09.023

Janjua S, Shahid M, Fakhir-i-Abbas (2013) Phytochemical analysis and *in vitro* antibacterial activity of root peel extract of *Raphanus sativus* L. var niger. Adv Med Plant Res 1:1–7

Egbuna C, Awuchi CG, Kushwaha G, Rudrapal M, Patrick-Iwuanyanwu KC, Singh O, Odoh UE, Khan J, Jeevanandam J, Kumarasamy S, Chukwube VO, Narayanan M, Palai S, Găman MA, Uche CZ, Ogaji DS, Ezeofor NJ, Mtewa AG, Patrick-Iwuanyanwu CC, Kesh SS, Shivamallu C, Saravanan K, Tijjani H, Akram M, Ifemeje JC, Olisah MC, Chikwendu CJ (2021) Bioactive compounds effective against type 2 diabetes mellitus: a systematic review. Curr Top Med Chem 21(12):1067–1095. https://doi.org/10.2174/1568026621666210509161059

Khan J, Deb PK, Priya S, Medina KD, Devi R, Walode SG, Rudrapal M (2021) Dietary flavonoids: cardioprotective potential with antioxidant effects and their pharmacokinetic, toxicological and therapeutic concerns. Molecules 26:4021. https://doi.org/10.3390/molecules26134021

Egbuna C, Hassan S (eds) (2021) Dietary phytochemicals: a source of novel bioactive compounds for the treatment of obesity, cancer and diabetes. Springer, Cham. https://doi.org/10.1007/978-3-030-72999-8

Genovese S, Fiorito S, Locatelli M, Carlucci G, Epifano F (2014) Analysis of biologically active oxyprenylated ferulic acid derivatives in citrus fruits. Plant Foods Hum Nutr 69:255–260. https://doi.org/10.1007/s11130-014-0427-8

Epifano F, Genovese S (2013) Recent acquisitions on naturally occurring oxyprenylated secondary plant metabolites. In: Brahmachari G (ed) Chemistry and pharmacology of naturally occurring bioactive compounds, 1st edn. CRC Press, Boca Raton, pp 239–257. https://doi.org/10.1201/b13867

Saravanan K, Mallikarjuna Aradhya S (2011) Potential nutraceutical food beverage with antioxidant properties from banana plant bio-waste (pseudostem and rhizome). Food Funct 2:603–610. https://doi.org/10.1039/c1fo10071h

Montelongo RG, Lobo MG, Gonalez M (2010) Antioxidant activity in banana peel extract: testing extraction conditions and related bioactive compounds. Food Chem 119:1030–1039. https://doi.org/10.1016/j.foodchem.2009.08.012

Kołodziejczyk K, Markowski J, Kosmala M, Krol B, Płocharski W (2007) Apple pomace as a potential source of nutraceutical products. Pol J Food Nutr Sci 57:291–295

Bhushan S, Kalia K, Sharma M, Singh Band Ahuja PS (2008) Processing of apple pomace for bioactive molecules. Crit Rev Biotechnol 28:285–296. https://doi.org/10.1080/07388550802368895

Fromm M, Bayha S, Carle R, Kammerer DR (2012) Characterization and quantitation of low and high molecular weight phenolic compounds in apple seeds. J Agric Food Chem 60:1232–1242. https://doi.org/10.1021/jf204623d

Walia MK, Rawat S, Bhushan Y, Padwada S, Singh B (2014) Fatty acid composition, physicochemical properties, antioxidant and cytotoxic activity of apple seed oil obtained from apple pomace. J Sci Food Agric 94:929–934. https://doi.org/10.1002/jsfa.6337

Arogba SS (2000) Mango (*Mangifera indica*) kernel: chromatographic analysis of the tannin, and stability study of the associated polyphenol oxidase activity. J Food Compos Anal 13:149–156. https://doi.org/10.1006/jfca.1999.0838

Martínez R, Torres P, Meneses MA, Figueroa JG, Pérez-Álvarez JA, Viuda-Martos M (2012) Chemical, technological and *in vitro* antioxidant properties of mango, guava, pineapple and passion fruit dietary fibre concentrate. Food Chem 135:1520–1526. https://doi.org/10.1016/j.foodchem.2012.05.057

Serna-Cock L, García-Gonzales E, Torres-León C (2016) Agro-industrial potential of the mango peel based on its nutritional and functional properties. Food Rev Int 32:364–376. https://doi.org/10.1080/87559129.2015.1094815

Maier T, Schieber T, Kammerer DR, Carle R (2009) Residues of grape (*Vitis vinifera* L.) seed oil production as a valuable source of phenolic antioxidants. Food Chem 112:551–559. https://doi.org/10.1016/j.foodchem.2008.06.005

Abdrabba S, Hussein S (2015) Chemical composition of pulp, seed and peel of red grape from Libya. Global J Sci Res 3(2):6–11. https://doi.org/10.12691/jfnr-7-2-7

Manso T, Gallardo B, Salvá A, Guerra-Rivas C, Mantecón AR, Lavín P, de la Fuente MA (2016) Influence of dietary grape pomace combined with linseed oil on fatty acid profile and milk composition. J Dairy Sci 99:1111–1120. https://doi.org/10.3168/jds.2015-9981

Castro-Vargas HI, Baumann W, Ferreira SR, Parada-Alfonso F (2019) Valorization of papaya (*Carica papaya* L.) agroindustrial waste through the recovery of phenolic antioxidants by supercritical fluid extraction. J Food Sci Technol 56(6):3055–3066. https://doi.org/10.1007/s13197-019-03795-6

Liu L, Xie B, Cao S, Yang E, Xu X, Guo S (2007) A-type procyanidins from *Litchi chinensis* pericarp with antioxidant activity. Food Chem 105:1446–1451. https://doi.org/10.1016/j.foodchem.2007.05.022

Nagendra PK, Yang B, Yang S, Chen Y, Zhao M, Ashraf M, Jiang Y (2009) Identification of phenolic compounds and appraisal of antioxidant and antityrosinase activities from litchi (*Litchi sinensis* Sonn.) seeds. Food Chem 116:1–7. https://doi.org/10.1016/j.foodchem.2009.01.079

Li S, Chen L, Yang T, Wu Q, Lv Z, Xie B, Sun Z (2013) Increasing antioxidant activity of procyanidin extracts from the pericarp of *Litchi chinensis* processing waste by two probiotic bacteria bioconversions. J Agric Food Chem 61:2506–2512. https://doi.org/10.1021/jf305213e

Deng GF, Xu XR, Guo YJ, Xia EQ, Li S, Wu S, Chen F, Ling WH, Li HB (2012) Determination of antioxidant property and their lipophilic and hydrophilic phenolic contents in cereal grains. J Funct Foods 4:906–914. https://doi.org/10.1016/j.jff.2012.06.008

Han NM, May CY (2010) Determination of antioxidants in oil palm leaves (*Elaeis guineensis*). Am J Appl Sci 7:1243–1247. https://doi.org/10.3844/ajassp.2010.1243.1247

Han NM, May CY (2012) Determination of antioxidants in oil palm empty fruit bunches. Am J Appl Sci 9:1862–1867. https://doi.org/10.3844/ajassp.2012.1862.1867

Ofori-Boateng C, Lee K (2013) Sustainable utilization of oil palm wastes for bioactive phytochemicals for the benefit of the oil palm and nutraceutical industries. Phytochem Rev 12(1):173–190. https://doi.org/10.1007/s11101-013-9270-z

Akhtar S, Ismail T, Fraternale D, Sestili P (2015) Pomegranate peel and peel extracts: chemistry and food features. Food Chem 174:417–425. https://doi.org/10.1016/j.foodchem.2014.11.035

Orgil O, Schwartz E, Baruch L, Matityahu I, Mahajna J, Amir R (2014) The antioxidative and antiproliferative potential of non-edible organs of the pomegranate fruit and tree. LWT Food Sci Technol 58:571–577. https://doi.org/10.1016/j.lwt.2014.03.030

Varzakas T, Zakynthinos G, Verpoort F (2016) Plant food residues as a source of nutraceuticals and functional foods. Foods (Basel, Switzerland) 5(4):88. https://doi.org/10.3390/foods5040088

Maimunah MA, Norhashila H, Samsuzana AA, Ola L (2020) Pineapple (*Ananas comosus*): a comprehensive review of nutritional values, volatile compounds, health benefits, and potential food products. Food Res Int 137:109675. https://doi.org/10.1016/j.foodres.2020.109675

Lima RS, Ferreira SRS, Vitali L, Block JM (2018) May the superfruit red guava and its processing waste be a potential ingredient in functional foods? Food Res Int 115:451–459. https://doi.org/10.1016/j.foodres.2018.10.053

Angulo-López JE, Flores-Gallegos AC, Torres-León C, Ramírez-Guzmán KN, Martínez GA, Aguilar CN (2021) Guava (*Psidium guajava* L.) fruit and valorization of industrialization by-products. Processes 9:1075. https://doi.org/10.3390/pr9061075

Strati IF, Oreopoulou V (2014) Recovery of carotenoids from tomato processing by-products—a review. Food Res Int 65:311–321. https://doi.org/10.1016/j.foodres.2014.09.032

Zeyada NN, Zeitoum MAM, Barbary OM (2008) Utilization of some vegetables and fruit waste as natural antioxidants. Alexandria J Food Sci Technol 5:1–11

Kallel F, Driss D, Chaari F, Belghith L, Bouaziz F, Ghorbel R, Chaabouni SE (2014) Garlic (*Allium sativum* L.) husk waste as a potential source of phenolic compounds: influence of extracting solvents on its antimicrobial and antioxidant properties. Ind Crops Prod 62:34–41. https://doi.org/10.1016/j.indcrop.2014.07.047

Chantaro P, Devahastin S, Chiewchan N (2008) Production of antioxidant high dietary fiber powder from carrot peels. LWT Food Sci Technol 41(10):1987–1994. https://doi.org/10.1016/j.lwt.2007.11.013

Wadhwa M, Bakshi MPS (2013) Utilization of fruit and vegetable wastes as livestock feed and as substrates for generation of other value-added products. Rap Publication 4: 1–67. E-ISBN 978-92-5-107632-3 (pdf)

Sagar NA, Pareek S, Sharma S, Yahia EM, Lobo MG (2018) Fruit and vegetable waste: bioactive compounds, their extraction, and possible utilization. Compr Rev Food Sci Food Saf 17(3):512–531. https://doi.org/10.1111/1541-4337.12330

Sainvitu P, Nott K, Richard G, Blacker C, Jerome C, Wathelet JP, Deleu M (2012) Structure, properties and obtention routes of flaxseed lignan secoisolariciresinol: a review biotechnology. Agron Soc Environ 16:115–124

Oliveira VB, Yamada LT, Fagg CW, Brandão MG (2012) Native foods from Brazilian biodiversity as a source of bioactive compounds. Food Res Int 48(1):170–179. https://doi.org/10.1016/j.foodres.2012.03.011

Kaur B, Chakraborty D, Kaur G (2013) Biotransformation of rice bran to ferulic acid by pediococcal isolates. Appl Biochem Biotechnol 170:854–867. https://doi.org/10.1007/s12010-013-0223-1

Wang X, Yan JY, Wang YW, Huang MT, Ho CT, Huang Q (2008) Enhancing anti-inflammation activity of curcumin through O/W nano-emulsions. Food Chem 108:419–424. https://doi.org/10.1016/j.foodchem.2007.10.086

Chapter 2
Potato Peels as Source of Nutraceutics

Barbara Sawicka, Dominika Skiba, and Piotr Barbaś

Abbreviations

CIP	International Potato Center
DNP	1,8-dinitroprene
FAO	Food and Agriculture Organization
IFPRI	International Food Policy Research Institute
IG	Glycaemic index
PP	Potato peel
PPF	Potato peel fiber
SSF	Solid state fermentation
USDA	United States Department of Agriculture

2.1 Introduction

Barbara Sawicka

Potato (*Solanum tuberosum* L.) is one of the main crops in the world. According to the Food and Agriculture Organization, the annual potato production in 2019 was

B. Sawicka (✉)
Department of Plant Production and Commodities Science, University of Life Sciences in Lublin, Akademicka, Lublin, Poland
e-mail: barbara.sawicka@up.lublin.pl

D. Skiba · P. Barbaś
Department Agronomy of Potato, Plant Breeding and Acclimatization Institute – National Research Institute, Jadwisin Branch, Jadwisin, Poland
e-mail: dominika.skiba@up.lublin.pl; p.barbas@ihar.edu.pl

© The Author(s), under exclusive license to Springer Nature Switzerland AG 2022
C. Egbuna et al. (eds.), *Food and Agricultural Byproducts as Important Source of Valuable Nutraceuticals*, https://doi.org/10.1007/978-3-030-98760-2_2

over 330 million tonnes [1]. Developing countries will have higher rates of growth in potato production and productivity than developed countries, according to estimates published by the International Food Policy Research Institute (IFPRI) and the International Potato Center (CIP). According to the growth rates, the total potato production in the world in 2020 will amount to 404 million tonnes. Worldwide demand for potatoes for food, processing and animal feed is expected to increase by around 40% over the same period. There are several challenges to achieving these goals, namely a growing population, declining arable land, declining water availability, increased food demand, increased environmental degradation, decreased input efficiency and adverse climate change. This species plays a key role in the food security of developing countries. As much as 50% of the global demand for food energy is covered by potatoes, rice, wheat, and corn. World potato consumption averaged 35 kg/person/year in 2019, with regional differences [1].

Trends in potato consumption indicate that in Europe potato consumption is steadily declining, while in Africa and Asia, between 1961 and 2019, there was an increase of ~70% [2]. Due to the fact that the biodiversity of potatoes is huge the genus Solanum has more than 2000 species, of which more than 170 produce tubers [3]. As a result of human activity, more than 10,000 varieties of this species have been created so far [4]. Within the species *Solanum tuberosum*, there are two subspecies: ssp. tuberosum and ssp. andigena. In the world the first subspecies is of importance mainly, photo periodically passive or as a long-day plant. *Solanum tuberosum* ssp. tuberosum is widely cultivated and used in Europe and North America, while the cultivation of *Solanum tuberosum* ssp. andigena is limited only to South America, while *Solanum phureya* is used in breeding for interspecies crosses [4]. Thus, the potato variety is one of the main factors influencing the nutrient content. Potato tubers produce a minimal amount of fat (0.1–0.5%), they are a rich source of starch, vitamins, especially vitamins C and B6, and can also provide a large e amount of fiber, macro- and microelements and many bioactive compounds. Potato is included in nutritional guidelines (FBDG) in many countries of the world [1, 5, 6]. The agri-food industry is one of the most important industries, especially for food safety in the world. It produces a large mass of by-products, such as organic waste, which must be managed in an appropriate manner, not only to avoid environmental pollution, but also to contribute to economic growth through the use of by-products [6–11]. Each year, nearly a million tonnes of food are wasted due to a poor food supply chain, improper storage, and delays in transport [12]. In 2018, the Department of Food and Agriculture of the United Nations (FAO, Quebec City, QC, Canada) estimated that 33% of human food is wasted every year [13].

Potato peel as a by-product of the food processing industry can be a cheap, valuable, and affordable starting material for the production of economically important products, including dietary fiber, biopolymers, biofilms, natural antioxidants, and natural food additives [6, 7, 9].

This chapter summarizes the work on the use of potato skins for food and non-food purposes, including the extraction, verification and utilization of bioactive ingredients, nutritional aspects, biotechnological application, in livestock feed and other uses of potato skins as the main industrial waste of potato processing.

2.2 Nutritional Values of Potatoes

Barbara Sawicka

Potatoes are an important part of the South American Indian diet more than 2000 BC [14]. The Indians were also forerunners in the processing of potatoes and their use as valuable food for humans. They even obtained the potato extract by freezing it (thanks to the low temperatures in mountainous regions) and drying the tubers afterwards. This product was called chumo, it could be stored longer and eaten instead of bread [15]. The quality of potato tubers is primarily related to their chemical composition. The most important feature of the chemical composition of potato tubers is the dry matter content. Its amount depends on the genetic characteristics of the varieties and environmental conditions. The dry matter content of potato tubers ranges from 17.0 to 25.7%, depending on the cultivar [16].

Potatoes have a high glycaemic index [GI], which means that after consuming 50 g of potato, glucose levels can increase by as much as 95% [17, 18]. As a result, potatoes are digested more slowly, which reduces the increase in blood sugar levels [19]. On the other hand, a positive effect is the fact that the consumption of a potato with a high index provides the body with the necessary glucose and energy in a short time [17, 18]. Potato starch sticks together during thermal treatment. After cooling of the heat-treated products containing the glued starch, it is retrograded, and this part of the starch is not digested. This is called resistant starch. It consists of starch and its degradation products not absorbed in the small intestine. Resistant starch in the large intestine is fermented by probiotic bacteria of the genera *Bifidobacterium* and *Lactobacillus* [19]. The so-called insoluble non-starch substances that are components of cell walls, such as cellulose, hemicellulose, lignin, etc., i.e., dietary fiber (2.0–2.3%) [20]. It is resistant to the action of digestive enzymes and has no energy value, but it is essential in food as it facilitates the access of digestive enzymes. It also improves intestinal peristalsis and adsorbs bile acids and toxic heavy metals. Some of its ingredients act as a prebiotic for probiotic microorganisms in the large intestine. Fiber regulates the speed of digestion and contributes to the feeling of fullness. It can also move food through the intestines too quickly, so that less minerals are absorbed through the food. Fiber can also induce gas and cramps, especially when fiber intake increases overnight [19]. For example, potato puree slows down the digestion rate, reduces the amount of stool in the intestines, and allows the intestines to rest. According to USDA (United States Department of Agriculture) [21], 1 serving of potatoes, without the skin, contains 3.2 g of dietary fiber, slightly more than half of the fiber that is obtained from a large baked potato with the skin intact [22]. The protein content in potato tubers is on average only about 2%. According to USDA [2018], in 100 g of raw potato tubers, the protein content is 2.14 g, and when cooked "in jackets" –2.86 g [17]. Potato protein is characterized by a very high nutritional value, because it contains all the essential amino acids in the right amounts and proportions. It is the only vegetable protein with a value comparable to proteins of animal origin [20, 23]. Potato protein is unique, it combines favourable textural parameters with the lack of gluten and lactose, as well as allergenic properties [20, 23, 24]. The American Society for

Potato Research has recognized potato as a food with excellent protein quality, offering a number of possibilities for its use. It is located in the peel and in the layer immediately below it, so 50–95% of it is removed by peeling [20, 23].

Phenolic compounds are the dominant antioxidants in food, moreover, they enhance the action of other antioxidants by protecting vitamin C and β-carotene and contributing to the enhancement of vitamin E [17].

The main phenolic compound of potatoes (chlorogenic acid) is not resistant to thermal cooking. During cooking, its losses amount to about 65%, while baking destroys it completely. In the potato plant, phenolic compounds play a major role in its resistance to diseases, inhibiting the action of many pathogens and transforming into suberin deposited in the cell walls of the damaged tuber, constituting a barrier against pathogens [24].

The energy value of 100 g of potato tubers or processed potato products depends on the method of preparing the potato and amounts to: 9 kcal for young potatoes, 550 kcal for each—for crisps [17]. Potato tubers boiled in water, sautéed, baked, or mashed, have a much lower energy value than cereal seeds, legumes, and their processing products. The caloric value of cooked tubers does not exceed 90 kcal in 100 g of the product. Baked tubers (93 kcal per 100 g) and boiled, without skin (86 kcal per 100 g) have a higher energy value than raw (79 kcal per 100 g) and cooked in jackets (78 kcal per 100 g) [11, 17, 21]. The nutritional value of potato tubers also depends on the thickness of the peeling, as well as the method and time of preparing the tubers for consumption [11].

The main phenolic compound of potatoes (chlorogenic acid) is not resistant to thermal cooking. During cooking, its losses amount to about 65%, while baking destroys it completely. In the potato plant, phenolic compounds play a major role in its resistance to diseases, inhibiting the action of many pathogens and transforming into suberin deposited in the cell walls of the damaged tuber, constituting a barrier against pathogens [14].

The energy value of 100 g of potato tubers or processed potato products depends on the method of preparing the potato and amounts to: 9 kcal for young potatoes, 550 kcal for each—for crisps [17]. Potato tubers boiled in water, sautéed, baked, or mashed, have a much lower energy value than cereal seeds, legumes, and their processing products. The caloric value of cooked tubers does not exceed 90 kcal in 100 g of the product. Baked tubers (93 kcal per 100 g) and boiled, without skin (86 kcal per 100 g) have a higher energy value than raw (79 kcal per 100 g) and cooked in jackets (78 kcal per 100 g) [11, 17, 21]. The nutritional value of potato tubers also depends on the thickness of the peeling, as well as the method and time of preparing the tubers for consumption [11].

2.3 The Composition of the Potato Peel

Barbara Sawicka

Potato peel (PP) without technological treatment is a concentrated source of minerals, dietary fiber, phenolic compounds, anthocyanins, and bioactive compounds. Therefore, the nutraceutical potential of potato skins, especially those with a low glycoalkaloids content, is of increasing interest, not only from nutritionists, but also from the pharmaceutical and cosmetic industries. Hence, the recycling and disposal of waste generated in the potato industry is very important. High humidity, fibrous nature, dark color, musty smell, and bitter taste are the main limitations for the effective use and storage of potato skins. Consequently, the potato skins should be dried in place [25]. Traditionally, potato peel waste is usually used for the production of low-value animal feed [26], for the production of fertilizers or as a raw material for the production of biogas. As a result, a significant amount of nutrients and bioactive compounds with antioxidant, antibacterial, apoptotic, chemopreventive and anti-inflammatory properties are wasted [27]. However, the focal point of this chapter is still proper waste management with the practical feasibility of processing [8, 25]. Knowing the physico-chemical properties of potato peelings can help develop an environmentally friendly approach to their use. The various compounds are present in the skin of a potato [16, 28, 29]. The composition, variety, and color of the skin of potatoes vary depending on the geographical areas of cultivation [15, 30]. Potato peel contains polyphenols and phenolic acids, which are responsible for its antioxidant activity, and fatty acids and lipids have antibacterial properties [31, 32]. Potato peel also contains starch (25%), non-starch polysaccharides (30%), protein (18%), acid-soluble and insoluble lignin (20%), lipids (1%) and ash (6%) of their dry weight [33, 34]. The lipid fraction includes long chain fatty acids, alcohols, triglycerides, and sterol esters. There are also lignin units in the potato cell walls [27, 34]. Potato peel is rich mainly in starch (52% dry weight), the content of fermented reducing sugar is limited to 0.6% of their dry weight. Potato peel fermentation is not possible, hence pre-hydrolysis (enzymatic or acidic) of carbohydrates is required to increase the content of fermented reducing sugar [28]. Basic Elemental analysis of the potato peel showed that it contained (in% dry weight): C (43.78 ± 0.15), H (5.96 ± 0.12), N (4.06 ± 0.01) and O (46.21 ± 0.28). The C/N ratio of the potato peel is 10.7, and the pH is 6.5 [9]. The calorific value of the potato peel is 17.37 ± 0.38 (MJ/kg) [20]. Various phenolic acids such as gallic acid (58.6–63.0 mg/100 g), protocatechic acid (216.0–256.0 mg/100 g), vanillic acid (43.0–48.0 mg/100 g), caffeic acid (278.0–296.0 mg/100 g), chlorogenic acid (753.0–821.3 mg/100 g), *p*-hydroxybenzoic acid (82.0–87.0 mg/100 g) and *p*-coumaric acid (41.8–45.6 mg/100 g) are found in the potato skin extract [22, 27].

Helal et al. [34] when evaluating potato skin extracts in terms of efficiency, antioxidant activity, total phenolics and total flavonoids, they obtained the best extracts due to their effectiveness as an antioxidant and antimicrobial agent. The composition of the dried potato skins indicated that total carbohydrates (65.5%) were the main component, while lipids (2.6%) were of minor importance. Potato peel methanol extract at a concentration of 80% (PPME) had a higher extract yield (10.4%), total phenols (3.8 mg GAE g^{-1} DW), flavonoids (0.13 mg rutin g^{-1} DW) and antioxidants (80.9%) than other solvent extracts. Vanillin (176.6 μg L^{-1}) and hesperdin (6058.5 μg L^{-1}) turned out to be the most abundant phenolic and flavonoid compounds in PPME, respectively. The results of the tests of physical properties, combustion and emission of waste substances and potato peel pellets carried out by Halel et al. [34] shows that they are most used for biofuels. Potato peel has a higher fat content than peeled potatoes (0.10%) [28]. Helal et al. [34] obtained a similar composition of fatty acids, and these were characterized by a higher content of (polyunsaturated fatty acids).

2.4 Peeling Methods and Method Efficiency

Barbara Sawicka

The method of peeling tubers is a critical factor influencing the chemical composition and suitability for further use of the potato peel [34, 35]. As with the steaming method used to produce dried potatoes, more dietary fiber and less starch are obtained compared to the potato peel obtained by the rubbing method used in the potato crisp industries [36]. Extruded and non-extruded potato skins were also used as a source of dietary fiber, which replaces wheat bran and cellulose, due to the ability to bind with carcinogenic benzopyrene in in vitro digestion [24, 37]. It turned out that un-extruded, steam potato skins showed a higher binding potential of more benzopyrene than potato skins extruded at 110 °C with 30% humidity. The binding itself was found to be lower in the wearable potato peel as compared to the steamed peel. The steam peeled potato skins were then extruded at 110–15 °C; at 30–35% humidity, which was related to the increased content of lignin and dietary fiber with a lower content of starch at the same time. In the extruded grated potato skin, the lignin content was reduced, but the dietary fiber content was unchanged, and the content of soluble non-starch polysaccharides was increased in both peeling cases. Thus, it has been proved that the insoluble fiber obtained from steamed extruded skins, giving a greater amount of glucose, gives the possibility of the formation of resistant starch. On the other hand, peeled potato peel gives approx. 63% of fiber (hemicellulose, lignin, cellulose, and pectin substances) in terms of dry matter. The fractions obtained in this way give 7.7% of ash, 3.4% of pectin, 14.7% of protein, 2.2% of cellulose, 66.8% of starch, 0.9% of reducing sugars and 1.4% of total soluble sugars [24, 38].

2.5 Potato Skins as a Source of Dietary Fiber

Piotr Barbaś

2.5.1 Pharmaceutical Use

Dietary fiber is a bulking medium, improving intestinal mobility and stool hydration [4]. Dietary fiber, in a broad sense, includes various carbohydrates, hemicelluloses, cellulose, pectin's, lignin, gums, etc. which also have pharmaceutical use. Camire et al. [37] demonstrated the insolubility of the fiber extracted from the potato skin and its ability to bind bile acids during an in vitro study, which is one of the key mechanisms by which a dietary fiber source can lower plasma cholesterol. Dietary fiber extracted from potato skin shows a hypocholesterolemic effect in rats fed with potato skin. Rats showed 40% lower plasma cholesterol. High levels of low-density lipoproteins (LDL), cholesterol, high triglycerides, dyslipidaemia contributed to the agammaegation of platelets, increasing the risk of hypertension [39] and cardiovascular diseases (CVD) [34, 40]. The fiber contained in potato peel also contributes to lowering blood glucose levels in diabetic patients.

Dietary fiber improves the absorption of other sugars by emptying the digestive system and stimulating the insulin response. As a result, it can be promoted as an antidiabetic drug extracted from potato skin with reduced preparation costs and efficient use. Dietary fiber from potato skin is believed to play some protective role against carcinogenesis and mutagenesis through a variety of mechanisms, including binding to mutagenic and carcinogenic materials, increasing stool volume and water absorption, shortening intestinal transit time, and lowering stool pH through fermentation via the intestinal microflora [40].

Moreover, it has been proven that potato skins are an environmentally friendly and economical choice to be used as adsorbents in pharmaceutical wastewater treatment [41]. There are some toxic compounds, such as environmental pollutants from pharmaceutical preparations, that eventually end up in drinking water, groundwater, and surface water as a result of inadequate wastewater treatment. Here, the most useful adsorption material with a higher adsorption capacity for various pollutants (metals and dyes, etc.) is activated carbon [40].

Potato peel waste was also used to produce coal samples after hydrothermal treatment. Then, with slight modification with oxidizing agents and KOH activation, the above-mentioned substances were used to remove pharmaceutical compounds from wastewater [34, 41].

2.5.2 Application in the Bakery Industry

Fiber obtained from potato skins in combination with wheat flour was also used in baking. The share of fiber from the peels was almost 50%. Potato Peel Fiber (PPF) can serve as a cheap raw material with better physical and chemical properties compared to wheat bran. It has a better water holding capacity with a low starch content, but this may be accompanied by some quality defects such as a musty smell which can be removed with extruded potato skins [39]. Pectin polysaccharides appear to be the soluble component responsible for keeping DNP in solution. Competition between soluble and insoluble fiber components can have serious implications for the availability and distribution of hydrophobic mutagens in the gastrointestinal tract when added to the incubation medium in the absence of cell walls. One theory explaining the protective effect of some forms of dietary fiber against colon cancer is that certain mutagens and/or the cancer promoter are adsorbed on these dietary fibers, making mutagens and/or the cancer promoter less accessible to intestinal mucosa cells [42]. Al-Weshahy and Rao [5] showed a significant effect of potato peel in the production of biscuits using the proportion of potato peel, respectively, replacing wheat flour with an acceptable sensory result.

2.6 Biotechnological Application

Barbara Sawicka

Potato skins can be used as a cheap carrier for the production of alkaline protease and alpha-amylase enzyme with potential application in detergents [43], as it was found that potato skins have antimicrobial activity, and potato skin extract is effective against gram-negative bacteria [44]. Potato peel extract has also been tested for antimicrobial activity and has been shown to be effective against bacteria as well as some species of fungi. Compared to the antibiotic (streptomycin), the effect of potato skin extract had a significant effect on *Clavibacter michigenensis* and *Pseudomonas aeruginosa* [45].

Potato peel produces various extracellular hydrolytic enzymes with Bacillus subtilis. However, in terms of activity and quantity, the highest productivity was achieved in studies with *Bacillus amyloliquifaciens* grown on a potato skin substrate. The mannose enzyme produced is known in the industrial sector for the bioconversion of agricultural industrial residues by hydrolysis of the axial chain of heteromannans (softwood hemicellulose) [46]. Potato skins were tested in various industrial processes, they were useful in the extraction of vegetable oil from legumes, because they reduced the viscosity of the extracts during the production of soluble coffee and oligosaccharides, and they were also used in the paper and textile industries [47].

Starch waste from potato peel is also a good raw material for industrial fermentation. The production of polygalacturonases, proteases and amylases extracted from the potato peel waste substrate with Bacillus subtilis has proven to be used on an industrial scale [48]. They were also used in the microbiological processing of fruit and vegetable waste in order to produce essential enzymes and organic acids [49].

2.6.1 *Production of Biogas*

From potato peel waste, biogas (CH_4) is produced by anaerobic fermentation. The method is simple and cheap and can be carried out in the same area. From 1 ton of starch waste, 250 m^3 of biogas can be produced [48, 50]. Potential of anaerobically digested potato peelings after lactic fermentation and their suitability for methane production was investigated. The remains after fermentation of potato skins turned out to be highly productive and showed a maximum cumulative methane production of 273 dm kg^{-1}. Furthermore, these authors initiated a biogas fermentation system based on potato peel residues known as mixed fermentation with good results. In turn, Sanaei-Moghadamet et al. [51] based on potato waste, they produced 2.8 kWh per (kgVS) of energy. However, after the fermentation residue of potato peel increased from 6.4 to over 9.0%, the production of methane was reduced [33].

2.6.2 Lactic Acid Production

Lactic acid, widely used in the pharmaceutical, food and cosmetic industries, is synthesized by fermenting carbohydrates with fungi and bacteria [52]. Thus, potato peels can be used to produce lactic acid through anaerobic fermentation by inoculating potato peel waste with a mixed culture in a sequential batch reactor. Was used, inter alia, gelatinized potato peel waste with a range of 30–50 g L^{-1} DM to optimize production efficiency [33]. Moreover, the extensive use of potato peel waste in the food industry, as well as its use in the pharmaceutical, cosmetic, and other sectors, demonstrate their important use and industrial potential. The production of lactic acid from a zero-value raw material such as potato peel can significantly reduce the production costs of butyric acid, which is essential in many branches of the pharmaceutical and cosmetic industries.

2.6.3 Production of the Enzyme Polyphenol Oxidase

The waste potato peel is also a suitable source of the enzyme polyphenol oxidase. A two-phase aqueous system consisting of a potassium phosphate buffer and polyethylene glycol can be used to extract and purify polyphenol oxidase from potato peel waste. It was found that after partial purification of polyphenol oxidase from the aqueous biphasic system and further purification of the enzyme by gel permeation chromatography which increases the purity by more than 12 times, the aqueous biphasic system proved to be a smart option to obtain purified polyphenol oxidase enzyme from potato peel waste [48]. Niphadkar et al. [53] extracted polyphenol oxidase from potato skins mixed with a phosphate buffer of various pH values and concentrations. Under optimal experimental conditions, the maximum specific activity of polyphenol oxidase was obtained at the level of 3573 units/mg of protein.

2.6.4 Production of Other Enzymes

Potato peel (PP) can be successfully used as a substrate for microbial growth in solid state fermentation (SSF) and thus in the synthesis of by-products such as enzymes. Solid State Fermentation (SSF) is one of the microbiological culture systems. The main assumption of this process is the use of a solid raw material as a substrate for microbial cells and a source of water and necessary building and nutrients [34, 54]. Plant biomass meets the above requirements due to the relatively high content of both moisture and organic carbon in the form of, among others, from cellulose, hemicellulose, or pectin's [54]. Mannanases have been used in many industrial processes, incl. Reducing the viscosity of extracts synthesized in oligosaccharides and instant coffee, as well as in the extraction of vegetable oil from legumes. Potato peel has also been used as a carbon source for mannanase production using *Bacillus amylolique faciens*. The maximum activity of mannanase was obtained [55–57]. Amylases were the first enzymes commercially produced by microorganisms with a wide range of applications in the food, baking, feed, detergent, and other industries. Amylases are produced by Bacillus subtilis, BS9, PB1, Aspergillus niger and Bacillus licheniformis grown on potato skins [34, 55–57]. High enzymatic activity was obtained by the strains Bacillus licheniformis and Bacillus subtilis [57]. The protease obtained from Bacillus subtilis DM-04 using potato peel as a substrate shows an enzymatic activity of 400.0 ± 13.0 (U/gd). It is mainly used in the detergent industry as an additive. The production of protease by SSF lowers the total cost of production as it does not require additional supplementation of carbon and co-nitrogen [43]. Cellulolytic (cellulase, xylanase, carboxymethylcellulose and filter papers) and extracellular (α-amylase, protease, polygalacturatelyase) enzymes can also be produced using Aspergillus niger and Bacillus subtilis. These enzymes showed good enzymatic activity. Polyphenol oxidase is used to remove reactive textile dyes and phenolic pollutants from wastewater and contaminated soil [34, 54, 58].

2.6.5 Bio-fertilizers

Potato waste in the form of peels was effectively used for the synthesis of bio-fertilizers. Potato peel still contains large amounts of protein and starch, which, when decomposed by soil microorganisms, can produce fertilizers with a high nitrogen content [59]. Vermicompost produced on the basis of PP and earthworms (*Pheretima elongate*) was characterized by a higher number of bacteria than in the surrounding soil environment [49]. Slurry from a biogas plant operating on the basis of PP turned out to be a very useful biological fertilizer for use on soils requiring increased nutrient abundance [60]. Similar to potato skins and legume skins, neem leaves were fermented in water for 45 days to synthesize bio-fertilizers. Its application showed general vegetative growth and improvement of physicochemical properties of fruits and vegetables. The content of N, P and K increased accordingly and was improved by mixing the residues obtained from ethanol production in the solid state fermentation process with microorganisms such as: *Anabaena variabilis, Azotobacter chroococcum, Azospirillum lipoferum, Cylindrospermum muscicol, Fischerella muscicola, Nostoc muscorum, fertilissima* [61].

2.6.6 Biofuels, Biohydrogen and Biochar

The demand for the development of alternative energy sources is increasing due to the reduction of oil reserves and the climate policy. Pyrolysis is one of the tools enabling access to liquid fuels, such as bio-oils or bioethanol [62]. Bio-oils are more environmentally friendly compared to petroleum sources. Bio-oils include phenols, aldehydes, organic acids, ketones, esters, furans, alcohols, alkenes, and various oxidized complexes. The main idea behind environmentally friendly fuel is to turn waste into more valuable energy sources. In this way, waste potato peel can be converted into biofuels by pyrolysis. Liang et al. [63] found that waste after fermentation of potato skins will be used as a biofuel. Likewise, another biofuel, i.e., bioethanol, can also be recovered using potato skins due to their high starch content. Ethanol fermentation occurs as a result of the saccharification process. i.e., converting starch into fermentable sugars. As an alternative to petroleum fuels, bioethanol is considered an important energy source [54, 64].

Hydrogen has been considered as an important energy source because its combustion generates a significant amount of energy per unit mass. Burning hydrogen releases environmentally friendly gases, compared to oil sources which emit greenhouse gases into the environment. Recently, more attention has been paid to the production of biohydrogen gas using agricultural waste and microbes [65]. The combination of photofermentation followed by sequential dark fermentation results in greater biohydrogen production by potato skin waste compared to any other fermentation process [54].

From the fermentation of potato peels and their residues, biochar and bio-oil synthesized using screw pyrolysis at the temperature of 450 °C are obtained. Bio-oils produced from potato waste and their fermentation residues showed a higher level of suberin and lipid components and better absorption properties, compared to bio-oils produced by hybrid pyrolysis of poplars [63].

2.6.7 Production Bio-sorbents

Recently, potato peel waste is used to produce bio-sorbents, which are used to treat wastewater and various other sources of pollution, i.e., pigments, dyes, heavy metals, etc. An updated list was provided for in particular potato skin waste, white can be subjected to several treatments in order to obtain materials with specific adsorbate properties. Interesting achievements in this area are presented in Table 2.1.

2.7 Sources of Antioxidants

Dominika Skiba

2.7.1 Potato Skins as a Source of Antioxidants in the Food Industry

Due to its high phenolic content, potato peel has gained recognition as a natural antioxidant and has been proven to be 10 times greater than the flesh of about 50% of all polyphenols present in potato tubers. Potential uses of potato peel as an antioxidant were investigated in food, as synthetic antioxidants (butylated hydroxyanisole) and BHT (butylated hydroxytoluene) were typically used to prevent oxidation.

Table 2.1 List of sorbents and treatment of potato peel waste

Adsorbate	Processing of potato skins	Maximum adsorbent capacity
Cr ions (VI)	Untreated	1.97 mg/g
Fe	Untreated	2.17 μg/g i 2.91 mg/g
Co (II)	Pyrolysis and phosphoric acid	–
Cu (II)	Untreated	84.74 mg/g
Ni (II)	Untreated	2702 mg/g
Cr (IV)	Hydrochloric acid	3.28 mg/g

Source: own on the base [34, 54, 66]

Al-Weshahy and Rao [5] found them toxic as they can disrupt the normal levels of liver enzymes or cause liver edema. There is an urgent need for natural antioxidants, and this was the first attempt to do so from potato skin waste by extracting natural antioxidants and enriching them in food systems.

Potato peel methanol extract contains a higher level of total polyphenols than the aqueous extract. 90% of the phenolic compounds present in potato skins were composed of chlorogenic acid having three major isomers including cryptochlorogenic acid, chlorogenic acid and neochlorogenic acid [67]. Potato peels were analysed for total polyphenols and antioxidant capacity. A high content of total polyphenols was found (177 mg/100 g DM). Thus, a strong positive relationship was demonstrated between the total polyphenol content in potato skin extracts and their antioxidant power.

Freeze-dried potato peel extract provides maximum antioxidant power at pH 5–6, less activity at neutral and alkaline pH. Moreover, freeze-dried potato peel extracts, heated at over 80 °C for 30 min, reacted with a decrease in the antioxidant potential. Potato peel extract turned out to have antioxidant potential in preventing lipid oxidation in meat, conserved radiation [68]. Potato peel extract delays lipid oxidation when added to meat prior to the irradiation process. Moreover, it has been found that potato peel extract, as a natural antioxidant, prevents the degradation of lipids in food. It has been proven that potato peel extract can be safely used as a natural antioxidant inhibiting lipid oxidation in fats and oils and in other food products [34, 69].

2.7.2 Pharmaceutical Uses

In vitro studies suggest that potato peel can be used as an ingredient in functional or healthy foods to mitigate oxidative potential. Potato skins affect the antioxidant status as well as the glycaemic index in streptozotocin-induced diabetes in male rats [70]. The addition of powdered potato peel to the rats' diets reduced both kidney and liver overgrowth while simultaneously regulating serum AST (aspartate aminotransferase) and ALT (alanine aminotransferase) in induced diabetic rats. Serum enzymes, including alanine aminotransferase and aspartate aminotransferase, are used to assess the level of liver dysfunction in acute hepatitis with liver damage. Antioxidant potential of potato skin extracts as a defence against oxidative damage to erythrocytes in vitro by observing morphological and structural changes in the cell membrane [71].

The phenols present in the potato peel extract consist of protocatechuic acid, coffee acid, gallic acid and chlorogenic acid. Potato peel extract at a concentration of 2.5 mg/mL inhibits 80–85% of lipid peroxidation. Under experimental conditions, potato skins delay the structural changes induced by H_2O_2 in rat red blood cells. Potato peel extract was found to be an important protector of the membrane of human erythrocytes from oxidative damage due to its strong antioxidant activity. The liver is more susceptible to the absorption of xenobiotics and drugs in high concentrations. The detoxification reactions in both phases increase the hydrophilicity of the substrate by metabolizing xenobiotics. Drug metabolizing enzymes detoxify xenobiotics, but on the other hand they also increase or bioactivate the toxicity of others [72]. Bioactivation makes the liver vulnerable to damage from toxic substances. Potato peel antioxidants have shown promising therapeutic agents against liver damage. Singh and Rajini [71] investigated the protective effect of potato peel extract against induced liver damage. Oral administration of the extract reduced liver damage.

2.8 Discussion

Barbara Sawicka

The Food and Agriculture Organization of the United Nations (FAO, Quebec City, QC, Canada) estimated in 2018 that 33% of human food is wasted each year [13]. The attention paid to the problem of food waste, understood as the inedible by-products of food production practices, is constantly increasing due to the economic, social, and environmental factors associated with the management of these materials. Food waste factors can be at any level between production, harvesting, distribution of food raw materials, processing, packaging, storage, and consumption. Food waste has a very harmful effect on the environment. The idea of "Zero Waste" is an effective way to solve the problems associated with food waste, so that all food or by-products are reused or recycled. This topic fits well with the management of potato skins as a by-product in potato food processing. The first step to better control food waste, food distribution and redistribution of unsold and unnecessary food is a waste hierarchy such as composting, bio combination or similar [25, 34, 73]. Agri-industrial waste is considered to be one of the most preferred sources of renewable energy because it can provide sustainable energy and is environmentally friendly. The use of agricultural waste can also significantly reduce environmental pollution. Therefore, it is important to use waste, thus contributing to a circular economy. One of the ways to use waste or plant mixtures of agricultural origin is to transform them into solid fuels (pellets, briquettes). However, obtaining high-quality pellets is difficult as it is influenced by many factors such as density, humidity, calorific value, harmful emissions. All these indicators must meet specific requirements and standards [10, 74].

Studies with various biomasses, such as tree bark, wheat, and rapeseed straw [75], pumpkin seeds, grape seeds, olive seeds, mixtures of peel and fruit peel, waste from the palm industry [10], waste biomass from rapeseed meal, maize straw, barley straw [75, 76], potato peelings and many more. The most important properties of granules are their durability, which can be influenced by humidity, size, binders, granulation conditions and temperature. One of the binders for production may be faba bean, which contributes to the sustainability of cultivation systems through their ability to biologically bind N_2, which leads to a reduction in the occurrence of diseases, weeds, and pests [77]. Pellets produced from horse field waste and potato skins have good properties as biofuels [77]. A lot of waste is also generated in the processing of potatoes. When preparing potato products, approximately 35–46% of the recycled pulp becomes waste. It was established that in Europe, the volume of potato pulp as the main by-product in the production of potato starch may reach one million tonnes per year [78]. The cause is potato waste without proper handling and many diseases such as viruses (PVY, PLRV), potato blight, ring rot, etc. Potatoes contain a variety of nutrients: up to 24% starch, 2% vegetable protein, enzymes, various nutrients, organic and mineral salts. The high content of starch allows the pulp to be treated as an additive to the pre-granulation binder for plant materials characterized by a lower susceptibility to compaction. Potato pulp and mixtures of potato pulp with other vegetable wastes, such as buckwheat hull, oat bran and others has been proven that the granules have high energy properties, and the heat of combustion of potato pulp pellets in dry matter is over 16 MJ kg^{-1}, and their calorific value over 15 MJ kg^{-1} [78]. When developing new solid biofuels, their physical and chemical indicators should be carefully analysed. They are important for the transport, storage, and combustion of biofuels. Before the granulation process, it is necessary to properly prepare the materials by cutting and grinding in order to obtain a finer fraction of the materials, because the density of biofuels depends on their quality [77]. When developing new biofuels (liquid or solid), a compromise is sought between heat combustion and harmful emissions, which encourages the search for fuels that can provide the required calorific value at lower emissions [77, 78].

Kumar and Srinivasa Rao [73], based on the results of the EU REFRESH project, consider food by-processing in Europe and potential food safety hazards to consider when assessing suitability for human consumption. They are compared with the nutritional benefits offered by these products and their potential uses in food supply chains. In general, it is investigated whether the leftover fruit and vegetables, e.g., waste from potato processing, are safe, sustainable and, at the same time, nutritiously valuable [34, 73]. Valorisation of by-products from food processing is only possible on a large scale if consumers find it safe and acceptable practice. Extracting valuable compounds from by-products and using them in the preparation of functional foods can be a way to gain consumer acceptance. In addition, the current EU food safety legislation does not sufficiently consider the by-products of food processing. A way to fill this regulatory gap could be the adoption of private food safety standards that have shown a propensity for sustainability issues in food supply

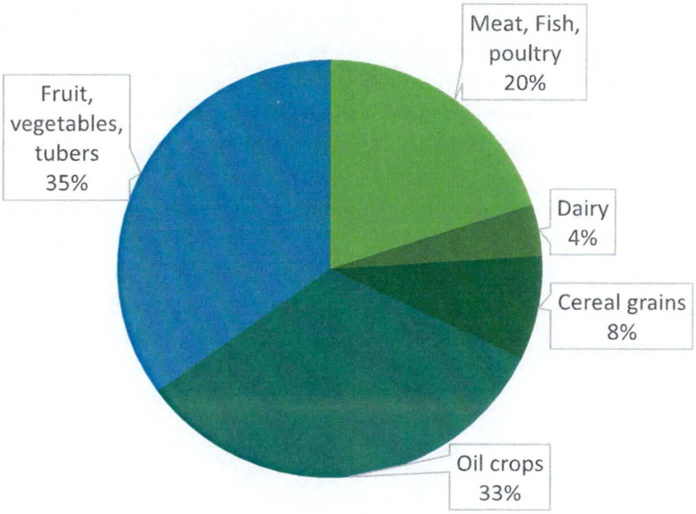

Fig. 2.1 Food waste estimates per sub-sector at the processing stage. (Source: on based on FAO [1])

chains. Finally, it has shown that it is indeed possible and feasible for certain food processing by-products to be valorised while ensuring sustainability, food safety and nutritional value. Extracting valuable compounds from by-products and using them in the preparation of functional foods can be a way to gain consumer acceptance [73].

Food processing is one of the most important industries in the world because it is associated with food safety. However, by-products of such industrial activities, which are mainly organic material, must be handled appropriately to avoid environmental hazards. The sanitation and disposal of food processing by-products is a priority, there should be many approaches, including recycling such ingredients and using them in several food and/or non-food applications [73]. However, the most important in this approach are environmental safety, stability, and economic feasibility [6, 26, 27]. Industrial processing generates annually from 70 to 140 thousand jobs worldwide. Tons of potato skins [9]. In the sectoral structure, the share of by-products of individual sectors of the agri-food industry is shown in Fig. 2.1.

2.9 Conclusion

Barbara Sawicka

While implementing the mission of "Feeding the Word", which is to engage in the production of responsible food for the ever-growing world population, not only the nutritional value, but also the healing and energy properties of the potato should be considered. Potato tubers should be treated not only as functional, pro-health, gluten-free and non-allergenic food, but also as a medicinal, energy raw material, useful in the food, pharmaceutical, cosmetic, fodder and energy processing industries. In order to stimulate the economy of developing countries, the correct and safe use of agricultural waste and the agri-food industry must be effective not only in increasing economic expansion, but above all in reducing the poverty zone. Therefore, potato peel waste can be used to synthesize many compounds such as: biosorbent, biohydrogen, biogas and bio-fertilizer, enzymes, lactic acid, etc., which can serve as a basis for developing links between industry and agriculture and sharing new ideas technologies and future directions of sustainable economic growth. Potato peel, a zero-value material in the potato processing industry, is increasing day by day with the increasing consumption of processed potato products, thereby helping to stimulate the economic sector, especially in developing countries.

References

Barbara Sawicka

1. FAO (2020) Chapter 3. Food security and nutrition. In: Statistical yearbook. World food and agriculture. https://www.fao.org/3/cb1329en/online/cb1329en.html. Accessed 27 Dec 2021
2. Wijesinha-Bettoni R, Mouille B (2019) Potatoes' contribution to global food security, nutrition, and a healthy diet. Am J Potato Res 96:139–149
3. Bhushan S, Kalia K, Sharma M et al (2008) Processing of apple pomace for bioactive molecules. Crit Rev Biotechnol 28:285–296. https://doi.org/10.1080/07388550802368895
4. Bilal M, Asgher M, Iqbal HMN, Zhang HB (2017) Biotransformation of lignocellulosic materials into value-added products a review. Int J Biol Macromol 98:447–458
5. Al-Weshahy A, Rao V (2012) Potato peel as a source of important phytochemical antioxidants of nutraceuticals and their role in human health—a review. IntechOpen, London. https://doi.org/10.5772/30459. Corpus ID: 53129157
6. Joshi A, Sethi S, Arora B, Azizi AF, Thippeswamy BS (2020) Potato peel composition and utilization. In: Potato. Springer Nature, Berlin, ISBN: 978-981-15-7661-4. https://doi.org/10.1007/978-981-15-7662-1_13
7. Chiellini E, Cinelli P, Chiellini F et al (2004) Environmentally degradable bio-based polymeric blends and composites. Macromol Biosci 4:218–231. https://doi.org/10.1002/mabi.200300126
8. Paleologou I, Vasiliou A, Grigorakis S et al (2016) Optimisation of a green ultrasound-assisted extraction process for potato peel (*Solanum tuberosum*) polyphenols using bio-solvents and response surface methodology. Biomass Convers Bior 6:289–299. https://doi.org/10.1007/s13399-015-0181-7
9. Chang K (2019). Polyphenol antioxidants from potato peels: extraction optimization and application to stabilizing lipid oxidation in foods. In: Proceedings of the national conference on undergraduate research (NCUR), New York, NY, USA
10. Javed A, Ahmad A, Tahir A, Shabbir U, Nouman M, Hameed A (2019) Potato peel waste—its nutraceutical, industrial and biotechnological applications. AIMS Agric Food 4(3):807–823. https://doi.org/10.3934/agrfood.2019.3.807
11. Raigond P, Parmar V, Thakur A, Lal MK, Changan SS, Kumar D, Dutt S, Singh B (2021) Composition of different carbohydrate fractions in potatoes: effect of cooking and cooling. Starch 73(7–8):2100015. https://doi.org/10.1002/star.202100015
12. Caldeira C, De Laurentiis V, Corrado S, Van Holsteijn F, Sala S (2019) Quantification of food waste per product group along the food supply chain in the European Union: a mass flow analysis. Resour Conserv Recycl 149:479–488
13. Abiad MG, Meho LI (2018) Food loss and food waste research in the Arab world a systematic

review. Food Secur 10:311–322
14. Spooner DM, Hijmans RJ (2001) Potato systematics and germplans collecting, 1989–2000. Am J Potato Res 78(2001):237–268
15. Spooner DM, van Treuren RR, de Vicente MC (2005) Molecular markers for germplasm and Genebank Management, Technical Bulletin 10. International Plant Genetic Resources Institute, Rome, p 136
16. Sepelev I, Galoburda R (2015) Industrial potato peel waste application in food production: a review. Res Rural Dev 1:130–136
17. Leszczyński W (2012) Żywieniowa wartość ziemniaka i przetworów ziemniaczanych (Przegląd literatury). Biuletyn IHAR 266:5–20. (in Polish)
18. Foster-Powell K, Holt SHA, Brand-Miller JC (2002) International table of glycemic index and glycemic load values. Am J Clin Nutr 76(2002):5–56
19. Sawicka B, Johar KSR, Sood PP, Gupta PD (2017) Imbalance of gut microbiota induces cancer. J Cell Tissue Res 17(2):6073–6084. (Available online at www.tcrjournals.com) ISSN: 0973-0028, E-ISSN: 0974-091
20. Pszczółkowski P, Sawicka B (2016) Wartość odżywcza białka wybranych odmian ziemniaka [w:] Bioprodukty – pozyskiwanie, właściwości i zastosowanie w produkcji żywności, Red. G. Lewandowicz, J. Le Than-Blicharz. Wydział Nauk o Żywności i Żywieniu Uniwersytet Przyrodniczy w Poznaniu, pp 56–64. ISBN 978-83-7160-835-3. (in Polish)
21. USDA National Nutrient Database for Standard Reference (2012) http://www.ndb.nal.usda.gov. Accessed 15 Mar 2021
22. Ulmius M (2011) Characterisation of dietary fibre properties to optimise the effects on human metabolism and the transcriptome. In: Biomedical nutrition, pure and applied biochemistry. Lund University, Lund, p 74. ISBN: 978-91-7422-262-3
23. Harkema J (2015) Potato proteins "free from" texture & nutrition. In: Conference Solanic® potato proteins—free from food expo, Barcelona, 4–5 Juni, 2015
24. Nemś A, Miedzianka J, Pęksa A, Kita A (2015) Zawartość związków prozdrowotnych w ziemniakach odmian o różnej barwie miąższu. Bromatologia Chemia Toksykologia 47(3):473–478
25. Minajeva A, Jasinskas A, Domeika R, Vaiciukevičius E, Lemanas E, Bielski S (2021) The study of the faba bean waste and potato peels recycling for pellet production and usage for energy conversion. Energies 14:2954. https://doi.org/10.3390/en14102954
26. Nelson M (2010) Utilization and application of wet potato processing coproducts for finishing cattle. J Anim Sci 88:E133–E142. https://doi.org/10.2527/jas.2009-2502
27. Liang S, McDonald AG (2014) Chemical and thermal characterization of potato peel waste and its fermentation residue as potential resources for biofuel and bioproducts production. J Agric Food Chem 62:8421–8429. https://doi.org/10.1021/jf5019406
28. Arapoglou D, Varzakas T, Vlyssides A et al (2010) Ethanol production from potato peel waste (PPW). Waste Manag 30:1898–1902. https://doi.org/10.1016/j.wasman.2010.04.017
29. Mohdaly AAA, Hassanien MFR, Mahmoud A et al (2013) Phenolics extracted from potato, sugar beet, and sesame processing by-products. Int J Food Prop 16:1148–1168. https://doi.org/10.1080/10942912.2011.578318
30. Pathak PD, Mandavgane SA, Kulkarni BD et al (2015) Fruit peel waste as a novel low-cost bio adsorbent. Rev Chem Eng 31:361–381
31. Maldonado AFS, Mudge E, Gänzle MG et al (2014) Extraction and fractionation of phenolic acids and glycoalkaloids from potato peels using acidified water/ethanol-based solvents. Food Res Int 65:27–34. https://doi.org/10.1016/j.foodres.2014.06.018
32. Jeddou KB, Chaari F, Maktouf S et al (2016) Structural, functional, and antioxidant properties of water-soluble polysaccharides from potatoes peels. Food Chem 205:97–105
33. Liang S, McDonald AG (2015) Anaerobic digestion of pre-fermented potato peel wastes for methane production. Waste Manag 46:197–200. https://doi.org/10.1016/j.wasman.2015.09.029
34. Helal M, El-Adewale TA, El-Battaglia E, El-Bedawey AA, Youssef SM (2020) Evaluation of potato peel extract as a source of antioxidant and antimicrobial substances. Menoufia J Food Dairy Sci 5:79–90. https://doi.org/10.21608/mjfds.2020.113467
35. Onyeneho SN, Hettiarachchy NS (1993) Antioxidant activity, fatty acids, and phenolic acids

compositions of potato peels. J Sci Food Agric 62:345–350
36. Camire ME, Violette D, Dougherty MP et al (1997) Potato peel dietary fiber composition: effects of peeling and extrusion cooking processes. J Agric Food Chem 45:1404–1408
37. Camire ME, Zhao J, Violette DA et al (1993) In vitro binding of bile acids by extruded potato peels. J Agric Food Chem 41:2391–2394. https://doi.org/10.1021/jf00036a033
38. Mahmood A, Greenman J, Scragg A et al (1998) Orange and potato peel extracts: analysis and use as Bacillus substrates for the production of extracellular enzymes in continuous culture. Enzyme Microb Technol 22:130–137
39. Alonso A, Beunza JJ, Bes-Rastrollo M et al (2006) Vegetable protein and fiber from cereal are inversely associated with the risk of hypertension in a Spanish cohort. Arch Med Res 37:778–786. https://doi.org/10.1016/j.arcmed.2006.01.007
40. Erkkilä AT, Lichtenstein AH (2006) Fiber and cardiovascular disease risk: how strong is the evidence. J Cardiovasc Nurs 21:3–8. https://doi.org/10.1097/00005082-200,601,000-00003
41. Kyzas GZ, Deliyanni EA (2015) Modified activated carbons from potato peels as green environmental-friendly adsorbents for the treatment of pharmaceutical effluents. Chem Eng Res Des 97:135–144
42. Harris PJ, Roberton AM, Hollands HJ, Ferguson LR (1991) Adsorption of a hydrophobic mutagen to dietary fibre from the skin and flesh of potato tubers. Mutat Res 260(2):203–213. https://doi.org/10.1016/0165-1218(91)90009-B
43. Mukherjee AK, Adhikari H, Rai SK et al (2008) Production of alkaline protease by a thermophilic Bacillus subtilis under solid-state fermentation (SSF) condition using imperata cylindrica grass and potato peel as low-cost medium: characterization and application of enzyme in detergent formulation. Biochem Eng J 3(9):353–361. https://doi.org/10.1016/j.bej.2007.09.017
44. Sharma N, Tiwari DP, Singh SK et al (2014) The efficiency appraisal for removal of malachite green by potato peel and neem bark: isotherm and kinetic studies. Int J Chem Environ Eng 5:83–88
45. Prasad AGD, Pushpa HN (2007) Antimicrobial activity of potato peel waste. Asian J Microbiol Biotechnol Environ Sci 9:559–561
46. Sharma D, Ansari MJ, Gupta S et al (2015) Structural characterization and antimicrobial activity of a biosurfactant obtained from Bacillus pumilus DSVP18 grown on potato peels. Jundishapur J Microbiol 8:e21257
47. Saldaña MD, Valdivieso-Ramirez CS (2015) Pressurized fluid systems: phytochemical production from biomass. J Supercrit Fluid 96:228–244. https://doi.org/10.1016/j.supflu.2014.09.037
48. Arapoglou D, Vlyssides A, Varzakas T et al (2009) Alternative ways for potato industries waste utilization. In: Proceedings of the 11th international conference on environmental science and technology, Chaina, Crete, Greece, pp 3–5
49. Pandit N, Ahmad N, Maheshwari S et al (2012) Vermicomposting biotechnology an eco-loving approach for recycling of solid organic wastes into valuable biofertilizers. J Biofertil Biopesticide 3:1–8
50. Fu BR, Sun Y, Wang YX et al (2013) Construction of biogas fermentation system based on feedstock of potato residue. J Northeast Agric Univ 44:102–106
51. Sanaei-Moghadam A, Abbaspour-Fard MH, Aghel H et al (2014) Enhancement of biogas production by co-digestion of potato pulp with cow manure in a CSTR system. Appl Biochem Biotechnol 173:1858–1869. https://doi.org/10.1007/s12010-014-0972-5
52. Datta R, Henry M (2006) Lactic acid: recent advances in products, processes, and technologies—a review. J Chem Technol Biotechnol 81:1119–1129
53. Niphadkar SS, Vetal MD, Rathod VK et al (2015) Purification and characterization of polyphenol oxidase from waste potato peel by aqueous two-phase extraction. Prep Biochem Biotechnol 45:632–649. https://doi.org/10.1080/10826068.2014.940970
54. Zasłona H, Duraj P, Trusek-Hołownia A (2015) *Penicillium chrysogenum* culture conditions for the production of polygalacturonase from waste material. Eng Chem Apparatus 4:213–215
55. Mabrouk ME, El Ashwani AM (2008) Production of 946-mannanase by *Bacillus amyloliquefaciens* 10A1 cultured on potato peels. Afr J Biotechnol 7:1123–1128
56. Dos Santos TC, Gomes DPP, Bonomo RCF et al (2012) Optimization of solid state fermentation of potato peel for the production of cellulolytic enzymes. Food Chem 133:1299–1304. https://doi.org/10.1016/j.foodchem.2011.11.115

57. Jadhav SA, Kataria PK, Bhise KK et al (2013) Amylase production from potato and banana peel waste. Int J Curr Microbiol Appl Sci 2:410–414
58. Mukherjee S, Bandyopadhayay B, Basak B et al (2012) An improved method of optimizing the extraction of polyphenol oxidase from potato (Solanum tuberosum L.) Peel. Not Sci Biol 4:98–107. https://doi.org/10.15835/nsb417186
59. Priyanga K, Reji A, Bhagat JK et al (2016) Production of organic manure from potato peel waste. Int J Chem Tech Res 9:845–847
60. Muhondwa J, Martienssen M, Burkhardt M et al (2015) Feasibility of anaerobic digestion of potato peels for biogas as mitigation of greenhouse gases emission potential. Int J Environ Res 9:481–488
61. Chintagunta AD, Jacob S, Banerjee R et al (2016) Integrated bioethanol and bio manure production from potato waste. Waste Manag 49:320–325. https://doi.org/10.1016/j.wasman.2015.08
62. Izmirlioglu G, Demirci A (2012) Ethanol production from waste potato mash by using Saccharomyces cerevisiae. Appl Sci 2:738–753. https://doi.org/10.3390/app2040738
63. Liang S, Han Y, Wei L et al (2015) Production and characterization of bio-oil and bio-char from pyrolysis of potato peel wastes. Biomass Convers Biorefinery 5:237–246. https://doi.org/10.1007/s13399-014-0130-x
64. Afsar N, Özgür E, Gürgan M et al (2011) Hydrogen productivity of photosynthetic bacteria on dark fermenter effluent of potato steam peels hydrolysate. Int J Hydrogen Energy 36:432–438
65. Panagiotopoulos IA, Karaoglanoglou LS, Koullas DP et al (2015) Technical suitability mapping of feedstocks for biological hydrogen production. J Clean Prod 102:521–528. https://doi.org/10.1016/j.jclepro.2015.04.055
66. Jasinskas A, Minajeva A, Šarauskis E, Romaneckas K, Kimbirauskiene R, Pedišius N (2020) Recycling and utilization of faba bean harvesting and threshing waste for bioenergy. Renew Energy 162:257–266
67. Rodríguez-Meizoso I, Marin FR, Herrero M et al (2006) Subcritical water extraction of nutraceuticals with antioxidant activity from oregano. Chemical and functional characterization. J Pharm Biomed 41:1560–1565
68. Formanek Z, Lynch A, Galvin K et al (2003) Combined effects of irradiation and the use of natural antioxidants on the shelf-life stability of overwrapped minced beef. Meat Sci 63:433–440. https://doi.org/10.1016/S0309-1740(02)00063-3
69. Rommi K, Rahikainen J, Vartiainen J et al (2016) Potato peeling costreams as raw materials for biopolymer film preparation. J Appl Polym Sci 133:1–11
70. El Bohi KM, Hashimoto Y, Muzandu K et al (2009) Protective effect of *Pleurotus cornucopiae* mushroom extract on carbon tetrachloride-induced hepatotoxicity. J Vet Res 57:109–118
71. Singh N, Rajini P (2008) Antioxidant-mediated protective effect of potato peel extract in erythrocytes against oxidative damage. Chem Biol Interact 173:97–104. https://doi.org/10.1016/j.cbi.2008.03.008
72. Mohammed NMS, Salim HAM (2017) Adsorption of Cr (VI) ion from aqueous solutions by solid waste of potato peels. Sci J Uni Zakho 5:254–258. https://doi.org/10.25271/2017.5.3.392
73. Kumar A, Srinivasa Rao P (2020) Optimization of pulsed-mode ultrasound assisted extraction of bioactive compounds from pomegranate peel using response surface methodology. Food Measure 14:3493–3507. https://doi.org/10.1007/s11694-020-00597-9
74. Ramadan MF (2016) Potato lipids. In: Advances in potato chemistry and technology. Elsevier, Amsterdam
75. Gil MV, Oulego P, Casal MD, Pevida C, Pis JJ, Rubiera F (2010) (2010) Mechanical durability and combustion characteristics of pellets from biomass blends. Bioresour Technol 101:8859–8867
76. Ishii K, Furuichi T (2014) Influence of moisture content, particle size and forming temperature on productivity and quality of rice straw pellets. Waste Manag 34:2621–2626
77. Jensen ES, Peoples MB, Hauggaard-Nielsen H (2010) Faba bean in cropping systems. Field Crop Res 115:203–216
78. Obidziński S (2014) Palletization of biomass waste with potato pulp content. Int Agrophysics 28:85–91

Chapter 3
Red Beet Pomace as a Source of Nutraceuticals

Muhammad Afzaal, Farhan Saeed, Aftab Ahmed, Muhammad Armghan Khalid, Fakhar Islam, Ali Ikram, Muzzamal Hussain, Faisal Fareed, and Waqas Anjum

3.1 Introduction

Dietary management should always be approached to boost immunity. Vegetables are regarded as a vibrant segment of a healthy diet because they are an incredible resource of phytonutrients and various other nutraceuticals components. They have a significant influence on the health system and can give medical advantages such as disease treatment and prevention [1].

Recently, the root veggie *Beta vulgaris rubra* often called red beetroot distinguished out among minimally processed vegetables due to its higher nutritional value and usability. Beetroot is currently harvested in numerous nations across the world, is usually consumed as part of a healthy diet, and is extensively utilized in food production [2]. Red beetroot is started as the tenth most antioxidant-dense vegetable containing bioactive compounds like carotenoids, phenolic compounds, flavonoids, nitrate, and betalain, saponins, etc. Beetroot consumption has the therapeutic benefit that may result in better clinical outcomes for quite a several diseases like atherosclerosis, diabetes, dementia, aging, hypertension, cancer, liver diseases [3]. Red beet is utilized in several ways, and a substantial percentage of it is exploited in the industrial manufacture of pickles. A modest amount of the root vegetable is used to make beet juice [4]. During the processing of beet-root for the preparation of ready-to-drink, a significant quantity of vegetable slush remains after filtering of

M. Afzaal (✉) · F. Saeed · M. A. Khalid · F. Islam · A. Ikram · M. Hussain · F. Fareed · W. Anjum
Department of Food Sciences, Government College University Faisalabad, Faisalabad, Pakistan
e-mail: muhammadafzaal@gcuf.edu.pk

A. Ahmed
Department of Nutritional Sciences, Government College University Faisalabad, Faisalabad, Pakistan

© The Author(s), under exclusive license to Springer Nature Switzerland AG 2022

C. Egbuna et al. (eds.), *Food and Agricultural Byproducts as Important Source of Valuable Nutraceuticals*, https://doi.org/10.1007/978-3-030-98760-2_3

juice and it is accessible as a by-product of food and beverages [5]. Surplus by-products, which are often used for animal food and the manufacture of fiber or biomass, are one of the most serious issues in the food industry [6]. These compounds might be employed in the pharmaceutical sector and as food additives as suppliers of color-giving substances and bio-active antioxidants [7].

Beetroot pomace (peel, crown, flesh) is commonly seen as a waste product from the beetroot processing industry, despite the fact that it contains greater levels of betalains and phytonutrients that could be used in medical and food purposes. The reason is that during the process of dejuicing a vast number of secondary metabolites as well as dietary fiber components are not converted into liquid form and persist in the pomace upon pressing. There is an urgent need to explore various techniques for repurposing this waste or extracting betalains and phytochemicals at a low cost [8]. Red beet pomace has many functional qualities because of rich in bioactive compounds and high fiber content; that can give intriguing technological characteristics for the food sector, like enrichment of cookies, pasta, cakes [9]. Betalain is widely utilized in modern food systems They're one of the most well-known natural colorants and one of the first to be developed for use in food production [10]. Natural colorants can be regarded as a substitute for artificial colorant and it is generally regarded as safe substance for substance. Beetroot pomace enhances the acceptance of food products because of its superior taste flavor, and color [11]. It's a bountiful, low-cost resource of natural components with functional potential. It is crucial to utilize co-products before they become trash to enhance the eco-sustainability of the food processing sector and result in the production of valuable products which gives benefits to both consumers as well as the juice industry to earn profit [12].

3.2 Nutraceuticals

Nutraceuticals have sparked a lot of attention because of their assumed safety and possible nutritional and medicinal benefits [1]. The notion of nutraceuticals was raised from a survey conducted in the United Kingdom, France, and Germany which found that consumers value food more than exercise or genetic factors in obtaining excellent health. It may be able to minimize or eliminate the requirement for traditional medicines by utilizing nutraceuticals, decreasing the risk of side effects [13]. A nutraceutical can be described simply as food (or portion of a diet) that has health and medical advantages, such as illness prevention and therapy (Fig. 3.1) [1, 14]. According to Kalia et al. [15], nutraceuticals were classified into many subcategories, including antioxidants, dietary fiber, polyphenols, polyunsaturated fatty acids, spices, prebiotics, and probiotics.

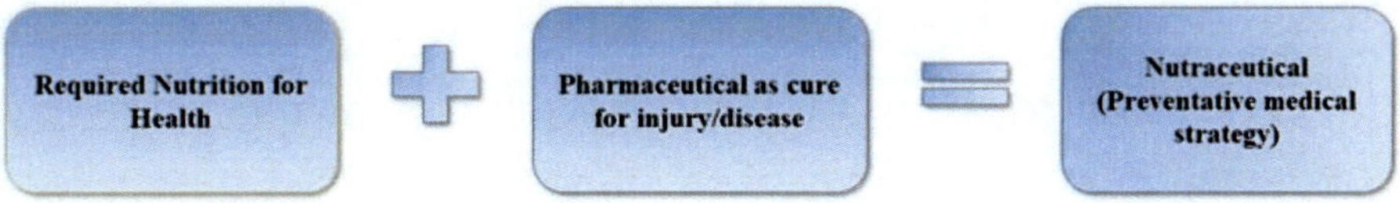

Fig. 3.1 Nutraceutical concept [13]

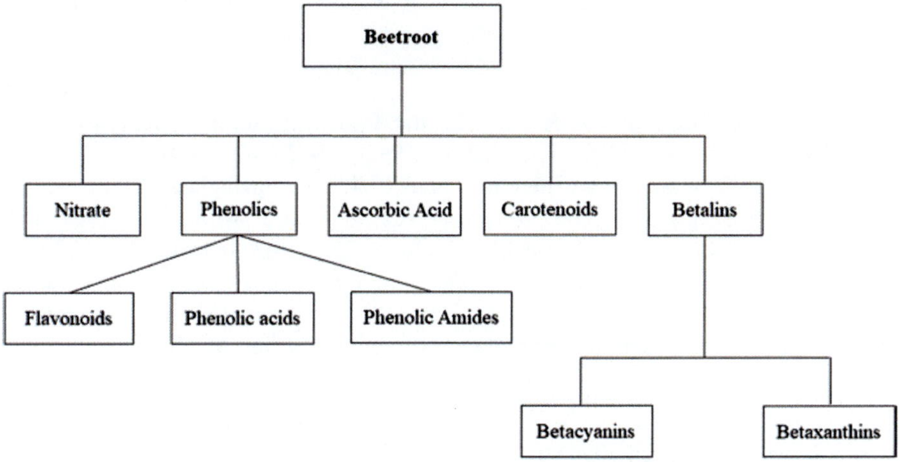

Fig. 3.2 Flow chart of potential bio-active components in beetroot [13, 14]

3.3 Nutritional Composition

Red beet (*Beta vulgaris*) is a renowned natural food having a wide range of applications and nutritional attributes. Beetroot is a popular root vegetable that is high in carbs, lipids, protein, minerals, and various functional components having numerous health benefits. Beetroot is a great supplier of magnesium, potassium, manganese, copper, and iron. Moreover, it contains a lot of antioxidants and vitamins such as A, B (complex), and C (Fig. 3.2). The beetroot pomace that is disposed of is a promising source of betalains and phenols as well as dietary fiber, especially the soluble fiber that's why it is highly recommended [16]. Beetroot has a good array of nutrients that are shown in Fig. 3.2.

3.3.1 Macronutrients

Beetroot comprises carbs (9.96 g/100 g) like fructose, starch, glucose, sucrose, as well as protein (1.68 g/100 g) and fat (0.18 g/100 g). Leaves, on the other hand, include 5 g of carbohydrates per 100 g, 14.8 mg/100 g of protein as well as starch 4.5 g per 100 g [17]. Essential and non-essential amino acids are abundant in

Table 3.1 Percentage composition of red-beetroot by USDA [19]

Composition	Source	Amount
Energy (kcal)	Root	43
Carbohydrate	Root	9.56
Fat	Root	0.17
Protein	Root	1.61
Total sugar	Root	6.76
Fiber	Root	2.8

beetroot. Each 100 g of edible part has the following amino acids: isoleucine (0.048 g), leucine (0.068 g), tryptophan (0.019 g), methionine (0.018 g), tyrosine (0.038 g), cysteine (0.019 g), glutamic acid (0.428 g), phenylalanine (0.046 g), valine (0.056 g), glycine (0.031 g), valine (0.056 g) and proline (0.042 g) [18]. Every 100 g of the edible part, beetroot includes 25 mg of phytosterol, 0.027 g of total saturated fatty acids (SFA's), 0.060 g of total polyunsaturated fatty acids (PUFA's), and 0.032 g of total monounsaturated fatty acids (MUFA's). Table 3.1 depicts the percentage composition of red beetroot according to USDA [19].

3.3.2 Micronutrients

Micronutrients include both minerals and vitamins are very essential for the growth and maintenance of the body. Beetroot encompasses a wide variety of minerals (Na, Fe, Zn, P, Mg, Ca, Mn, etc.) and vitamin C, A, K, B complex. The number of micronutrients in raw red-beet are listed in Table 3.2 as listed.

3.4 Bioactive Profile

The pomace of beetroot is a promising source of natural antioxidant molecules for use as a dietary or food antioxidant. According to postharvest treatments, intraspecific variability, and edaphic parameters, the pigmented part of the beetroot ranged from 0.4 to 2.0% of the dry mass.

3.4.1 Betalains

Beet-root is indeed a potent supplier of water-soluble nitrogenous color, known as betalains. Betalain is the principal bioactive chemical found in red beets. Betalains are found in red beetroot pomace between 0.75 and 3.75 mg/g of dry weight. A

Table 3.2 Micronutrients in raw beet [3, 19, 20]

	Values	
Beet compounds	Leaves	Tuber
Water content (g)	91 ± 4	91.3 ± 4.3
Vitamin A (IU)	6.33	—
Vitamin C (mg)	30	7.2 ± 2.5
Thiamine (mg)	—	0.31
Riboflavin (mg)	—	0.27
Niacin (mg)	0.40	0.33
Folate (μg)	15	109
Vitamin B_6 (mg)	0.1	0.067
Potassium (mg)	762	325 ± 4.5
Calcium (mg)	117	16 ± 3.5
Sodium (mg)	226	78.0 ± 5
Phosphorus (mg)	41	40
Magnesium (mg)	70	23 ± 2
Zinc (mg)	0.38	0.365 ± 0.015
Manganese (mg)	0.391	0.359 ± 0.04
Copper (mg)	0.19	0.07
Zeaxanthin + lutein (μg)	1.50	0
Lycopene (μg)	0	30 ± 0.3
β-carotene (μg)	11.6	0
Iron (mg)	2.6	0.8

secondary metabolite of L-tyrosine, betalains are found in red beetroot. In terms of antioxidant capability, betalain is ten times stronger than tocopherol and three times more effective than catechin. There have been two primary categories of betalains: yellow betaxanthins and red betacyanins. Betacyanins, in which the balsamic acid is compacted with cyclo-dihydroxyphenylalanine, and betaxanthins, which are ammonium derivatives of betalamic acid with different amines [21].

The peel typically contains the majority of betalains, accounting for up to 54% of the total, with the crown (32%) as well as flesh (14%). However, the phenolic composition of the peel contains L-tryptophane as well as betacyanin and betaxanthins, as well as derivatives of the cyclodopa glucosides. Betaxanthins are yellow, with a maximum wavelength (lm) of around 480 nm, regardless of the origin of the amino acid. Betacyanins on the other hand are purple and have an absorbance spectrum focused on lm = 540 nm [20]. However, the level of phytoconstituents in red beetroot largely depends on the processing method, the best being free drying (Fig. 3.3).

Fig. 3.3 Freeze drying of pomace from red beetroot [22]

3.4.2 *Phenolic Compounds*

Beetroot contains a large number of phenolic substances. This has been observed that beetroot contains 50–60 μmol/g dry weight of phenolic acids [23]. Pomace, the byproduct of the processing of beet juice, is extreme in antioxidants (betalains and phenolic compounds). Total phenolic compounds in red beetroot pomace were 1.87–11.98 mg GAE/g of dry weight. Red beet pomace contains phenolic chemicals such as vanillic, ferulic, caffeic, protocatehuic, and *p*-hydroxybenzoic acids. The content of phenolics decreases in the following order: flesh (13%), a crown (37%), and peel (50%) [24].

3.4.3 *Flavonoids*

Red beetroot pomace contains total flavonoids 54.844 mg/100 g. In beetroot, the most prevalent flavonoids include betavulgarin, betagarin, cochlophilin A, and dihydroisorhamnetinas. Two flavanones, betavulgarin and betagarin were identified from beetroot leaves. It was shown that the rutin, quercetin and 4′-hydroxy-5-methoxy-6,7-methylenedioxy flavanone were present in the *B. vulgaris* perennis ethyl acetate fractions [8].

3.4.4 *Carotenoids*

Carotenoids, commonly known as tetraterpenoids, are organic pigments that belong to the carotene family. Beetroot contains a significant level of carotenoids, which are powerful antioxidants. Because the carotenoids in red beetroot are in such small concentrations, they are not typical. The carotenoids in red beet are lutein and b-carotene, both of which are powerful antioxidants that protect against cancer [3]. Carotenoids are antioxidants that can help you avoid illness and boost your immune system. Carotenoids, in particular, may effectively shield your eyes from the harmful effects of blue light and lower your incidence of macular impairment later age [1, 14].

3.4.5 Saponins

Saponins, which are triterpene or steroid glycosides, are mostly found in legumes and roots like beetroot, sugar beet, tea, oats. They have a bitter taste, surfactant action [14]. Biologically active molecules, saponins, are derived from plants to fight pathogens as well as herbivores. There are approximately 11 triterpene saponins in *B. vulgaris*, according to early studies. Almost all saponins are comprised of oleanolic-acid derivatives [3]. According to the study, 26 triterpene saponins were identified in beetroots, 17 of which had never been described before and 7 of which were novel. The concentration of saponins is 8.22 mg/100 mL beetroot juice. Saponins lessen cancer risks, lower blood lipids, and reduce blood glucose sensitivity. Diet rich in saponin can avert tooth decay and platelet coagulation, treat hypercalciuria in humans, and function as an antidote to acute heavy metal poisoning [25, 26].

3.4.6 Dietary Fiber

A huge volume of pomace generated by juice companies is a low-cost supplier of fiber. Dietary fiber is an essential component of the diet that provides a variety of health aspects and aids in the prevention of many illnesses. Dietary fiber is beneficial for human wellbeing because it is resistant to hydrolysis by a person's alimentary enzymes; it ferments completely or partially in the large intestine; and mostly comprises cellulose, oligosaccharides, lignin, pectin, waxes, and gums [27]. Plant-based foods have been linked to high levels of dietary fiber. A healthy person should take (20–35)g of dietary fiber each day. The deficiency of dietary fiber in the diet is linked to several diseases like cardiovascular disease, diverticulosis, constipation, and cancer. Dietary fiber in beet pomace is approximately 62.75%. However, beetroot pomace is an innovative concept for fiber enrichment in food products due to its improved functionality because of the well-adjusted ratio of soluble and insoluble fiber, improved hydration characteristics, improved fermentability, and existence of phytochemicals [28].

3.4.7 Antioxidant Activity

Natural food antioxidants are becoming increasingly popular as a substitute for synthetic food antioxidants. The antioxidant composition of beetroot juice is superior to that of cranberry, apple, citrus fruits, and yellow passion fruit due to its greater levels of total phenolics (1169 mg GAE/L), flavonoids (925 mg catechin equivalent/L), and pigments (854 mg/L). Aglycone-betanidine and betanin have been testified to have great antioxidant activity and are beneficial in preventing lipid peroxidation [29, 30]. The total antioxidant potential of beetroot juice (80.5%), beetroot powder (95.3%), beetroot chips (95.7%), and cooked beetroot (85.8%) was documented by Vasconcellos et al. [26].

3.5 Health Promoting Potential

Sophisticated lifestyles have acquired many converts in the field of public health which leads to serious chronic diseases like cancer, diabetes, heart disease. As a result, incorporating fruits and vegetables into one's daily diet has become critical. Numerous scientific studies have showcased that beetroot consumption has physiological benefits that may optimize clinical outcomes for a range of illnesses [31].

3.5.1 Bioavailability

A dietary component must be available in vivo, which means that after consumption, its bioactive substances are assimilated via the digestive system and become present in sufficient quantities in the plasma to be used by cells [30]. Moreover, to achieve a systemic circulatory system and perform a certain beneficial role, food ingredients must keep their chemical composition across multiple digestion steps, each of which poses a significant metabolic dilemma for the molecule and influences the extent and rate of absorption [32, 33].

In this context, both inorganic nitrate as well as betalains, the principal active constitutes of beetroot, have been regarded as a viable source of bioactive compounds in humans [34]. Kanner et al. [35] found 0.5–0.9% of the consumed betacyanins (isobetanin, betanin) in participants' urine 12 h after drinking 300 mL of beet juice. This suggests that, even in minute doses, betacyanins may be absorbed by individuals. They also discovered that the highest urine exclusion ratio of betacyanains (absorption indicator) occurred 2–4 h after consumption, although there was a substantial amount of inter-individual heterogeneity within such a time frame [35].

3.5.2 *Oxidative Stress*

Beetroot supplementation may be an effective method for strengthening intrinsic antioxidant defenses, therefore protecting biomolecules from oxidative stress. Under usual metabolic circumstances, the physiological state of a cell is regarded to be in a phase of redox equilibrium, or in other terms, an equilibrium occurs among (pro-oxidant) oxidizing and (antioxidant) reducing agents [36]. RONS plays a critical role in an extensive range of physiological and metabolic activities at very low concentrations, including expression of a gene, apoptosis, cell proliferation, and muscular contraction. Excessive vulnerability of cells to extracellularly derived RONS (xenobiotic, UV rays) or intracellularly produced RONS (inflammation, abnormal cellular metabolism) can interrupt the antioxidant capacity of cells, creating an imbalance in redox cell function, that further leads to the situation known as oxidative stress.

According to Tesorier et al. [37], several in vitro investigations have demonstrated that betalain pigments, in particular, shield cellular components from oxidative damage. Beetroot possesses many highly biologically active phenolics, including caffeic acid, epicatechin, rutin all of which are good antioxidants [2]. Likewise, nitrite and other NO sources, such as red beet, have been proven to limit radical formation and aggressively scavenge potentially harmful RONS like hydrogen peroxide. Implying that nitrate may have antioxidant properties [38, 39]. The scientists also discovered that betanin, the utmost prevalent betalain present in beetroot (300–600 mg kg^{-1}), was the more efficient lipid peroxidation inhibitor. The strong antioxidant activity of betanin appears to be due to its extraordinary electron-providing ability and capacity to overcome extremely reactive radicals attacking cell membranes [35].

3.5.3 *Inflammation*

In most cases, inflammation is viewed as a good mechanism that governs our inherent reaction to physiological stressors such as infection, trauma, and another organism that might inflict harm to the body and disturb homeostasis [40]. Immunological response may nonetheless have unfavorable effects on the host. Failure to eliminate the intruding component and renew adequate immune systems can result in persistent inflammation and long-term cell malfunction [41]. Betalains and beetroot extracts have been discovered to be effective anti-inflammatory agents. Their anti-inflammatory actions appear to be mediated in part by interacting with pro-inflammatory communication pathways [42]. Beetroot's anti-inflammatory and antioxidant properties have sparked curiosity in its possible application in illnesses defined by abnormal immune cellular function. Betacyanin extracts have also been shown to have chemopreventive effects in skin, lung, and liver cancerous cells in animal studies, and more latterly, in human skin, breast, pancreatic and prostrate cancerous cells [41, 43].

The antioxidant and anti-inflammatory properties of beetroot have prompted interest in its potential application in diseases characterized by aberrant immune cell activity. Indeed, persistent inflammation has been related to the development of malignant tumors, and research shows that betalain extracts produced from beetroot may help to attenuate these effects [44].

3.5.4 Endothelial Function

Nitrate from beetroot is metabolized to nitrite, which can then be decreased further to create NO [45]. The endothelium regulates vascular homeostasis by regulating thrombotic action, vascular tone, platelet function, and the sensitive balance between vasoconstricting agents, (thromboxane, endothelin) and the discharge of vasodilating substance (prostacyclin, NO) [46].

As NO regulates numerous endothelial actions, a decrease in NO bioavailability with age has been identified as the primary cause of endothelial dysfunction [47]. Endothelial dysfunction and has been acknowledged as a key potential cause for numerous cardiovascular diseases and therefore has been linked to the progression of atherosclerosis and hypertension. Likewise, beetroot rich in nitrate also contain bioactive components (i.e., caffeic acid, betalains) which have a role in the management of cardiovascular disease [48, 49].

3.5.5 Cognitive Function

Impaired cerebral circulation of blood is one of the main pathogenic processes that accompany the deterioration of cognitive function with aging [50]. Indeed, a decline or diminishment in cerebral perfusion with age has been linked to several neurological diseases associated with impaired cognitive capacity, including clinical dementia, brain injury, and Alzheimer's disease [51]. A disturbance in neurovascular performance, which is mediated in part by decreased NO activity, is one of the primary triggers and adverse outcomes for the start and progression of cerebral hypoperfusion. As a result, a NO originator like beet may be able to improve cerebral blood flow and challenge cognitive function impairments [52]. According to Gilchrist et al. [53], elderly (67 years) type-2-diabetics who received 250 mL of beet juice (nitrate: 7.5 mmol) for 2 weeks had substantial progress in simple response time when compared with the control group. Other cognitive tasks involving decision-making, quick processing, morphology, and spatial memory, on the other hand, revealed no impact.

3.6 Application of Red Beet Pomace in the Food and Beverage Industry

Red beet pomace has many functional qualities and that's why it has acquired popularity in the food and beverage sector (Table 3.3) and is used as a food colorant, preservative, and in the production of a variety of functional food items.

Table 3.3 Application in food product

Food product	Inclusion of betalains	Level of addition (w/w)	Outcomes	References
Baked items	Beet-root powder	2–10%	Wheat dough with good hydration and farinographic qualities, as well as physical and sensory features of cooked rolls	[5]
Biscuits	Extract of beetroot	2.5%	Acceptability, bioactivities, and nutritional values have all been improved	[54]
Pasta	Beetroot	150 g/kg	The inclusion of beetroot may increase the quality of the pasta and produce an appealing color on the pasta	[55]
Noodles	Red beet pulp	10–40%	Noodles with 30% beet pulp had the best cooking, sensorial and nutritional characteristics	[3]
Meat	Red beet powder	0.15%	Beetroot might be used to partly substitute nitrites in meat processes, and it has high antioxidant activity as well as benefits for longer shelf life	[56]
Fish	Peel extract	Red beet peel extracts in 0.1% (w/v) ice	Red-beetroot peel extract (0.1%) might increase the storage life of trout while also improving the chemical and sensorial quality of the fish	[57]
Sausages	Betalain powder derived from beetroot	33, 45, 56 ppm	The inclusion of betalain in sausages is depicted to be more resistant to light susceptibility throughout storage than the color of sausages using nitrite-nitrate salts	[58]
Candies	Beetroot pomace	9.24%	Enhanced the candies' phytochemical characteristics	[59]
Yogurt	Incorporate beetroot powder	2%	Best sensory acceptability, rheological characteristics, and physicochemical properties	[60]
Beverages	Beetroot	4 g/L	Better health	[61]

3.6.1 Natural Colourant

Consumers are gradually avoiding synthetic colors as they become more aware of the benefits of consuming healthier foods. Now, the food, as well as beverages industry, chiefly focuses on natural color [62]. Natural colorants are non-toxic (safe to use), nutritive, less polluting, and produce soft and delicate effects. Because of the large cost, food manufacturers are looking for more cost-effective ways to get natural colorants, such as industrial byproducts from food industries. In this aspect, beetroot pomace has attained prominence since it is an excellent source of betalain, which is a commonly used colorant in the food sector. Betanin (E162), [63] is used as a food colorant in ice cream, yogurt, and other goods. The colorant derived from beetroot waste remained stable for 20 days [9]. Furthermore, the addition of betalain to a product improves its nutritional and nutraceutical qualities [64, 65].

3.6.2 Natural Preservative

To increase product quality and shelf life, several preservatives and methods are utilized [66]. Synthetic nitrite and nitrate are widely used in the food processing sector because these ingredients not only act as potent antibacterial agents against pathogenic organisms that deteriorate the product but also improve the sensory qualities. In essence, they can be harmful and dramatically increase health issues, which motivates food manufacturers to pay more attention to natural preservatives [67]. Nitrates and nitrites are now naturally generated from beetroot pomace. The use of beetroot pomace is not only rich in nitrate/nitrite but also contains numerous useful components that in addition to preserving the food quality also added functional characteristics [68]. Biodegradable packaging containing an antibacterial ingredient is extremely prevalent. The inclusion of beetroot extract increases edible packaging's antibacterial effectiveness [69, 70].

3.6.3 Functional Foodstuff

Beetroot pomace is a valuable source of phytonutrients as well as dietary fiber, and it has fascinating technical qualities for the food sector, such as enrichment of cookies, cakes, and pasta [9]. Due to increased nutritional knowledge, bakery goods (cookies, cakes, and biscuits) are now classified as junk food that provides empty calories. As a result, such popular snack foods must be nutritionally upgraded by value addition, so that these are filled with fiber and vital nutrients. Most pomace powders also serve as natural colorants, flavors, and aromas, without interfering with the physiognomies of the items into which they are incorporated [71, 72].

Beetroot pomace powders can be employed as a low-cost, non-caloric bulking ingredient in place of flour, sugar, or fat since they increase food functioning by increasing water and oil retention and improving emulsion constancy [73]. The crude fiber content of beetroot pomace powder is about 62.75%. The addition of beetroot pomace powder improved the fiber content of cookies. The addition of beetroot pomace at a sufficient amount increases crude fiber, moisture, protein, and ash while decreasing carbohydrate content [16].

3.7 Future Perspective

Producing more food for a rising population while simultaneously addressing environmental issues is a key task that the world population confronts in the coming decades. Each year, vast amounts of agricultural by-products are generated as a result of the food processing operations. Similarly, the beetroot processing industry produced a large volume of pomace, which was either discarded or used as animal feed. However, the proportions of phytochemicals and other phenolics in the outer skin and pulp/pomace of red beetroot are usually much greater than in their respective edible tissues, indicating that these wastes and residues might be viable sources for isolating bio-active substances. The antioxidants (phytochemicals and polyphenols) or other bioactive substances from the source have antioxidative, anticancer, antimicrobial, and immune-modulatory properties. Beetroot pomace is a natural and cost-effective source of colorants, flavoring, dietary fiber, protein, antioxidants, and antimicrobials, that may be utilized as a source of natural food additives in the food industry. This approach assists processing industries in lowering their treatment costs and potentially earning extra profit from what was formerly deemed garbage, likewise, improving their productivity.

The synergistic efforts of environmentally friendly product preservation, waste minimization throughout the production process, and use of byproducts would significantly minimize waste while also raising the environmental profile of the food processing sector. This recommends that the by-product from produce (i.e., beetroot) should be used as a possible source for functional food components, natural antioxidants, and antibacterial compounds, as well as further processed into therapeutic functional food items rather than being thrown as trash.

3.8 Conclusion

This chapter provides a thorough summary of numerous data to illustrate the nutraceutical capability of beetroot pomace and its application in foods. Beetroot pomace, a byproduct of industrial food processing, was found to be a virtuous source of natural substances, particularly phenolic compounds and betacyanins, with substantial antioxidant property, that can be characterized by high proportions of the major constituent or to the synergy between these components. The usage of beetroot pomace is gaining popularity in the food industry since it is a low-cost source of dietary fiber and bioactive chemicals, which will aid in the valorization of foodstuff and may be utilized to create new 'functional products.' As a result, the fruit production industry may profit from effective waste management.

References

1. Egbuna C, Tupas GD (eds) (2020) Functional foods and nutraceuticals—bioactive components, formulations and innovations, 1st edn. Springer Nature, Cham. 646 pp
2. Georgiev VG, Weber J, Kneschke EM, Denev PN, Bley T, Pavlov AI (2010) Antioxidant activity and phenolic content of betalain extracts from intact plants and hairy root cultures of the red beetroot Beta vulgaris cv. Detroit dark red. Plant Foods Hum Nutr 65(2):105–111
3. Chhikara N, Kushwaha K, Sharma P, Gat Y, Panghal A (2019) Bioactive compounds of beetroot and utilization in food processing industry: a critical review. Food Chem 272:192–200
4. Yashwant K (2015) Beetroot: a super food. IJESTA 1:20–26
5. Kohajdová Z, Karovičová J, Kuchtová V, Lauková M (2018) Utilisation of beetroot powder for bakery applications. Chem Pap 72(6):1507–1515
6. Kushwaha R, Kumar V, Vyas G, Kaur J (2018) Optimization of different variable for eco-friendly extraction of betalains and phytochemicals from beetroot pomace. Waste Biomass Valoriz 9(9):1485–1494
7. Tumbas Šaponjac V, Čanadanović-Brunet J, Ćetković G, Jakišić M, Djilas S, Vulić J, Stajčić S (2016) Encapsulation of beetroot pomace extract: RSM optimization, storage and gastrointestinal stability. Molecules 21(5):584
8. Maraie NK, Abdul-Jalil TZ, Alhamdany AT, Janabi HA (2014) Phytochemical study of the Iraqi beta vulgaris leaves and its clinical applications for the treatment of different dermatological diseases. World J Pharm Pharm Sci 3:5–19
9. Dalla Costa AP, Hermes VS, de Oliveira RA, Flôres SH (2017) Minimally processed beetroot waste as an alternative source to obtain functional ingredients. J Food Sci Technol 54(7):2050–2058
10. Azeredo HM (2009) Betalains: properties, sources, applications, and stability—a review. Int J Food Sci Technol 44(12):2365–2376
11. Slavov A, Karagyozov V, Denev P, Kratchanova M, Kratchanov C (2013) Antioxidant activity of red beet juices obtained after microwave and thermal pretreatments. Czech J Food Sci 31(2):139–147
12. Varzakas T, Zakynthinos G, Verpoort F (2016) Plant food residues as a source of nutraceuticals and functional foods. Foods 5(4):88
13. Chintale Ashwini G, Kadam Vaishali S, Sakhare Ram S, Birajdar Ganesh O, Nalwad DN (2013) Role of nutraceuticals in various diseases: a comprehensive review. Int J Res Pharm Chem 3:290–299
14. Egbuna C, Mishra AP, Goyal MR (eds) (2021) Preparation of phytopharmaceuticals for the management of disorders. Academic, New York. https://doi.org/10.1016/C2019-0-01420-3
15. Kalia AN (2005) Textbook of industrial pharmacology. CBS Publisher and Distributor, New Delhi, pp 204–208
16. Sahni P, Shere DM (2016) Physico-chemical and sensory characteristics of beet root pomace powder incorporated fibre rich cookies. Int J Food Ferment Technol 6(2):309–315
17. Agarwal K, Varma R (2014) Biochemical screening of beetroot leaves. Int J Pharm Sci Rev Res 1(23):127–134

18. Nemzer B, Pietrzkowski Z, Spórna A, Stalica P, Thresher W, Michałowski T, Wybraniec S (2011) Betalainic and nutritional profiles of pigment-enriched red beet root (Beta vulgaris L.) dried extracts. Food Chem 127(1):42–53
19. USDA (2019) USDA national nutrient database for standard reference. https://ndb.nal.usda.gov/ndb/foods/show/2849?fg=&man=&lfacet=&format=&count=&max=25&offset=&sort=&qlookup=beet
20. Ceclu L, Nistor OV (2020) Red beetroot: composition and health effects—a review. J Nutr Med Diet Care 6:43
21. Clifford T, Howatson G, West DJ, Stevenson EJ (2015) The potential benefits of red beetroot supplementation in health and disease. Nutrients 7(4):2801–2822
22. Janiszewska-Turak E, Rybak K, Grzybowska E, Konopka E, Witrowa-Rajchert D (2021) The influence of different pretreatment methods on color and pigment change in beetroot products. Molecules 26(12):3683
23. Hidalgo A, Brandolini A, Čanadanović-Brunet J, Ćetković G, Šaponjac VT (2018) Microencapsulates and extracts from red beetroot pomace modify antioxidant capacity, heat damage and colour of pseudocereals-enriched einkorn water biscuits. Food Chem 268:40–48
24. Čanadanović-Brunet JM, Savatović SS, Ćetković GS, Vulić JJ, Djilas SM, Markov SL, Cvetković DD (2011) Antioxidant and antimicrobial activities of beet root pomace extracts. Czech J Food Sci 29(6):575–585
25. Güçlü-Üstündağ Ö, Mazza G (2007) Saponins: properties, applications and processing. Crit Rev Food Sci Nutr 47(3):231–258
26. Vasconcellos J, Conte-Junior C, Silva D, Pierucci AP, Paschoalin V, Alvares TS (2016) Comparison of total antioxidant potential, and total phenolic, nitrate, sugar, and organic acid contents in beetroot juice, chips, powder, and cooked beetroot. Food Sci Biotechnol 25(1):79–84
27. Sahni P, Shere DM (2018) Utilization of fruit and vegetable pomace as functional ingredient in bakery products: a review. Asian J Dairy Food Res 37(3):202–211
28. Parveen H, Bajpai A, Bhatia S, Singh S (2017) Analysis of biscuits enriched with fibre by incorporating carrot and beetroot pomace powder. Indian J Nutr Diet 54(4):403–413
29. Kathiravan T, Nadanasabapathi S, Kumar R (2014) Standardization of process condition in batch thermal pasteurization and its effect on antioxidant, pigment and microbial inactivation of ready to drink (RTD) beetroot (Beta vulgaris L.) juice. Int Food Res J 21(4):1305
30. Wootton-Beard PC, Ryan L (2011) A beetroot juice shot is a significant and convenient source of bioaccessible antioxidants. J Funct Foods 3(4):329–334
31. Sawicki T, Martinez-Villaluenga C, Frias J, Wiczkowski W, Peñas E, Bączek N, Zieliński H (2019) The effect of processing and in vitro digestion on the betalain profile and ACE inhibition activity of red beetroot products. J Funct Foods 55:229–237
32. Rein MJ, Renouf M, Cruz-Hernandez C, Actis-Goretta L, Thakkar SK, da Silva PM (2013) Bioavailability of bioactive food compounds: a challenging journey to bioefficacy. Br J Clin Pharmacol 75(3):588–602
33. Ting Y, Jiang Y, Ho CT, Huang Q (2014) Common delivery systems for enhancing in vivo bioavailability and biological efficacy of nutraceuticals. J Funct Foods 7:112–128
34. van Velzen AG, Sips AJ, Schothorst RC, Lambers AC, Meulenbelt J (2008) The oral bioavailability of nitrate from nitrate-rich vegetables in humans. Toxicol Lett 181(3):177–181
35. Kanner J, Harel S, Granit R (2001) Betalains a new class of dietary cationized antioxidants. J Agric Food Chem 49(11):5178–5185
36. Powers SK, Jackson MJ (2008) Exercise-induced oxidative stress: cellular mechanisms and impact on muscle force production. Physiol Rev 88(4):1243–1276
37. Tesoriere L, Fazzari M, Angileri F, Gentile C, Livrea MA (2008) In vitro digestion of betalainic foods. Stability and bioaccessibility of betaxanthins and betacyanins and antioxidative potential of food digesta. J Agric Food Chem 56(22):10487–10492
38. Lundberg JO, Carlström M, Larsen FJ, Weitzberg E (2011) Roles of dietary inorganic nitrate in cardiovascular health and disease. Cardiovasc Res 89(3):525–532

39. Wink DA, Hines HB, Cheng RY, Switzer CH, Flores-Santana W, Vitek MP, Ridnour LA, Colton CA (2011) Nitric oxide and redox mechanisms in the immune response. J Leukoc Biol 89(6):873–891
40. Monteiro R, Azevedo I (2010) Chronic inflammation in obesity and the metabolic syndrome. Mediat Inflamm 2010:289645
41. Kundu JK, Surh YJ (2012) Emerging avenues linking inflammation and cancer. Free Radic Biol Med 52(9):2013–2037
42. El Gamal AA, AlSaid MS, Raish M, Al-Sohaibani M, Al-Massarani SM, Ahmad A, Hefnawy M, Al-Yahya M, Basoudan OA, Rafatullah S (2014) Beetroot (Beta vulgaris L.) extract ameliorates gentamicin-induced nephrotoxicity associated oxidative stress, inflammation, and apoptosis in rodent model. Mediators Inflamm 2014:983952
43. Kapadia GJ, Rao GS, Ramachandran C, Iida A, Suzuki N, Tokuda H (2013) Synergistic cytotoxicity of red beetroot (Beta vulgaris L.) extract with doxorubicin in human pancreatic, breast and prostate cancer cell lines. J Complement Integr Med 10(1):113–122
44. Das S, Williams DS, Das A, Kukreja RC (2013) Beet root juice promotes apoptosis in oncogenic MDA-MB-231 cells while protecting cardiomyocytes under doxorubicin treatment. J Exp Second Sci 2:1–6
45. Hobbs DA, Kaffa N, George TW, Methven L, Lovegrove JA (2012) Blood pressure-lowering effects of beetroot juice and novel beetroot-enriched bread products in normotensive male subjects. Br J Nutr 108(11):2066–2074
46. Davignon J, Ganz P (2004) Role of endothelial dysfunction in atherosclerosis. Circulation 109(23 Suppl 1):III-27
47. Lidder S, Webb AJ (2013) Vascular effects of dietary nitrate (as found in green leafy vegetables and beetroot) via the nitrate-nitrite-nitric oxide pathway. Br J Clin Pharmacol 75(3):677–696
48. Joris PJ, Mensink RP (2013) Beetroot juice improves in overweight and slightly obese men postprandial endothelial function after consumption of a mixed meal. Atherosclerosis 231(1):78–83
49. Machha A, Schechter AN (2011) Dietary nitrite and nitrate: a review of potential mechanisms of cardiovascular benefits. Eur J Nutr 50(5):293–303
50. Bond V, Curry BH, Adams RG, Asadi MS, Millis RM, Haddad GE (2013) Effects of dietary nitrates on systemic and cerebrovascular hemodynamics. Cardiol Res Pract 2013:435629
51. Poels MM, Ikram MA, Vernooij MW, Krestin GP, Hofman A, Messen WJ, Van der Lugt A, Breteler MM (2008) Total cerebral blood flow in relation to cognitive function: the Rotterdam Scan Study. J Cereb Blood Flow Metab 28(10):1652–1655
52. Presley TD, Morgan AR, Bechtold E, Clodfelter W, Dove RW, Jennings JM, Kraft RA, King SB, Laurienti PJ, Rejeski WJ, Burdette JH (2011) Acute effect of a high nitrate diet on brain perfusion in older adults. Nitric Oxide 24(1):34–42
53. Gilchrist M, Winyard PG, Fulford J, Anning C, Shore AC, Benjamin N (2014) Dietary nitrate supplementation improves reaction time in type 2 diabetes: development and application of a novel nitrate-depleted beetroot juice placebo. Nitric Oxide 40:67–74
54. Amnah MA (2013) Nutritional, sensory and biological study of biscuits fortified with red beet roots. Life Sci J 10(3):1579–1584
55. Rekha MN, Chauhan AS, Prabhasankar P, Ramteke RS, Rao GV (2013) Influence of vegetable purees on quality attributes of pastas made from bread wheat (T. aestivum). CyTA J Food 11(2):142–149
56. Xiao CG, Tan LL, Zhu PP, Yang HJ, Chen D, Lu WJ, Ren FZ, Guo HY, Ge SY, Tang HG, Chen LH (2019) The coloring effect and antioxidant effect of red beet in minced pork. J Chin Inst Food Sci Technol 19:111–121
57. Yavuzer E, Özogul F, Özogul Y (2020) Impact of icing with potato, sweet potato, sugar beet, and red beet peel extract on the sensory, chemical, and microbiological changes of rainbow trout (Oncorhynchus mykiss) fillets stored at (3±1° C). Aquac Int 28(1):187–197
58. van Elbe JH, Klement JT, Amundson CH, Cassens RG, Lindsay RC (1974) Evaluation of betalain pigments as sausage colorants. J Food Sci 39:128–132

59. Kumar V, Kushwaha R, Goyal A, Tanwar B, Kaur J (2018) Process optimization for the preparation of antioxidant rich ginger candy using beetroot pomace extract. Food Chem 245:168–177
60. Dabija A, Codină GG, Ropciuc S, Stroe SG (2019) Studies regarding the production of a novel yogurt using some local plant raw materials. J Food Process Preserv 43(6):e13826
61. Butt MS, Randhawa MA, Shahid M (2019) Nephroprotective effects of red beetroot-based beverages against gentamicin-induced renal stress. J Food Biochem 43(7):e12873
62. Mortensen A (2006) Carotenoids and other pigments as natural colorants. Pure Appl Chem 78(8):1477–1491
63. Stintzing FC, Carle R (2007) Betalains–emerging prospects for food scientists. Trends Food Sci Technol 18(10):514–525
64. Vidal PJ, López-Nicolás JM, Gandía-Herrero F, García-Carmona F (2014) Inactivation of lipoxygenase and cyclooxygenase by natural betalains and semi-synthetic analogues. Food Chem 154:246–254
65. Zielińska-Przyjemska M, Olejnik A, Dobrowolska-Zachwieja A, Grajek W (2009) In vitro effects of beetroot juice and chips on oxidative metabolism and apoptosis in neutrophils from obese individuals. Phytother Res 23(1):49–55
66. Ozaki MM, Munekata PE, de Souza Lopes A, do Nascimento MdS, Pateiro M, Lorenzo JM, Pollonio MA (2020) Using chitosan and radish powder to improve stability of fermented cooked sausages. Meat Sci 167:108165
67. Petkoska AT, Daniloski D, D'Cunha NM, Naumovski N, Broach AT (2021) Edible packaging: sustainable solutions and novel trends in food packaging. Food Res Int 140:109981
68. Jin SK, Choi JS, Moon SS, Jeong JY, Kim GD (2014) The assessment of red beet as a natural colorant, and evaluation of quality properties of emulsified pork sausage containing red beet powder during cold storage. Korean J Food Sci Anim Resour 34(4):472
69. Domínguez R, Munekata PE, Pateiro M, Maggiolino A, Bohrer B, Lorenzo JM (2020) Red beetroot. A potential source of natural additives for the meat industry. Appl Sci 10(23):8340
70. Tran TN, Athanassiou A, Basit A, Bayer IS (2017) Starch-based bio-elastomers functionalized with red beetroot natural antioxidant. Food Chem 216:324–333
71. Chauhan S, Rajput H (2018) Production of gluten free and high fiber cookies using beet root waste powder and wheat flour husk. Pharma Innov J 7(11):556
72. Manopriya M, Arivuchudar R (2019) Cakes enriched with Beta vulgaris pomace powder: development, sensory evaluation, proximate composition and total antioxidant evaluation. Adalya J 8(11):688–694
73. Elleuch M, Bedigian D, Roiseux O, Besbes S, Blecker C, Attia H (2011) Dietary fibre and fibre-rich by-products of food processing: characterisation, technological functionality and commercial applications: a review. Food Chem 124(2):411–421

Chapter 4
Mango Peels as a Source of Nutraceuticals

Intan Soraya Che Sulaiman, Azham Mohamad, and Isharudin Md. Isa

4.1 Introduction

Mango (*Mangifera indica* L.), known as the king of fruits is among popular fruit production in the world market [1, 2]. Considered native to India and Southeast Asia [3], mango is a fruit of excellent antioxidant properties due to its good potential as a free radical scavenger [4, 5]. Mango fruit and its by-product contain an abundant source of beneficial compounds such as polyphenols, carotenoids, dietary fiber and vitamin E [3, 5, 6]. Generally, processing industries produce juice or nectar, pulp and jams, then discard a large amount of peel, seed, and seed kernel (stony pit) as bio-wastes [7, 8]. About 15–20% of the total weight of fresh mango represents by its peel [9, 10]. Likewise, peel wastes from the fruit may contain beneficial properties similar to that generally found in fruit [11]. As one of the significant by-products of mango fruit, the peels rich in health-enhancing constituents, particularly phenolic compounds, can be incorporated into nutraceutical, pharmaceutical, and functional food products [2, 11]. Figure 4.1 shows different features of mango.

Scientific researches reported that mango peel contains mangiferin, pectin, anthocyanins, β-carotene, gallic acid, galloyl glucose, and lutein [12–14]. Comparative studies on mango peel indicate that higher total polyphenol content is found in the ripened than the unripe peel [3, 5]. Other studies by Patiño-Rodríguez

I. S. Che Sulaiman
Faculty of Applied Sciences, Universiti Teknologi Mara, Arau, Perlis, Malaysia

A. Mohamad (✉)
Centre of Foundation Studies for Agricultural Science, Universiti Putra Malaysia, Serdang, Selangor, Malaysia
e-mail: azham@upm.edu.my

I. Md. Isa
Department of Land Management, Faculty of Agriculture, Universiti Putra Malaysia, Serdang, Selangor, Malaysia

© The Author(s), under exclusive license to Springer Nature Switzerland AG 2022

C. Egbuna et al. (eds.), *Food and Agricultural Byproducts as Important Source of Valuable Nutraceuticals*, https://doi.org/10.1007/978-3-030-98760-2_4

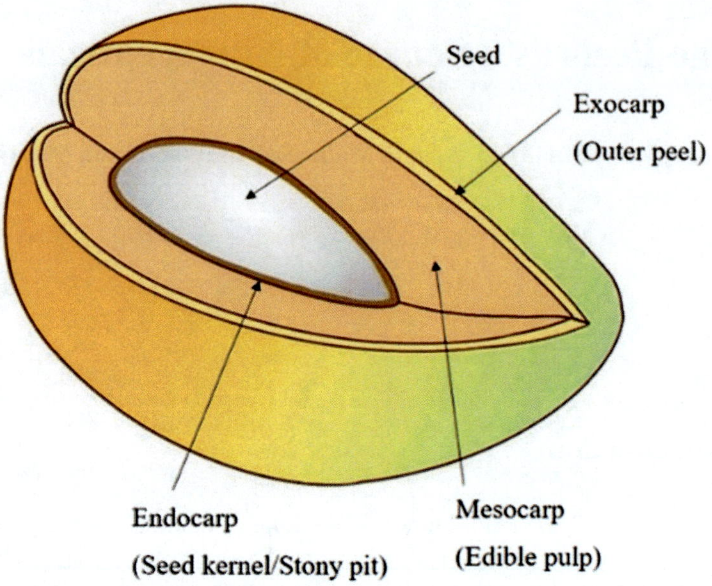

Fig. 4.1 Detail of different fractions of mango

et al. reported that unripe mango pulp had a higher polyphenol content than unripe mango peel [15]. Because mango is a climacteric fruit, the appearance of mango in terms of size, color, taste, and quality can be influenced by temperature [1, 5]. Generally, the quality and quantity of bioactive composition in mango peel variety may vary significantly due to soil condition, maturity stage, geographical site of growing, cultivar, farming practices and processing conditions [5, 16]. A study conducted by Alañón and co-workers reported that three different cultivars of mango peel from the Spanish region (Keitt, Osteen and Kent) showed higher phenolic content than the edible fraction [17]. Furthermore, Borrás-Enríquez and co-workers reported varied phenolic constituents values between cultivar of mangoes with Manililla mango peel recorded the highest values compared to Kent, Tommy Atkins and Manila [18].

Fruit processing industries generate by-products such as peel that retain high levels of underutilized bioactive substances [8]. Due to poor waste management, these uncommercialized resources were largely discarded into landfills, posing an environmental burden [2, 8, 19]. Hence, the reutilizing peel may reduce overflowing landfills and reveal their potential for therapeutic and nutraceuticals purposes [11, 20]. Stephen de Felice originally coined the term "nutraceuticals" in 1989, where he described the term as a hybrid category of products lying between "Nutrition" and "Pharmaceutical" [21, 22]. The term includes all food products that could provide nutrition and health benefits, along with deterrence and treatment of diseases. Nutraceuticals are not required for general practitioner prescription, work on general or multiple targeted actions, and contain less concentration of active compounds. While, pharmaceuticals products aim for a specific target of the action, need for prescription and involve a thorough drug evaluation process [21, 23, 24].

4.2 Bioactive Constituents in Mango Peel

Peel and seed are significant mango by-products processing industries, accounting for approximately 35–60% of the fresh fruit weight [17, 25]. Mango peel has higher dietary fiber [7] and carbohydrates than mango pulp and seed [5]. As mentioned earlier, cultivar variability, maturation stages and employed extraction method widely attributed to bioactive availability, particularly phenolic compounds in mango peel [16, 26]. A plant biosynthesizes phenolic metabolites to adapt to environmental stressors such as a limitation on nutrient supply, temperature, water and light [24]. Because of its protective function in plants, these secondary metabolites are widely distributed in the outermost layer of the fruit (outer peel). Accumulation of pectins and sugars in mango peel acts as a barrier and poses a challenge for extracting polyphenolic compounds [27]. These phenolic compounds include phenolic acids, mono- and di-galloyl compounds, benzophenones, gallotannins and flavonoids [17].

Phenolics in plants can be classified into three groups: free, esterified, and bound phenolics. Free and esterified phenolics are phenolic compounds conjugated to low molecular mass components and sugars, extracted with solvolytic solvents such as ethanol, methanol, acetone and water. While, bound phenolics are released by alkaline or acid hydrolysis. Alkali treatment can break the ester bonds of the cell wall structure to release bound phenolics, whereas acid hydrolysis may break glycoside bounds to free the aglycones [28, 29]. Hence, selecting appropriate solvent extraction is crucial to ensure the complete extraction of the targeted phenolic compounds. In a study conducted by López-Cobo and co-workers, mango peels of all tested cultivars from Spain demonstrated a higher amount of free phenolic compounds than other mango fractions. The authors also reported that free phenolic compounds in mango peel were found at the highest concentration in Sensación cultivar than Osteen and Keitt. Concerning bound phenolic compounds, the highest concentrations were demonstrated in seed, followed by peel, seed kernel and pulp [26]. The bioactive constituents quantified in this study are citric acid, galloylglucose isomer I, methyl-digallate ester isomer, gallic acid, protocatechuic acid and ellagic acid, which are found to be predominant in all cultivars tested. Table 4.1 summarizes the bioactive constituents in mango peel. As one of the primary phenolic acids existing in mango fruit, it is reported that gallic acid is also present in other mango peel cultivars (Ataulfo, Kensington Pride, Tommy Atkins, Haden and Kent) from Mexico, Australia, and Ecuador [2, 8, 14, 16, 34]. However, the concentration of gallic acid diminished during the ripe to an overripe period for Keitt, Osteen and Kent cultivars [17]. Concerning the influence of maturation stages on mango peel composition, Alañón et al. study the changes of bioactive constituents during the ripening process for three cultivars from Spain [17]. Based on the findings, galloyl

Table 4.1 Summary of the bioactive constituents in mango peel

Cultivar	Origin	Maturity stage	Bioactive constituent	Extraction solvent/ratio (v/v)	Extraction method	Analyzer	Reference
Keitt	Spain	Ripe	Citric acid, 5-galloylquinic acid, galloylglucose isomer I, digalloylglucose, methylgallate, methyl-digallate ester isomer, gallic acid, galloylglucose, vanillic acid, protocatechuic acid, ellagic acid, p-hydroxybenzoic acid isomer I, cyanidin 3-*O*-β-D-galactopyranoside, 7-*O*-methylcyanidin 3-*O*-β-D-galactopyranoside, heptadecadienylresorcinol, heptadecenylresorcinol	(a) Methanol/ water (80:20) (free polar fraction) (b) Diethyl ether/ ethyl acetate (1:1) and methanol/ water (1:1) (bound polar fraction) (c) Dichloromethane (alk(en) ylresorcinols fraction)	Freeze-dried and ultrasonic	HPLC-DAD-ESI-QTOF-MS and GC-QTOF-MS	[26]
Osteen			Citric acid, galloylglucose isomer I, methylgallate, quercetin glucoside, methyl-digallate ester isomer, gallic acid, galloylglucose, p-hydroxybenzoic acid isomer I, protocatechuic acid, ellagic acid, cyanidin 3-*O*-β-D-galactopyranoside, 7-*O*-methylcyanidin 3-*O*-β-D-galactopyranoside. Heptadecadienylresorcinol, heptadecenylresorcinol				
Sensación			Citric acid, galloylglucose isomer I, methylgallate, quercetin glucoside, quercetin galactoside, methyl-digallate ester isomer, gallic acid, p-hydroxybenzoic acid isomer I, protocatechuic acid, ellagic acid, cyanidin 3-*O*-β-D-galactopyranoside, 7-*O*-methylcyanidin 3-*O*-β-D-galactopyranoside, heptadecadienylresorcinol, heptadecenylresorcinol				

Keitt	Spain	Green	Hexagalloyl glucose, galloyl glucose, digalloylquinic acid, gallic acid	Methanol/water (80:20)	Sonication	HPLC-DAD-Q-TOF-MS	[17]
		Ripe	Hexagalloyl glucose, galloyl glucose, digalloyl glucose, digalloylquinic acid, gallic acid				
		Overripe	Gallotannins, galloyl glucose, digalloyl glucose, catechin, gallic acid				
Osteen		Green	Gallotannins, galloyl glucose, digallic acid, gallic acid				
		Ripe	Gallotannins, galloyl glucose, digallic acid, gallic acid				
		Overripe	Gallotannins, galloyl glucose, digallic acid, catechin, gallic acid				
Kent		Green	Gallotannins, galloyl glucose, digalloylquinic acid, catechin, gallic acid				
		Ripe	Gallotannins, galloyl glucose, digalloylquinic acid, digalloyl glucose, gallic acid				
		Overripe	Gallotannins, galloyl glucose, digalloyl glucose, gallic acid				
Sugar	Colombia	NAD	β-Cryptoxanthin, lutein, β-carotene	Mixture of polar (acetone:methanol 2:1) and non-polar (hexane) solvents	Supercritical fluid	RP-UHPLC-DAD	[13]
Tainong and Jinhuang	China	Ripe	α-Carotene, lycopene, β-carotene, β-cryptoxanthin, zeaxanthin and lutein	Ethanol:hexane (3:2)	Liquid nitrogen	HPLC	[12]
Manililla	Mexico	Green	Mangiferin	Ethanol/water (50:50)	Ultrasound-assisted	UHPLC	[18]

(continued)

Table 4.1 (continued)

Cultivar	Origin	Maturity stage	Bioactive constituent	Extraction solvent/ratio (v/v)	Extraction method	Analyzer	Reference
Osteen	Spain	Ripe	3-Carene, eucarvone, citral, β-selienene, trans-β-carylophyllene, α-humulene, D-cadinene, and α-gurjunene, 2-pentylfural, furfural, β-damascenone, β-ionone	Water	HS–SPME	GC–MS	[30]
Ataulfo	Mexico	Ripe	Mangiferin, gallic acid, quercetin, ellagic acid, gallocatechin, catechin	NAD	Organic-aqueous extraction	HPLC-DAD/MS	[14]
Ataulfo	Mexico	Ripe	Galloyl glucose, methyl digallate ester, methyl gallate	Methanolic extraction	Lyophilized and sonication	HPLC	[29]
Tommy Atkins	Ecuador	NAD	Violaxanthin, lutein, β-carotene, α-tocopherol	Hexane/acetone (1:1) (carotenoids and α-tocopherol)	Lyophilized and sonication	RRLC	[31]
			Procyanidin, (epi)afzelechin-(epi)catechin dimer, mangiferin, unidentified anthocyanin, quercetin galactoside, quercetin glucoside, quercetin arabinopyranoside, quercetin arabinofuranoside, quercetin rhamnoside	Methanol 70% (phenolic compound)		HPLC–DAD–ESI–MS^n	
Kesar	India	Ripe	Mangiferin	Lactic acid/sodium acetate/water (3:1:4)	Microwave-assisted extraction	HPLC	[32]
Tommy Atkins	Brazil	NAD	Catechin, procyanidin A2, procyanidin B1, Kaempferol 3-glucoside, quercetin piranoside, quercetin 3-glucoside, rutin	Ethanol, ethyl acetate, hexane	Low-pressure extraction methods (Soxhlet, maceration, ultrasound-assisted extraction)	RP-HPLC/DAD/FD	[33]
			Procyanidin A2, catechin, cinnamic, trans-resveratrol, quercetin 3, rutin	CO_2	Supercritical fluid extraction		
			Catechin, procyanidin B1, trans-resveratrol, quercetin 3, rutin	CO_2 plus 5% ethanol			

Kensington Pride	Australia	Ripe	Lupeol, mangiferin, gallic acid, vanillic acid, caffeic acid, chlorogenic acid, ferulic acid	NAD	Freeze-dried	HPLC-DAD	[34]
Keitt	Australia	NAD	Gallic acid, 2-hydroxybenzoic acid, ellagic acid, ellagic acid glucoside, galloyl glucose, gallagic acid, cinnamic acid, 3-caffeoylquinic acid, chicoric acid, sinapine, verbascoside, *p*-coumaroyl tartaric acid, dihydroquercetin, procyanidin dimer B1, procyanidin trimer C1, hesperidin, isorhamnetin	30% ethanol	Lyophilization	LC-ESI-QTOF/MS	[2]
Kensington Pride			Gallic acid, 2-hydroxybenzoic acid, gallic acid 3-*O*-gallate, cinnamic acid, feruloyl tartaric acid, 3-caffeoylquinic acid, isoferulic acid, sinapic acid, xanthohumol, procyanidin dimer B1, myricetin, isorhamnetin				
Keitt and Kensington Pride	Australia	NAD	Gallic acid, protocatechuic acid, *p*-hydroxybenzoic acid, chlorogenic acid, caffeic acid, syringic acid, catechin, epicatechin gallate, quercetin-3-galactoside, quercetin-3-glucuronide, kaempferol-3-glucoside, quercetin, kaempferol			HPLC-PDA	
Ataulfo	Mexico	NAD	Galloyl glucose, gallic acid, maclurin-3-C-(2-*O*-galloyl)-ß-D-glucoside, mangiferin, ethyl gallate, mangiferin-6′-*O*-gallate, valoneic acid dilactone, ellagic acid, *m*-digallic acid methyl ester	Ethanol/water (1:1)	Freeze-dried	HPLC-DAD	[8]
Tommy Atkins and Haden	Ecuador	NAD	Mangiferin, iso mangiferin, quercetin 3-*O*-galactoside, quercetin 3-*O*-glucoside, quercetin 3-*O*-xyloside, quercetin 3-*O*-arabinoside, quercetin, kaempferol	80% acetone	Supercritical antisolvent extraction	HPLC	[27]

(continued)

Table 4.1 (continued)

Cultivar	Origin	Maturity stage	Bioactive constituent	Extraction solvent/ratio (v/v)	Extraction method	Analyzer	Reference
Kensington Pride	Australia	Ripe	Gallic acid, caftaric acid, chlorogenic acid, *p*-hydroxybenzoic acid, caffeic acid, syringic acid, coumaric acid, ferulic acid, sinapinic acid	70% ethanol	Freeze-dried	LC-ESI-QTOF-MS/MS and HPLC-PDA	[35]
Langra, Amrapali, Hamlet, Bombay No. 1	India	Ripe	Violaxanthin, β-carotene, luteoxanthin, cyanidin-3-*O*-monoglucoside, pelargonidin 3-(feruloyl) diglucoside, cyanidin-3-*O*-acetylglucoside	1% acidic methanol, Milli Q water	Liquid nitrogen	UPLC	[36]
Arka Anmol, Lazzat Baksh, Peach, Banganapalli			Lutein, β-carotene, cyanidin-3-*O*-(6-*O*-acetyl)-5-*O*-diglucoside, petunidin-3-*O*-glucoside, trans-cyanidin-3-*O*-(6-*O*-*p*-coumaryl)-glucoside				
Tommy Atkins, Gulabi, Lalmuni, Janardhan Pasand			Luteoxanthin, β-carotene, cyanidin-3-*O*-monoglucoside, pelargonidin 3-(acyl)glucoside, delphinidin-3-*O*-(6-*O*-acetyl)-glucoside, petunidin-3-*O*-glucoside, peonidin-3-*O*-glucoside				
Tommy Atkins	Ecuador	Ripe	Gallic acid, rutin, mangiferin, β-carotene	Acetone/water (70:30)	Lyophilized	HPLC-UV/VIS and UPLC-PDA	[16]
Haden			Gallic acid, rutin, β-carotene				
Kent			Gallic acid, rutin, β-carotene				

NAD not appropriately described

glucose is the most abundant compound in all tested cultivars at all maturation stages (green, ripe and overripe). The Green and ripe mango peel of Keitt and Osteen contain high concentrations of galloyl glucose compared to its overripe sample. A similar pattern was observed for galloyl glucose, whereas the concentration of the bioactive decreased throughout the process of ripening to the overripe stage. Previously, galloyl glucose was also reported in mango pulp, seed and seed kernel but at a low concentration [37].

Mangiferin is a bioactive compound mainly isolated from various parts of the mango tree, including fruit peels [38]. A study by Meneses et al. evaluated mangiferin in the antioxidant extract by supercritical antisolvent extraction in two cultivars (Tommy Atkins and Haden) from Ecuador [27]. The recovery percentage was about 90% of the initial phenolic compounds, and the product obtained was micronized particles with high bioavailability. Several researchers had reported that the concentration of mangiferin was peaked at the ripe stage and decreased when the ripening process continued [18, 39]. Anthocyanin is a prominent plant phenolic pigment widely distributed in the plant kingdom to cause red, blue or purple color in fruits and vegetables [40]. Compared to the color variation of mango peels, red colored mango cultivar (Tommy Atkins, Gulabi, Lalmuni and Janardhan Pasand) showed the highest anthocyanin content than yellow and green colored types [36]. This could be due to the highest content of cyanidin-3-*O*-monoglucoside, pelargonidin 3-(acyl)glucoside, delphinidin-3-*O*-(6-*O*-acetyl)-glucoside, petunidin-3-*O*-glucoside and peonidin-3-*O*-glucoside in the red types as compared to others [36].

Apart from phenolic sources, mango peel is also a rich source of carotenoid [12, 13]. Carotenoids in mango contribute to the peel and flesh colors and their related concentration changes concerning fruit maturity stages. Liang et al. reported the presence of carotenoid contents (α-carotene, lycopene, β-carotene, β-cryptoxanthin, zeaxanthin and lutein) in the peel of two mango cultivars Jinhuang and Tainong, from China [12]. Lutein and β-carotene were previously reported in ripe mango cultivar Sugar from Colombia and four cultivars (Arka Anmol, Lazzat Baksh, Peach, Banganapalli) from India [13, 36]. The ripened mango demonstrated a higher carotenoid concentration than the unripe one [5]. β-carotene is a precursor to vitamin A has recently been incorporated in food products to improve shelf-life, attributed to its nutritional properties such as colorant and antioxidant properties [41].

4.3 Pharmacological Effects of Mango Peel

Numerous reports have documented the bioactivities of mango peel, including its antioxidant [16, 18, 32], anti-diabetic [42, 43], antimicrobial [44], antiproliferative [45], hyperlipidemia and lipid peroxidation properties [46]. Figure 4.2 illustrates the pharmacological studies on mango peel.

Studies on the antioxidant properties of mango peel are well documented in the works of literature. Regular consumption of antioxidant sources showed remarkable effects by preventing oxidative stresses, a root cause of tissue damage and chronic

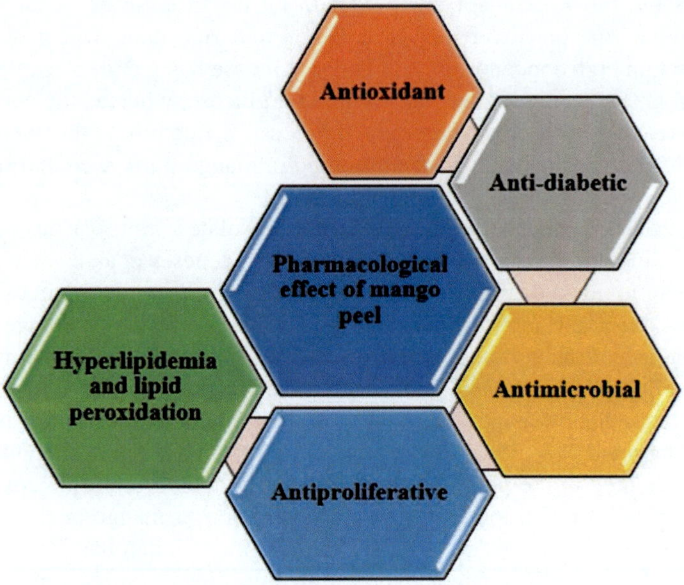

Fig. 4.2 Pharmacological studies on mango peel [9, 11, 16, 43–48]

diseases [11]. Recently, Marcillo-Parra and co-workers studied the antioxidant activity of mango peel in three different varieties (Tommy Atkins, Kent and Haden) from the Ecuadorian region [16]. The results showed significant differences among the varieties in antioxidant activity and total phenolic, flavonoid and carotenoid content. Haden cultivar had the highest total phenolic content, followed by Kent and Tommy Atkins. A correlation between the ABTS radical-scavenging activity and phenolic content was observed for all samples. Sánchez-Camargo et al. studied the effect of extraction conditions using microwave-assisted extraction on the antioxidant activity of mango peel from supercritical fluid extraction [9]. The antioxidant activity measured by trolox equivalent has shown to be influenced by microwave power and in correlation with a phenolic concentration in the extract. This antioxidant effect can be attributed to mangiferin, quercetin, and gallic acid quantified in the extract.

Diabetes is a metabolic disorder related to high glucose levels due to less production of insulin by the pancreas or reduced cell response to insulin [42]. Rodríguez-González studied the mechanisms associated with the anti-diabetic properties of mango juice by-products (peel and remnant pulp) [43]. The results showed that these by-products decreased serum glucose in streptozotocin-induced diabetic rats. Besides, the tested by-products closely mimic the insulin effects in 3T3-L1 adipocyte cells, enhancing Glut4, Irs1 and Pi3k expression. The finding suggests that the rich content of soluble fiber, carotenoids and polyphenols in mango peel and remnant pulp have contributed to its anti-diabetic properties. *In vitro* anti-diabetic properties of mango peel were studied by examining the inhibition of α-amylase and α-glucosidase activities of the peel extract [47]. From the results, mango peel extract demonstrated inhibition against α-amylase and α-glucosidase activities in a dose-dependent manner with IC_{50} values of 4 and 3.5 μg/mL, respectively. These results indicate the potential of extract as an anti-diabetic agent due to its ability to control blood glucose levels by inhibiting enzymes α-amylase and α-glucosidase. The absorption rate of glucose was reduced through delayed carbohydrate digestion and extended digestion time.

Further *in vivo* studies were conducted to determine the effect of mango peel extract on oral glucose tolerance, plasma insulin level and basic diabetic parameters in rats [47]. The blood glucose levels for three different doses of peel extracts (100, 150 and 200 mg/kg body weight) were recorded between 0 and 150 min. A significant reduction in glucose levels was observed in all treated groups with the higher reducing glucose level at 150 min. It is proposed that increased glucose tolerance levels were attributable to the mango peel's hypoglycemic activity and increased insulin secretion by time and concentration. Furthermore, mango peel extract in a dose-dependent manner significantly decreased fasting glucose level, fructosamine and glycated hemoglobin levels, while plasma insulin level increased when treated to streptozotocin-induced diabetic rats.

The antimicrobial activity of mango peel has been tested against several bacteria and fungi. Three different extracts of mango peel were evaluated against bacteria *Escherichia coli*, *Shigella species*, *Salmonella typhi* and *Enterobacter*, while its antifungal activity was tested against *Aspergillus niger*. In the study, acetone extract exhibited intense antibacterial activity against all tested bacterial within the diameter of inhibition zone 13–27 mm. However, no bactericidal effect was observed for aqueous extract. Aqueous and ethanol extracts demonstrated the highest antifungal activity against *Aspergillus niger* strain for the antifungal assay, while no antifungal activity was observed in acetone extract [44].

Free and bound (alkaline and acid releasable phenolics) polyphenols from mango Ataulfo peel have demonstrated the antiproliferative effectiveness against colon cancer cells [45]. A fraction of alkaline releasable phenolics in the study showed the most vigorous antiproliferative activity against human colon adenocarcinoma cells (IC_{50} = 138.2 μg/mL) and mouse connective cells (IC_{50} = 93.5 μg/mL). The quantification of gallic acid by UPLC-DAD revealed that alkaline releasable phenolics fraction contain the highest concentration of gallic acid and antioxidant activity compared to acid releasable phenolics fraction. In contrast, no gallic acid was detected in the free phenolics fraction. Because of gallic acid acting by a single electron transfer (SET) mechanism and a major compound in the bound polyphenols, these polyphenols' antioxidant and antiproliferative activity could be closely related to its SET mechanism.

Mango peel has been shown in *in-vivo* research to inhibit hyperlipidemia and lipid peroxidation induced by a hyperlipidemic diet in rats [46]. The experiments were carried out by including mango peels in a hyperlipidemic diet at 12% as a replacement basis. From the results, mango peels supplemented diet showed a decreasing concentration of serum thiobarbituric acid reactive substances (TBARS) value, indicating an increase of tested rats' resistance against oxidative stress. Because TBARS is a biomarker of oxidation in the body, decreasing the concentration of TBARS in blood demonstrates an improvement of antioxidant level in the samples tested. The authors also reported that the rat group supplemented with mango peels showed a reduction in liver cholesterol concentration (7.99 μmol/g) compared to the control (15.65 μmol/g). The tested group supplemented with mango peel displayed a reduction in glucose level and improvement in lipid profile than the control group. The soluble fiber content in mango peel could decrease glucose level by delayed gastric emptying rate and reduced macronutrient absorption [48].

4.4 Nutraceutical Applications

Over than a thousand varieties of mango are produced worldwide. They were greatly valued for their pleasant fragrance, great flavor and taste, and various functional properties because mango peel wastes contain similar beneficial properties generally found in the fruit. Many scientific research is being carried out on the formulation of nutraceutical products on mango peel such as food hydrocolloid [49, 50], food fortification [51, 52], edible film and coating [19, 53] and probiotic and prebiotic products [54]. The nutraceutical products based on mango peel are illustrated in Fig. 4.3.

Pectin is a food hydrocolloid which abundantly found in plant processing waste. In mango peels, pectin represents approximately 20–30% of the total peel weight of the fruit [49]. This compound can be called structural carbohydrates, acting as

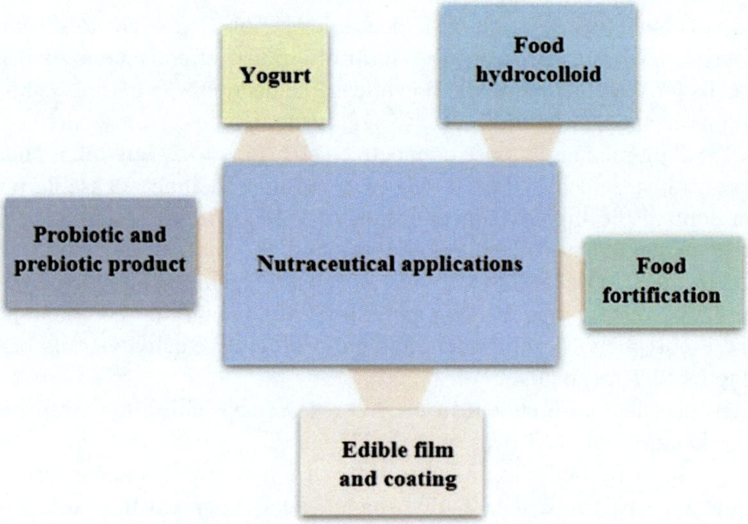

Fig. 4.3 Nutraceutical applications of mango peel. (Sources: Own study based on [19, 49, 52–58])

soluble dietary fiber in living plants. Pectin helps adherence and strengthens the plant tissues by limiting cell wall porosity [55]. Recently, pectin has been widely used in foods for antioxidant-fortified stabilizers in fruit drinks, gelling and thickening agents, fat replacers, drug delivery carriers, and acidified milk drinks [57]. Banerjee et al. [49] studied lemon juice's effect based on extraction of pectin from mango peel assisted by sonification. According to the authors, the highest pectin was obtained at 80 °C through sonication in the presence of lemon juice. Lemon juice serves as a natural acidifying agent to produce a low degree of esterification (DE) pectin (≤50%). Unlike commercial pectin with a higher DE value (>60%) that needs sugar for the gelation process, calcium is needed instead of gelling agent in low DE pectin. This nutraceutical property is essential for formulating low calorie and low-fat products.

Fortification of mango peel in biscuit formulation was reported by Salehi [51]. Adding 10% mango peel powder in the formulation yields satisfactory biscuits with mango flavor and improved nutritional value, including antioxidant properties. This is attributed to the content of dietary fiber, polyphenols, carotenoids and vitamins that increased water binding and holding capacity and thickening and gelling properties. Another work by Mayo-Mayo et al. [52] examined mango peel fortified tortilla chips to lower the glycemic index and enhance the functional value of the snack as prevention of health risks associated with obesity. The addition of mango peel to nixtamalized maize flour enhanced dietary fiber content and functional properties and reduced the glycemic index of tortilla chips. Low glycemic index food may have a fewer effect on blood sugar levels.

The application of mango peel as a promising biopolymer for film formulation or edible film for food packaging is attributed to their pectin composition. Pectin not only functions as a food hydrocolloid, but also as a film-forming agent [53]. Currently, thin film has been employed in food packaging to preserve the quality and freshness of the product, protect from environmental threats and improve shelf-life. Chaiwarit et al. [53] studied thin film fabricated from a combination of different polymers to improve film functionality and properties. The composite film fabricated with the low methoxyl pectin: mango peel pectin ratio of 1:2 with 40% (w/w) glycerol displayed appropriate tensile properties for pharmaceutical delivery. The resultant film displayed a higher percentage of elongation and Young's modulus values due to synergistic effects of the combination polymers. Another work by Torres-León et al. [19] investigated the formulation of edible coatings from mango peel and added with antioxidant extract of mango seed to improve the gas transfer rate of peach. Based on the findings, peach coated with a mixture of mango peel powder and antioxidant extract of mango seed kernel with glycerol showed significant reductions in ethylene and carbon dioxide production and consumed less oxygen than uncoated peaches. The production of ethylene during storage is associated with fruit senescence, thus delaying ethylene production and controlling gases transfer of CO_2 and O_2 are vital factors for controlling postharvest behavior and prolonging the freshness of the fruit.

Both prebiotics and probiotics have different roles that are important for human health [54]. By definition, probiotics are live microorganisms also known as gut microbiota, such as *Lactobacillus* and *Bifidobacterium* that live in the gastrointestinal tract to provide health benefits to the host. While, prebiotics are carbohydrates, mostly fiber that is non-digestible by humans and needs to digest by beneficial intestinal microbes (gut microbiota) for health-promoting benefits [54]. Recently, Hayayumi-Valdivia and co-workers [55] had investigated the effect of microencapsulation with the addition of mango peel powder on probiotics survival in ice cream. Results showed that the microencapsulation technique improved the viability of probiotics and formulation with 2% mango peel powder and achieved satisfactory results during the storage period (180 days). This is possibly because microencapsulated microorganisms of formulations take longer to decrease the logarithmic cycle during storage compared to free microorganisms of formulations with no encapsulation. The presence of mango peel powder increases the viability of *Lactobacillus acidophilus* and *Bifidobacterium lactis*, including the viscosity of the ice cream. However, it reduced its overrun and overall acceptability. A similar study reported by Ayar et al. suggested that dietary fiber from fruit processing by-products could be used in ice cream to enhance the survival of the probiotic strains with satisfactory sensory properties [58].

Perez-Chabela et al. [59] designed mango peel formulation as bioactive ingredients in functional yogurt. In the study, yogurt enriched with mango peel displayed lower syneresis and higher viscosity during the storage period than the control. Also, added mango peel formulation demonstrated higher titratable acidity and experienced a decrease in pH. A positive value of its prebiotic activity index scores (0.25) indicated the ability of this formulation to support the growth of probiotic microorganisms.

4.5 Conclusion

This chapter highlights the importance of mango peels as a promising source of nutraceuticals. Mango peel consists of various potential bioactive constituents such as polyphenols, carotenoids, pectin, and anthocyanin, which benefit humans to prevent and treat various diseases. Many factors significantly influence the mango peel composition, including cultivar, growing geographical site, maturity stage and many more. Also, choosing an appropriate extraction technique and an effective solvent is crucial because it significantly influences the extraction yield of compounds of interest.

References

1. Mandal V, Kundu S, Barman J, Adhikary R (2021) Harvesting strategy for different mango varieties based on comparative sugar and phenol contents. Proc Natl Acad Sci India Sect B Biol Sci 91(1):1–11. https://doi.org/10.1007/s40011-020-01188-w
2. Peng D, Zahid HF, Ajlouni S, Dunshea FR, Suleria HAR (2019) LC-ESI-QTOF/MS profiling of Australian mango peel by-product polyphenols and their potential antioxidant activities. Processes 7(10):764. https://www.mdpi.com/2227-9717/7/10/764
3. Lauricella M, Emanuele S, Calvaruso G, Giuliano M, D'Anneo A (2017) Multifaceted health benefits of Mangifera indica L. (mango): the inestimable value of orchards recently planted in Sicilian rural areas. Nutrients 9(5):525
4. Gentile C, Di Gregorio E, Di Stefano V, Mannino G, Perrone A, Avellone G et al (2019) Food quality and nutraceutical value of nine cultivars of mango (Mangifera indica L.) fruits grown in Mediterranean subtropical environment. Food Chem 277:471–479. https://doi.org/10.1016/j.foodchem.2018.10.109
5. Lebaka VR, Wee YJ, Ye W, Korivi M (2021) Nutritional composition and bioactive compounds in three different parts of mango fruit. Int J Environ Res Public Health 18(2):1–20
6. Ashok DA, Ravivarman J, Kayalvizhi DK (2020) Nutraceutical properties of recommended horticultural crops to develop human immune system against COVID-19. Int J Chem Stud 8(4):105–112
7. Wall-Medrano A, Olivas-Aguirre FJ, Ayala-Zavala JF, Domínguez-Avila JA, Gonzalez-Aguilar GA, Herrera-Cazares LA et al (2020) Health benefits of mango by-products. In: Food wastes and by-products. Wiley, New York, pp 159–191. https://onlinelibrary.wiley.com/doi/10.1002/9781119534167.ch6
8. de Ancos B, Sánchez-Moreno C, Zacarías L, Rodrigo MJ, Sáyago Ayerdí S, Blancas Benítez FJ et al (2018) Effects of two different drying methods (freeze-drying and hot air-drying) on the phenolic and carotenoid profile of 'Ataulfo' mango by-products. J Food Meas Charact 12(3):2145–2157. https://doi.org/10.1007/s11694-018-9830-4
9. Sánchez-Camargo AdP, Ballesteros-Vivas D, Buelvas-Puello LM, Martinez-Correa HA, Parada-Alfonso F, Cifuentes A et al (2021) Microwave-assisted extraction of phenolic compounds with antioxidant and anti-proliferative activities from supercritical CO2 pre-extracted mango peel as valorization strategy. LWT 137:110414. https://linkinghub.elsevier.com/retrieve/pii/S002364382031402X
10. Serna-Cock L, García-Gonzales E, Torres-León C (2016) Agro-industrial potential of the mango peel based on its nutritional and functional properties. Food Rev Int 32(4):364–376. http://www.tandfonline.com/doi/full/10.1080/87559129.2015.1094815
11. Dike CS, Orish CN, Nwokocha CR, Sikoki FD, Babatunde BB, Frazzoli C et al (2021) Phytowaste as nutraceuticals in boosting public health. Clin Phytosci 7(1):24
12. Liang M, Su X, Yang Z, Deng H, Yang Z, Liang R et al (2020) Carotenoid composition and expression of carotenogenic genes in the peel and pulp of commercial mango fruit cultivars. Sci Hortic (Amsterdam) 263:109072. https://doi.org/10.1016/j.scienta.2019.109072
13. Sánchez-Camargo AdP, Gutiérrez L, Vargas SM, Martinez-Correa HA, Parada-Alfonso F, Narváez-Cuenca C-E (2019) Valorisation of mango peel: proximate composition, supercritical fluid extraction of carotenoids, and application as an antioxidant additive for an edible oil. J Supercrit Fluids 152:104574. https://doi.org/10.1016/j.supflu.2019.104574
14. Bertha C, Alberto S-BJ, Tovar J, Sáyago-Ayerdi SG, Zamora-Gasga VM (2019) In vitro gastrointestinal digestion of mango by-product snacks: potential absorption of polyphenols and antioxidant capacity. Int J Food Sci Technol 54(11):3091–3098. https://onlinelibrary.wiley.com/doi/10.1111/ijfs.14224
15. Patiño-Rodríguez O, Bello-Pérez LA, Agama-Acevedo E, Pacheco-Vargas G (2020) Pulp and peel of unripe stenospermocarpic mango (Mangifera indica L. cv Ataulfo) as an alternative source of starch, polyphenols and dietary fibre. Food Res Int 138:109719. https://doi.org/10.1016/j.foodres.2020.109719

16. Marcillo-Parra V, Anaguano M, Molina M, Tupuna-Yerovi DS, Ruales J (2021) Characterization and quantification of bioactive compounds and antioxidant activity in three different varieties of mango (Mangifera indica L.) peel from the Ecuadorian region using HPLC-UV/VIS and UPLC-PDA. NFS J 23:1–7. https://doi.org/10.1016/j.nfs.2021.02.001
17. Alañón ME, Pimentel-Moral S, Arráez-Román D, Segura-Carretero A (2021) Profiling phenolic compounds in underutilized mango peel by-products from cultivars grown in Spanish subtropical climate over maturation course. Food Res Int 140:109852. https://doi.org/10.1016/j.foodres.2020.109852
18. Borrás-Enríquez AJ, Reyes-Ventura E, Villanueva-Rodríguez SJ, Moreno-Vilet L (2021) Effect of ultrasound-assisted extraction parameters on total polyphenols and its antioxidant activity from mango residues (Mangifera indica L. var. manililla). Separations 8(7):94
19. Torres-León C, Vicente AA, Flores-López ML, Rojas R, Serna-Cock L, Alvarez-Pérez OB et al (2018) Edible films and coatings based on mango (var. Ataulfo) by-products to improve gas transfer rate of peach. Food Sci Technol 97:624–631. https://doi.org/10.1016/j.lwt.2018.07.057
20. Gowe C (2015) Review on potential use of fruit and vegetables by-products as a valuable source of natural food additives. Food Sci Qual Manag 45:47–61. www.iiste.org
21. Maurya AP, Chauhan J, Yadav DK, Gangwar R, Maurya VK (2021) Nutraceuticals and their impact on human health. In: Preparation of phytopharmaceuticals for the management of disorders. Elsevier, Amsterdam, pp 229–254. https://doi.org/10.1016/B978-0-12-820284-5.00011-3
22. Télessy IG (2018) Nutraceuticals. In: The role of functional food security in global health. Academic, New York, pp 409–421
23. Kumar K (2020) Nutraceutical potential and utilization aspects of food industry by-products and wastes. In: Food industry wastes. Elsevier, Amsterdam, pp 89–111. https://linkinghub.elsevier.com/retrieve/pii/B978012817121900005X
24. Che Sulaiman IS, Mohamad A, Ahmed OH (2021) Murdannia loriformis: a review of ethnomedicinal uses, phytochemistry, pharmacology, contemporary application, and toxicology. Evid Based Complement Altern Med 2021:1–15. https://www.hindawi.com/journals/ecam/2021/9976202/
25. Jahurul MHA, Zaidul ISM, Ghafoor K, Al-Juhaimi FY, Nyam KL, Norulaini NAN et al (2015) Mango (Mangifera indica L.) by-products and their valuable components: a review. Food Chem 183:173–180. https://doi.org/10.1016/j.foodchem.2015.03.046
26. López-Cobo A, Verardo V, Diaz-de-Cerio E, Segura-Carretero A, Fernández-Gutiérrez A, Gómez-Caravaca AM (2017) Use of HPLC- and GC-QTOF to determine hydrophilic and lipophilic phenols in mango fruit (Mangifera indica L.) and its by-products. Food Res Int 100:423–434. https://doi.org/10.1016/j.foodres.2017.02.008
27. Meneses MA, Caputo G, Scognamiglio M, Reverchon E, Adami R (2015) Antioxidant phenolic compounds recovery from Mangifera indica L. by-products by supercritical antisolvent extraction. J Food Eng 163:45–53. https://doi.org/10.1016/j.jfoodeng.2015.04.025
28. Yu M, Yang L, Xue Q, Yin P, Sun L, Liu Y (2019) Comparison of free, esterified, and insoluble-bound phenolics and their bioactivities in three organs of Lonicera japonica and L. macranthoides. Molecules 24(5):970. https://www.mdpi.com/1420-3049/24/5/970
29. Pacheco-Ordaz R, Antunes-Ricardo M, Gutiérrez-Uribe J, González-Aguilar G (2018) Intestinal permeability and cellular antioxidant activity of phenolic compounds from mango (Mangifera indica cv. Ataulfo) peels. Int J Mol Sci 19(2):514. http://www.mdpi.com/1422-0067/19/2/514
30. Oliver-Simancas R, Díaz-Maroto MC, Pérez-Coello MS, Alañón ME (2020) Viability of pre-treatment drying methods on mango peel by-products to preserve flavouring active compounds for its revalorisation. J Food Eng 279:109953. https://linkinghub.elsevier.com/retrieve/pii/S0260877420300510
31. Ruales J, Baenas N, Moreno DA, Stinco CM, Meléndez-Martínez AJ, García-Ruiz A (2018) Biological active Ecuadorian mango 'Tommy Atkins' ingredients—an opportunity to reduce agrowaste. Nutrients 10(9):1138. http://www.mdpi.com/2072-6643/10/9/1138

32. Pal CBT, Jadeja GC (2020) Microwave-assisted extraction for recovery of polyphenolic antioxidants from ripe mango (Mangifera indica L.) peel using lactic acid/sodium acetate deep eutectic mixtures. Food Sci Technol Int 26(1):78–92. http://journals.sagepub.com/doi/10.1177/1082013219870010
33. Souza MEAO, Mezzomo N, Correa LC, Lima MS, Azevêdo LC, Ferreira SRS (2019) Recovery of antioxidant compounds from mango peel by green extraction processes. Int Food Res J 26(2):1845–1859
34. Vithana MDK, Singh Z, Johnson SK (2018) Cold storage temperatures and durations affect the concentrations of lupeol, mangiferin, phenolic acids and other health-promoting compounds in the pulp and peel of ripe mango fruit. Postharvest Biol Technol 139:91–98. https://doi.org/10.1016/j.postharvbio.2017.12.003
35. Suleria HAR, Barrow CJ, Dunshea FR (2020) Screening and characterization of phenolic compounds and their antioxidant capacity in different fruit peels. Foods 9(9):1206. https://www.mdpi.com/2304-8158/9/9/1206
36. Ranganath KG, Shivashankara KS, Roy TK, Dinesh MR, Geetha GA, Pavithra KCG et al (2018) Profiling of anthocyanins and carotenoids in fruit peel of different colored mango cultivars. J Food Sci Technol 55(11):4566–4577. https://doi.org/10.1007/s13197-018-3392-7
37. Alañón ME, Palomo I, Rodríguez L, Fuentes E, Arráez-Román D, Segura-Carretero A (2019) Antiplatelet activity of natural bioactive extracts from mango (Mangifera Indica L.) and its by-products. Antioxidants 8(11):517. https://www.mdpi.com/2076-3921/8/11/517
38. Du S, Liu H, Lei T, Xie X, Wang H, He X et al (2018) Mangiferin: an effective therapeutic agent against several disorders (review). Mol Med Rep 18:4775–4786
39. Ruiz-Montañez G, Ragazzo-Sánchez JA, Calderón-Santoyo M, Velázquez-De La Cruz G, Ramírez De León JA, Navarro-Ocaña A (2014) Evaluation of extraction methods for preparative scale obtention of mangiferin and lupeol from mango peels (Mangifera indica L.). Food Chem 159:267–272
40. Alvarez-Suarez JM, Cuadrado C, Redondo IB, Giampieri F, González-Paramás AM, Santos-Buelga C (2021) Novel approaches in anthocyanin research—plant fortification and bioavailability issues. Trends Food Sci Technol 117:92–105. https://linkinghub.elsevier.com/retrieve/pii/S0924224421000492
41. Socaci SA, Rugină DO, Diaconeasa ZM, Pop OL, Fărcaş AC, Păucean A et al (2017) Antioxidant compounds recovered from food wastes. In: Functional food—improve health through adequate food. InTech, Rijeka, pp 3–22. http://www.intechopen.com/books/functional-food-improve-health-through-adequate-food/antioxidant-compounds-recovered-from-food-wastes
42. Egbuna C, Awuchi CG, Kushwaha G, Rudrapal M, Patrick-Iwuanyanwu KC, Singh O, Odoh UE, Khan J, Jeevanandam J, Kumarasamy S, Chukwube VO, Narayanan M, Palai S, Găman MA, Uche CZ, Ogaji DS, Ezeofor NJ, Mtewa AG, Patrick-Iwuanyanwu CC, Kesh SS, Shivamallu C, Saravanan K, Tijjani H, Akram M, Ifemeje JC, Olisah MC, Chikwendu CJ (2021) Bioactive compounds effective against type 2 diabetes mellitus: a systematic review. Curr Top Med Chem 21(12):1067–1095. https://doi.org/10.2174/1568026621666210509161059. PMID: 33966619
43. Rodríguez-González S, Gutiérrez-Ruíz IM, Pérez-Ramírez IF, Mora O, Ramos-Gomez M, Reynoso-Camacho R (2017) Mechanisms related to the anti-diabetic properties of mango (Mangifera indica L.) juice by-product. J Funct Foods 37:190–199. https://linkinghub.elsevier.com/retrieve/pii/S1756464617304462
44. Thambi PA, John S, Lydia E, Iyer P, Monica SJ (2016) Antimicrobial efficacy of mango peel powder and formulation of recipes using mango peel powder (Mangifera indica L.). Int J Home Sci 2(2):155–161
45. Velderrain-Rodríguez G, Torres-Moreno H, Villegas-Ochoa M, Ayala-Zavala J, Robles-Zepeda R, Wall-Medrano A et al (2018) Gallic acid content and an antioxidant mechanism are responsible for the antiproliferative activity of 'Ataulfo' mango peel on LS180 cells. Molecules 23(3):695. http://www.mdpi.com/1420-3049/23/3/695

46. Umbreen H, Arshad MU, Noreen R, Aftab K (2020) Ameliorative effect of apple pomace and mango peels against hyperlipidemia and lipid peroxidation induced by hyperlipidemic diet. Sains Malaysiana 49(6):1273–1282. http://www.ukm.my/jsm/pdf_files/SM-PDF-49-6-2020/5.pdf
47. Gondi M, Prasada Rao UJS (2015) Ethanol extract of mango (Mangifera indica L.) peel inhibits α-amylase and α-glucosidase activities, and ameliorates diabetes related biochemical parameters in streptozotocin (STZ)-induced diabetic rats. J Food Sci Technol 52(12):7883–7893
48. Lattimer JM, Haub MD (2010) Effects of dietary fiber and its components on metabolic health. Nutrients. 2:1266–1289
49. Banerjee J, Vijayaraghavan R, Arora A, MacFarlane DR, Patti AF (2016) Lemon juice based extraction of pectin from mango peels: waste to wealth by sustainable approaches. ACS Sustain Chem Eng 4(11):5915–5920. https://pubs.acs.org/doi/10.1021/acssuschemeng.6b01342
50. Banerjee J, Singh R, Vijayaraghavan R, MacFarlane D, Patti AF, Arora A (2018) A hydrocolloid based biorefinery approach to the valorisation of mango peel waste. Food Hydrocoll 77:142–151. https://doi.org/10.1016/j.foodhyd.2017.09.029
51. Salehi F (2020) Recent applications of powdered fruits and vegetables as novel ingredients in biscuits: a review. Forum Nutr 45(1):1. https://nutrirejournal.biomedcentral.com/articles/10.1186/s41110-019-0103-8
52. Mayo-Mayo G, Navarrete-García A, Maldonado-Astudillo YI, Jiménez-Hernández J, Santiago-Ramos D, Arámbula-Villa G et al (2020) Addition of roselle and mango peel powder in tortilla chips: a strategy for increasing their functionality. J Food Meas Charact 14(3):1511–1519. https://doi.org/10.1007/s11694-020-00400-9
53. Chaiwarit T, Masavang S, Mahe J, Sommano S, Ruksiriwanich W, Brachais CH et al (2020) Mango (cv. Nam Dokmai) peel as a source of pectin and its potential use as a film-forming polymer. Food Hydrocoll 102:105611. https://doi.org/10.1016/j.foodhyd.2019.105611
54. Egbuna C, Tupas GD (eds) (2020) Functional foods and nutraceuticals—bioactive components, formulations and innovations, 1st edn. Springer Nature, Cham. 646 pp
55. Hayayumi-Valdivia M, Márquez-Villacorta LF, Pretell-Vásquez CC (2021) Effect of microencapsulation and mango peel powder on probiotics survival in ice cream. Braz J Food Technol 24:2–10. http://www.scielo.br/scielo.php?script=sci_arttext&pid=S1981-67232021000100418&tlng=en
56. Reichembach LH, de Oliveira L, Petkowicz C (2021) Pectins from alternative sources and uses beyond sweets and jellies: an overview. Food Hydrocoll 118:106824. https://doi.org/10.1016/j.foodhyd.2021.106824
57. Naqash F, Masoodi FA, Rather SA, Wani SM, Gani A (2017) Emerging concepts in the nutraceutical and functional properties of pectin—a review. Carbohydr Polym 168:227–239. https://doi.org/10.1016/j.carbpol.2017.03.058
58. Ayar A, Siçramaz H, Öztürk S, Öztürk YS (2018) Probiotic properties of ice creams produced with dietary fibres from by-products of the food industry. Int J Dairy Technol 71(1):174–182
59. Perez-Chabela MdL, Cebollon-Juarez A, Bosquez-Molina E, Totosaus A (2022) Mango peel flour and potato peel flour as bioactive ingredients in the formulation of functional yogurt. Food Sci Technol 42:e38220. http://www.scielo.br/scielo.php?script=sci_arttext&pid=S0101-20612021005010203&tlng=en

Chapter 5
Apple Pomace as a Source of Nutraceuticals

Shahira M. Ezzat, Maha Salama, Dina El Kersh, and Mohamed Salem

5.1 Introduction

Many wastes arising from the Food Supply Chain (FSC) constitute an environmental issue, causing severe socioeconomic impacts [1]. Food wastes can be classified into two major categories according to their origin: animal or fruit/vegetable wastes [2]. Animal wastes arise from dairy, meat, and fisheries processing, while, residues from vegetable and fruits processing produce the second category [3]. Logistic-related problems make the valorization of these wastes challenging [4].

Apple (*Malus domestica* Borkh., F. Rosaceae) is one of the earliest fruits known to humans. The consumption of apples is attributed to their high content of polyphenolic antioxidant compounds [5]. The world approximate production of apples in

S. M. Ezzat (✉)
Department of Pharmacognosy, Faculty of Pharmacy, Cairo University, Cairo, Egypt

Department of Pharmacognosy, Faculty of Pharmacy, October University for Modern Science and Arts (MSA), Giza, Egypt
e-mail: shahira.ezzat@pharma.cu.edu.eg; smelkomy@msa.eun.eg

M. Salama
Department of Pharmacognosy, Faculty of Pharmacy, Cairo University, Cairo, Egypt

Department of Pharmacognosy, Faculty of Pharmacy, The British University in Egypt, Cairo, Egypt
e-mail: maha.salama@pharma.cu.edu.eg; maha.salama@bue.edu.eg

D. El Kersh
Department of Pharmacognosy, Faculty of Pharmacy, The British University in Egypt, Cairo, Egypt
e-mail: dina.elkersh@bue.edu.eg

M. Salem
Department of Pharmacognosy, Faculty of Pharmacy, Menoufia University, Menoufia, Egypt
e-mail: mohamed.salem@phrm.menofia.edu.eg

© The Author(s), under exclusive license to Springer Nature Switzerland AG 2022
C. Egbuna et al. (eds.), *Food and Agricultural Byproducts as Important Source of Valuable Nutraceuticals*, https://doi.org/10.1007/978-3-030-98760-2_5

2019 reached 87 million tons, according to the Food and Agriculture Organization (FAO) [6]. Apple products, including juice, wine, jams and dried product, create about 25–30% by-products from skin, flesh, seeds and stems named apple pomace [7]. Apple pomace is the solid residue leftover after milling and pressing of apples during the making of apple juice, cider, or puree.

Millions of tons of apple pomace are generated annually that are collected as waste or used as animal feed. Waste from apple pomace represents a form of environmental pollution. The utilization of apple pomace as animal feed doesn't represent a protein-rich diet [8]. Additionally, the costs for apple pomace waste management is very high, while the recovery rate is relatively low [7, 9]. Composting of apple pomace creates secondary pollution due to greenhouse gasses production [7]. Moreover, composting creates surfaces for human disease vectors to breed, and it has the potential to contaminate ground waters [10]. Owing to the high content of dietary fibers and polyphenolic compounds, apple pomace has been included as an important nutraceutical as well as a functional ingredient in several food types [7, 9].

This chapter gives a clear overview of the possibilities for the utilization of apple pomace as a valuable nutraceutical. The chemical composition of apple pomace, its health benefits and safety are highlighted. The possibilities of using apple pomace as a functional ingredient in food products are discussed.

5.2 Chemical Profiling of Apple Pomace Extract

In apple juice manufacturing industry, the production of apple pomace as a waste by-product has attracted the attention for human use [11]. Apple pomace is made up of potentially health benefits ingredients classified as primary metabolites such as fatty acids, proteins and carbohydrates "simple and complex", fibres, minerals, and a myriad of secondary metabolites "triterpenoids and polyphenolics" [12] besides amygdalin, a cyanogenic glycoside is found predominantly in the seeds [13]. Apple pomace is comprised of heterogeneous parts mainly skin and minor parts of the flesh (95%), seeds (2–4%), and stems and calyx (1%). The variation of chemical profiling among apple pomace is controlled by variable factors *viz.* origin, variety, as well as type of processing [11]. Hereby in the following subsections, different classes of ingredients of apple pomace were discussed. The major identified phytochemicals as well as their analytical tool used for their identification and/or quantification have been illustrated in Fig. 5.1.

5.2.1 Primary Metabolites

Primary metabolites included within the apple pomace as proteins and fatty acids due to the presence of seeds raise its nutritional value [9]. **Unsaturated Fatty acids** (90%) in oil of apple seeds include mainly linoleic acid and oleic acids whereas

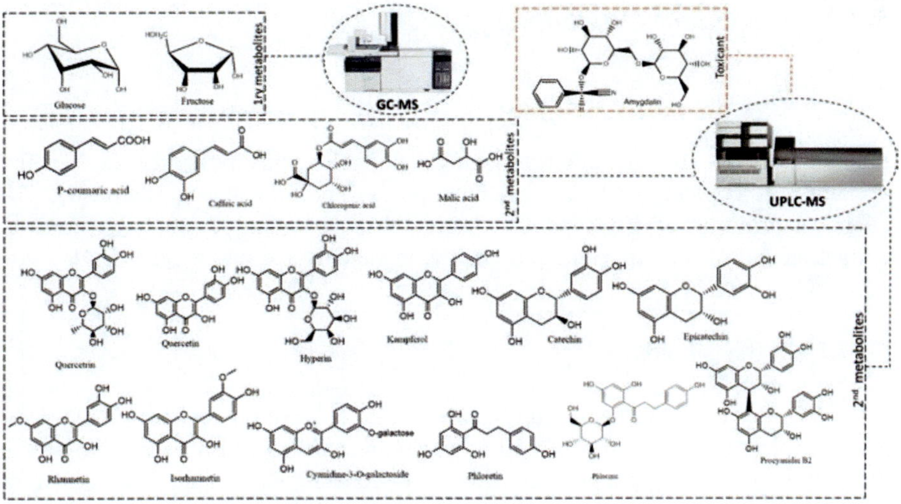

Fig. 5.1 Major metabolites in apple pomace

saturated fatty acids consist mainly of palmitic acid. **Tocopherols and phytosterols** also were detected in the apple seed oil [7]. **Simple carbohydrates** such as glucose and fructose have also been detected [12]. **Fibres** such as "pectin, cellulose and lignan" were found to be the major type of polysaccharides present in the apple pomace matrix. Its percentile ranged from 4.4 to 47.3 g fibres per 100 g pomace [12]. **Pectin** was found to be the major one (10–15%). It is a heteropolysaccharide that consists of galacturonic acid moieties plus neutral sugars as side chains. It is worthy of note that apple pomace and citrus peels are one of the major sources of pectin. Pectin quality varied as per the extraction method. As in Perussello et al. [11] pectin was extracted by different methods as conventional method "hot acidic solution", enzyme assisted technique, supercritical water method, microwave, ultrasound, electric field assisted; each method has its pros and cons aspects as listed in the review. Microwave, ultrasound and electric field assisted extraction yield pectin with unaltered chemical structure, better color due to better penetration to plant cells, whereas the conventional method is time consuming as it needs longer time for extraction. In enzymatic assisted extraction, selectivity was pronounced for pectin so better pectin extraction rate in less time contrary to conventional one. In supercritical water technique; better pectin solubility was found but this method is expensive [11]. Studies on the **minerals** contents in apple pomace has shown that potassium is the major micronutrient (398.4–880.2 mg/100 pomace) followed by sodium, phosphorus, and magnesium with variable levels. **Vitamin C and E** are also found in apple pomace ingredients which are well known being free radicle scavengers [12].

5.2.2 *Secondary Metabolites*

Secondary metabolites in functional foods are responsible for its health benefits. Many previous studies have identified different phytochemical classes in apple pomace extracts. They include triterpenoids, phenolic acids, polyphenolics *viz.* flavonoids, anthocyanins, dihydrochalcones [9], in addition to amygdalin [7]. The chemical profiling of apple pomace varies based on many factors *viz.* cultivar, extraction and squeezing techniques, extraction agents and temperature [14].

5.2.2.1 Triterpenoids

Triterpenoids class have been found to be a major class of phytochemicals in pomace of apples. Ursolic acid followed by oleanolic acid have also been identified in different varieties of apple pomace. Ursolic and oleanolic acids are well known for their antimicrobial as well as anti-inflammatory actions [9]. Skinner et al. [12] quantified the ursolic acid content in different skins of apple cultivars *viz.* Fuji, Gala, Smith, and Granny Smith. Smith apples skin proved to be the most enriched skin with ursolic acid (0.82 mg/cm^2) which could add to the pomace health benefits [12].

5.2.2.2 Phenolic Acids

In a previous study, the identification of **phenolic acids composition** was done by gas chromatography in Gala and Fuji cultivars apple pomace in Brazil [15]. The quantification of 16 different free, esterified, and insoluble bound phenolic acids was assessed which shows that benzoic acid, salicylic acid, *t*-cinnamic acid were found as major phenolic acids identified in both cultivars. It worth to mention Gala cultivars was enriched with more phenolic acids than that of Fuji one with total percentile 93.94 and 68.38 mg/g, respectively [15]. In a previous review article, mentioned a blend of major phenolics enriched in apple pomace exemplified in chlorogenic acid, protocatechuic acid, caffeic acid, *p*-coumaric acid, and ferulic acid which potentiate the antioxidant action of apple pomace [11].

5.2.2.3 Polyphenolics

A myriad of different flavonoids classes has been reported in many previous literatures of apple pomace viz. flavanols and flavonols as major flavonoid subclasses. Quercetin and its glycosides encountered the major flavonoid ingredient in apple pomace whereas the flavonoid content enriched in the skin so the pomace [12]. Fernandes et al. [16] identified the major flavonoids where quercetin was also the major aglycone in the pomace with its glycosides in agreement with Skinner et al. [12]. A blend of quercetin-*3-O*-hexosides and pentosides in parallel to quercetin-*3-O*-rutinoside and quercetin-*3-O*-deoxyhexoside were identified as well. Phloretin-*2-O*-glucoside was identified in apple pomace of royal Gala variety using ultra high performance-diode array detector coupled to mass spectroscopy (UHPLC-DAD-ESI-MS) [16]. Another literature recognized the polyphenolic profiling of apple pomace and identified different classes. Quercetin, Kaempferol, rhamnetin and isorhamnetin glycosides with epicatechin and procyanidin B2 phytochemicals. Dihydrochalcones as phloretin and phlorizin were also identified. Anthocyanins exemplified in cyanidin-*3-O*-galactoside was also assessed [9].

5.2.2.4 Amygdalin "Cyanogenic Glycoside"

Amygdalin, a cyanogenic glycoside presents in apple seeds, is reported as toxicant depending on its percentile to be metabolised by the human body. In apple pomace, the seeds accounts for 2–4% of total pomace content whereas the amygdalin content was quantified by maximum 4 mg/g still it is recommended that seeds better to be removed from the pomace of apples before further processing for the safety profile of the pomace [7]. In a previous study, amygdalin was quantified in seeds and prepared apple juice from 15 different varieties. Amygdalin content was ranged from 1 to 4 mg/g in all apple seeds varieties whereas the apple juices recorded much less lowered levels (0.01–0.08 mg/mL) due to the processing techniques of apple to prepare apple juices whereas this low percentile could be handled by human body [13].

5.3 Apple Pomace: Health Benefits and Safety

Many preclinical studies were performed on apple pomace which have proven that it can improve the lipid metabolism, gut health, body weight and glucose regulation in addition to its antioxidant effect [17].

A study done by Cho et al. found that rats fed high fat diet that consists of 8% lard and 7% soybean oil together with 10% apple pomace for 9 weeks could significantly decrease the body weight and percentage of body fat with the improved serum lipid profiles, reduced serum LDL-C and increased serum HDL-C. This effect is due to the high fiber content of the apple pomace which improved lipid absorption and metabolism [18]. Similarly, rats fed cholesterol diet supplemented with 5% apple pomace for 10 weeks had restored liver 3-hydroxy-3methyl-glutaryl-coenzyme A (HMG-CoA) reductase activity (a rate-controlling enzyme for de novo cholesterol synthesis), it also decreased conjugated dienes plasma levels, increased fractional catabolic rate of plasma cholesterol and reduced the erythrocyte antioxidant enzyme activity [19]. These results were in accordance with those obtained by Juśkiewicz et al. in which the erythrocyte SOD was significantly increased with a decrease in the levels of liver thiobarbituric acid reactive substances in rats fed on unprocessed apple pomace which are rich in polyphenols [20].

The study also showed that both apple pomace (unprocessed and processed) could improve gut health through a reduction in intestinal enzymes with no effect on nitrogen utilization. While the unprocessed apple pomace favorably improved antioxidant status [20]. Furthermore, Kosmala et al. reported that consumption of apple pomace could improve the health of the gastrointestinal tract with where the flavonoids present in the pomace could favorably alter the local interactions in the digestive tract [21].

Pigs fed on standard diet supplemented with 3.5% apple pomace for 6 weeks had improved villi breadth in the jejunum and ileum and decreased gut associated lymphoid tissue which indicated better absorption and decreased inflammations [22].

Ma et al. studied the effect of apple pomace in combination with rosemary extract on blood glucose in a model of fructose consumption–induced insulin resistance in rats. Where rats received fructose with 500 mg/kg of apple pomace and rosemary extract for 5 weeks had significantly lower fasting blood glucose, plasma insulin levels and improved insulin resistance than those receiving fructose alone through attenuating impaired CD36 cells and the GLUT-4 transporter [23].

A randomized, single-blind, 4 weeks crossover study made by Ravn-Haren et al. showed that apple pomace (22 mg/day) reduced lithocholic acid excretion, plasma medium- and short-chain acylcarnitines, primary bile acids, deydropiandrosterone sulphate, and lysophospholipids, which are associated with cholesterol transport from the liver. In this study, the apple pomace consumption could improve the cholesterol levels, gut microbial functionality and insulin sensitivity [24].

Concerning the safety of apple pomace, results of toxicity studies have shown that apple pomace is a safe feed additive for livestock as it provides a good source of nutrients with no adverse effects on milk production, protein digestion or pregnancy [25, 26].

However apple seeds contain the naturally toxic cyanogenic glycoside namely amygdalin, but animal studies showed that the amygdalin content of apple seeds don't have negative impact on the digestion markers or the levels of blood lipids. Rats fed on apple seeds for 14 days it had an improvement effect on the antioxidant status of rats, so it was concluded that it is safe to consume apple pomace that contains apple seeds [27].

The National Institutes of Health Toxicology Data Network recorded the lethal dose of HCN to be 50–300 mg. It was calculated that each gram of apple seeds can produce 0.6 mg of HCN, this means that 83–500 apple seeds which is equivalent to 800 g of apple pomace should be consumed for the acute cyanide toxicity to appear [28, 29].

It was also reported that the pesticide residues content in apples could be considered within the safe range for human consumption [30]. Therefore, apple pomace appears to be safe for human consumption. However, further studies are required to confirm that different pesticides and toxic compounds are not retained at toxic levels in apple pomace.

5.4 Pomace as Functional Ingredient in Food Products

Juice processing of apple results in the generation of pomace as by-product which consists of peel, stem, seeds, and pulp. Apple pomace is considered waste, disposing this waste exerts burden on the environment and might cause health hazards. Apple pomace is reported to be a rich source of health-benefitting nutritional bioactive materials as: carbohydrates, dietary fiber and minerals. In addition, to its high content of antioxidant compounds (polyphenols). Therefore, by few modifications' apple pomace could be added as an efficient ingredient in different food products, meanwhile, reusing of this waste in other beneficial aspects (waste management). Apple pomace could be incorporated to bakery products, meat products, confectionery and dairy products either to enhance their nutritional value or as flavouring or stabilizing agent. Moreover, dietary fibers present in apple pomace increased significantly the total fiber content with an added value of fortification to the final product and improving health benefits [7, 9, 12]. The most common application of apple pomace in food industry resides their high antioxidant ingredients that are added as preservatives in addition to their redox properties that prevents many chronic diseases associated with oxidative stress [31]. Though re-using pomace in food products is a step forward towards waste management and fruit waste valorization, but still this might lead to loss of the appropriateness due to lack of a comprehensive study of the physical and chemical structure of apple pomace that may affect the food quality.

5.5 Incorporating Apple Pomace in Food Products

Apple pomace is a good choice in confectionary products due to its high content of pectin. These products included jelly on adding apple pomace paste, or added in jams. This was based on stirring apple pomace at suitable pH $\simeq 3.2$ which led to an increase in the total phenolics and total flavonoids that consequently affects positively its anti-oxidant property, in addition to, the good flavour and acceptable appearance. On adding apple pomace pectin—using ultrasonic homogenizer and adding an enzyme—to cookies with the aim of replacing fats, this led to increasing the height of cookies, increasing the moisture content and decrease the hardness and fracture. Therefore, pectin could be an essential component to replace fat in bakery products [32].

Recent studies used apple pomace skin—as powder—as an alternative to wheat flour in muffin during the baking process, the percentage added started from 4 up to 30%. Promising results were obtained indicating an increase in the polyphenolic content and antioxidant activity [33]. Another study supported the addition of pomace powder to replace wheat flour in cookies and muffin. The resultant product was rich in fibers having dark colour. This dark colour was attributed to Maillard reaction—monosaccharides in the apple pomace—during heating undergo caramelization [34]. In the same aspect, addition of 5% apple pomace as a replacement to wheat flour in baking biscuits. The baked products showed hard consistency, while the sensory grade of the biscuit remains accepted [35]. Most efforts have been lately directed to incorporate apple pomace in diet—added as fibre supplement in bread—to improve the nutritive value of bread. Apple pomace was added in a concentration 11% to wheat bread, results showed that there was increase in weight from 3 to 7% Meanwhile, the volume was reduced accompanied by an increase in bread's hardness equivalent to the pomace added [31].

The concept that natural dyes are regarded as important alternatives for synthetic dyes placed apple pomace pigment as one of the lead substitutes for tartrazine (azo dye which cause skin allergy, angioedema and genotoxic). The process depends on the oxidation of the phenolic ingredients in pomace apple at pH $^{<}$5 to get the yellow colour (as tartrazine) or pH >6 to obtain the water soluble reddish orange colour that can be a substitute for curcumin which is water insoluble [36].

5.6 Apple Pomace in Gluten Free Food Products and Nutritional Improvement

Coeliac disease is an autoimmune system disease, where the immune system attacks the tissues on eating food gluten resulting damage of the GIT especially the small intestine. The symptoms range from diarrhea, to abdominal pain and bloating. This is also accompanied by fatigue due to iron deficiency anaemia or vitamin B_{12} deficiency anaemia. Accordingly, an unexpected weight loss occurs. Alternative, gluten-free food products are now available in the market targeting people suffering from coeliac disease to improve their health.

Apple pomace (10%) was added in producing gluten-free cake, although the cake obtained showed decreased volume, hard consistency, but the cake was highly rich in fibres with acceptable sensory grade [37]. On the other hand, apple incorporated in cookies decrease the glycemic index by 10% placing apple pomace with an added value glycemic index enriched product food [38]. Additionally, adding apple pomace 3–9% with brown rice in a gluten-free crackers led to an increase in mineral content with enhancement of the antioxidant activity: an increase in the total phenolic and total flavonoid content [39]. From that perspective, apple pomace could be regarded as an essential nutritional and functional component in bakery products.

5.7 Apple Pomace in Diary Food Products

Apple pomace fibre is now regarded as a crucial ingredient in fiber enriched food products. A fibre prepared by acid-alkali digestion method from apple pomace yielding from 2.5 to 10% fibre was added with *Lactobacillus acidophilus* to prepare yoghurt. The product showed an increase in fiber content up to 10%, decrease in acidity and fat content. In conclusion, yoghurts containing 5% fibre as a functional ingredient acquired best sensory grade as regards to colour and flavour [40]. Another approach for incorporating micronized apple pomace in yoghurt drinks as an emulsifying agent evidenced that the bioactive ingredients in apple pomace form emulsions with acceptable oxidative and texture characters [41, 42].

5.8 Apple Pomace in Meat Food Products

Replacement of chicken's and beef's meat by 10–20% apple pomace results in an increase in fibre and pectin contents in addition to enhancing the antioxidant activity, however there was a reduction in the firmness and the products acquired darker colours [34]. Notably, replacing meat chicken—to produce sausage or nuggets—with 3–6% apple pomace in sausage production improved the sensory grade of the product. This was manifested by an increase in the cooking yield as well as the stability of the emulsion with an added value of enriched fibre content, lower lipid oxidation during storage with negative effect on the microbial quality of the product [43]. In another study, a reduced fat chicken sausage was prepared by substituting 5–10% of pork fat by 1–2% fibre apple pomace. This prospective reputes the fibre extracted from apple pomace in controlling fat consumption [44].

5.9 Apple Pomace-Challenges and Limitations

Further researches are required to pursue the health risks that apple pomace might impose on consumption in functional food products. The first concern, is using apple seeds together with the skin and pulp; as seeds contain amygdalin that eventually produces cyanide gas that could be toxic. Fortunately, recent studies reported the safety of apple pomace in livestock feed as an additive also, pesticide concentrations are within safety thresholds for human consumption [12].

Another health hazard is pesticide residue in apple pomace, which was considered negligible and with low risk according to recent studies [12, 45]. With more focus and desire to increase and improve the yield and quality of apple crop, fungicides are added which impose another health risk on the apple pomace safety. However, as reported by EPA (The United States Environmental Protection Agency), these fungicides are regarded to be with low-risk [9, 46]. Based on the previous concerns, and due to continuous apprises in using pesticides or any fungicides, studies and quality control on apple pomace is a crucial step before incorporation in food stuffs.

Another important step in using apple pomace is the method of drying, to obtain the maximum bioactive compounds that could be of added value to the final product. In an advanced study investigating the effect of drying temperature and pressure on the antioxidant capacity and phenolic compounds of apple pomace extract.

Results evidenced the composition and function of apple pomace is highly affected by the drying method applied. Drying apple pomace at 60 °C applying conventional atmospheric oven affected positively the antioxidant activity. Moreover, the applied drying temperature controls the drying duration as well as the rate of water loss, and the antioxidant compounds present after drying [47].

5.10 Conclusion

As a by-product of apple industry, apple pomace has considerable amount of polyphenolic compounds, minerals and dietary fibers. The nutritional as well as the medicinal values of apple pomace highlight its use as a nutraceutical and fortification ingredient in food products, including plant food, meat and dairy products. The polyphenolic content of apple pomace increases the antioxidant value of food products, while, the high content of dietary fibres increases its nutritional profile. The possibility of direct use of apple pomace in food products may help reduce the chances of it pilling the environment as a waste.

References

1. Parfitt J, Barthel M, Macnaughton S (2010) Food waste within food supply chains: quantification and potential for change to 2050. Philos Trans R Soc B Biol Sci 365(1554):3065–3081
2. Garcia-Garcia G, Woolley E, Rahimifard S, Colwill J, White R, Needham L (2017) A methodology for sustainable management of food waste. Waste Biomass Valoriz 8(6):2209–2227
3. Jayathilakan K, Sultana K, Radhakrishna K, Bawa AS (2012) Utilization of byproducts and waste materials from meat, poultry and fish processing industries: a review. J Food Sci Technol 49(3):278–293
4. Esparza I, Jiménez-Moreno N, Bimbela F, Ancín-Azpilicueta C, Gandía LM (2020) Fruit and vegetable waste management: conventional and emerging approaches. J Environ Manag 265:110510
5. Francini A, Sebastiani L (2013) Phenolic compounds in apple (malus x domestica Borkh.): compounds characterization and stability during postharvest and after processing. Antioxidants (Basel) 2(3):181–193
6. Food and Agriculture Organization (FAO) (2019) FAOSTAT Statistical Database of the United Nation Food and Agriculture Organization (FAO) statistical division. Rome
7. Antonic B, Jancikova S, Dordevic D, Tremlova B (2020) Apple pomace as food fortification ingredient: a systematic review and meta-analysis. J Food Sci 85(10):2977–2985
8. Kammerer DR, Judith K, Regine V, Reinhold C (2014) Recovery of polyphenols from the by-products of plant food processing and application as valuable food ingredients. Food Res Int 65:2–12
9. Lyu F, Luiz SF, Azeredo DRP, Cruz AG, Ajlouni S, Ranadheera CS (2020) Apple pomace as a functional and healthy ingredient in food products: a review. Processes 8(3):319
10. Dhillon GS, Kaur S, Brar SK (2013) Perspective of apple processing wastes as low-cost substrates for bioproduction of high value products: a review. Renew Sust Energ Rev 27(C):789–805
11. Perussello CA, Zhang Z, Marzocchella A, Tiwari BK, Safety F (2017) Valorization of apple

pomace by extraction of valuable compounds. Compr Rev Food Sci 16(5):776–796
12. Skinner RC, Gigliotti JC, Ku K-M, Tou JC (2018) A comprehensive analysis of the composition, health benefits, and safety of apple pomace. Nutr Rev 76(12):893–909
13. Bolarinwa IF, Orfila C, Morgan MR (2015) Determination of amygdalin in apple seeds, fresh apples and processed apple juices. Food Chem 170:437–442
14. Krasnova I, Segliņa D (2019) Content of phenolic compounds and antioxidant activity in fresh apple, pomace and pomace water extract—effect of cultivar. Proc Latvian Acad Sci 73:513–518
15. Soares MC, Ribeiro ÉT, Kuskoski EM, Gonzaga LV, Lima A, Mancini Filho J et al (2008) Composition of phenolic acids content in apple (Malus sp) pomace. Semina: Ciências Agrárias 29(2):339–347
16. Fernandes PA, Ferreira SS, Bastos R, Ferreira I, Cruz MT, Pinto A et al (2019) Apple pomace extract as a sustainable food ingredient. Antioxidants 8(6):189
17. Bhushan S, Kalia K, Sharma M, Singh B, Ahuja PS (2008) Processing of apple pomace for bioactive molecules. Crit Rev Biotechnol 28(4):285–296
18. Cho K-D, Han C-K, Lee B-H (2013) Loss of body weight and fat and improved lipid profiles in obese rats fed apple pomace or apple juice concentrate. J Med Food 16(9):823–830
19. Bobek P, Ozdín L, Hromadova M (1998) The effect of dried tomato, grape and apple pomace on the cholesterol metabolism and antioxidative enzymatic system in rats with hypercholesterolemia. Food Nahrung 42(5):317–320
20. Juśkiewicz J, Żary-Sikorska E, Zduńczyk Z, Król B, Jarosławska J, Jurgoński A (2012) Effect of dietary supplementation with unprocessed and ethanol-extracted apple pomaces on caecal fermentation, antioxidant and blood biomarkers in rats. Br J Nutr 107(8):1138–1146
21. Kosmala M, Kołodziejczyk K, Zdunczyk Z, Juskiewicz J, Boros D (2011) Chemical composition of natural and polyphenol-free apple pomace and the effect of this dietary ingredient on intestinal fermentation and serum lipid parameters in rats. J Agric Food Chem 59(17):9177–9185
22. Sehm J, Lindermayer H, Dummer C, Treutter D, Pfaffl M (2007) The influence of polyphenol rich apple pomace or red-wine pomace diet on the gut morphology in weaning piglets. J Anim Physiol Anim Nutr 91(7–8):289–296
23. Ma P, Yao L, Lin X, Gu T, Rong X, Batey R et al (2016) A mixture of apple pomace and rosemary extract improves fructose consumption-induced insulin resistance in rats: modulation of sarcolemmal CD36 and glucose transporter-4. Am J Transl Res 8(9):3791
24. Ravn-Haren G, Dragsted LO, Buch-Andersen T, Jensen EN, Jensen RI, Németh-Balogh M et al (2013) Intake of whole apples or clear apple juice has contrasting effects on plasma lipids in healthy volunteers. Eur J Nutr 52(8):1875–1889
25. Abdollahzadeh F, Pirmohammadi R, Fatehi F, Bernousi I (2010) Effect of feeding ensiled mixed tomato and apple pomace on performance of Holstein dairy cows. Slovak J Anim Sci 43(1):31–35
26. Tiwari S, Narang M, Dubey M (2008) Effect of feeding apple pomace on milk yield and milk composition in crossbred (Red Sindhi x Jersey) cow. Livest Res Rural Dev 20(4):293–297
27. Opyd PM, Jurgoński A, Juśkiewicz J, Milala J, Zduńczyk Z, Król B (2017) Nutritional and health-related effects of a diet containing apple seed meal in rats: the case of amygdalin. Nutrients 9(10):1091
28. National Institutes of Health (2017) TOXNET: Toxicology Data Network, Hydrogen Cyanide National Institutes of Health Toxicology Data Network
29. Shalini R, Gupta D (2010) Utilization of pomace from apple processing industries: a review. J Food Sci Technol 47(4):365–371
30. EPA (US Environmental Protection Agency) (2000) Choosing a percentile of acute dietary exposure as a threshold of regulatory concern
31. Gómez M, Martinez MM (2018) Fruit and vegetable by-products as novel ingredients to improve the nutritional quality of baked goods. Crit Rev Food Sci Nutr 58(13):2119–2135
32. Min B, Bae IY, Lee HG, Yoo S-H, Lee S (2010) Utilization of pectin-enriched materials from apple pomace as a fat replacer in a model food system. Bioresour Technol 101(14):5414–5418

33. Rupasinghe HV, Wang L, Huber GM, Pitts NL (2008) Effect of baking on dietary fibre and phenolics of muffins incorporated with apple skin powder. Food Chem 107(3):1217–1224
34. Jung J, Cavender G, Zhao Y (2015) Impingement drying for preparing dried apple pomace flour and its fortification in bakery and meat products. J Food Sci Technol 52(9):5568–5578
35. Kohajdová Z, Karovičová J, Magala M, Kuchtová V (2014) Effect of apple pomace powder addition on farinographic properties of wheat dough and biscuits quality. Chem Pap 68(8):1059–1065
36. Fromm M, Loos HM, Bayha S, Carle R, Kammerer DR (2013) Recovery and characterisation of coloured phenolic preparations from apple seeds. Food Chem 136(3–4):1277–1287
37. Kırbaş Z, Kumcuoglu S, Tavman S (2019) Effects of apple, orange and carrot pomace powders on gluten-free batter rheology and cake properties. J Food Sci Technol 56(2):914–926
38. Alongi M, Melchior S, Anese M (2019) Reducing the glycemic index of short dough biscuits by using apple pomace as a functional ingredient. LWT 100:300–305
39. Mir SA, Bosco SJD, Shah MA, Santhalakshmy S, Mir MM (2017) Effect of apple pomace on quality characteristics of brown rice based cracker. J Saudi Soc Agric Sci 16(1):25–32
40. Issar K, Sharma P, Gupta A (2017) Utilization of apple pomace in the preparation of fiber-enriched acidophilus yoghurt. J Food Process Preserv 41(4):e13098
41. Lu Z, Ye F, Zhou G, Gao R, Qin D, Zhao G (2020) Micronized apple pomace as a novel emulsifier for food O/W pickering emulsion. Food Chem 330:127325
42. Wang X, Kristo E, LaPointe G (2020) Adding apple pomace as a functional ingredient in stirred-type yogurt and yogurt drinks. Food Hydrocoll 100:105453
43. Yadav S, Malik A, Pathera A, Islam RU, Sharma D (2016) Development of dietary fibre enriched chicken sausages by incorporating corn bran, dried apple pomace and dried tomato pomace. Nutr Food Sci 46(1):16–29
44. Choi Y-S, Kim Y-B, Hwang K-E, Song D-H, Ham Y-K, Kim H-W et al (2016) Effect of apple pomace fiber and pork fat levels on quality characteristics of uncured, reduced-fat chicken sausages. Poult Sci 95(6):1465–1471
45. Chen M, Tao L, McLean J, Lu C (2014) Quantitative analysis of neonicotinoid insecticide residues in foods: implication for dietary exposures. J Agric Food Chem 62(26):6082–6090
46. Liu S, Che Z, Chen G (2016) Multiple-fungicide resistance to carbendazim, diethofencarb, procymidone, and pyrimethanil in field isolates of Botrytis cinerea from tomato in Henan Province, China. Crop Protect 84:56–61
47. Gonelimali FD, Szabó-Nótin B, Máté M (2021) Optimal drying conditions for valorization of industrial apple pomace: potential source of food bioactive compounds. Prog Agric Eng Sci 17(S1):69–75

Chapter 6
Olive Pomace as a Source of Nutraceuticals

Selma Hamimed and Abdelwaheb Chatti

6.1 Introduction

The cultivation of new olive trees and increasing consumers demand have led to a steady increase in olive oil production. The oil is highly recognized as a healthy choice of a diet based on its rich composition of antioxidants and essential fatty acids [1]. The major olive oil producers include Spain (42%), Italy (17%), Greece (11%), Turkey (6%), Syria (6%), Tunisia (6%), Morocco (4%), Jordan (3%), and Lebanon (1.5%); while other smaller producers are Asia (15%), Africa (12%), and America (2%) [2]. The worldwide production of olive oil is considered to be 3.1 million tons per year [3].

The production of olive oil accounts for 20% of the oil itself, 50% of olive mill wastewater (OMW), and 30% of a pulpy waste or pomace, making the environmental impact of the byproducts a paramount cause worth investigating. OMW represents a hazardous environmental impact due to high chemical oxygen demand (COD) (80–200 g/L) and the presence of phenolic compounds (0.5–24 g/L); interestingly, Hamimed et al. [4] showed that the removal of phenolic compounds reduces the pollution charge of the OMW. On the other hand, the olive pomace is consisting of the remaining olive seed and pulp with a discharge of approximately 473,400–526,000 tons/year [5]. The absence of well-established treatment for olive pomace in the mills has created a wide gap being filled with the emerging valorization and sustainable methods for minimizing, preventing, and treating the olive pomace, to reduce the environmental hazards emanating from their uncontrolled discharge [6]. Moreover, olive pomace is considered an important source of waste

S. Hamimed (✉) · A. Chatti
Laboratory of Biochemistry and Molecular Biology, Faculty of Sciences of Bizerte, University of Carthage, Jarzouna, Tunisia
e-mail: alma.hamimed@fsb.rnu.tn

© The Author(s), under exclusive license to Springer Nature Switzerland AG 2022

C. Egbuna et al. (eds.), *Food and Agricultural Byproducts as Important Source of Valuable Nutraceuticals*, https://doi.org/10.1007/978-3-030-98760-2_6

biomass due to high organic matter, especially phenolic compounds, high energy content, and low ash content [7].

The total phenolic contents that could migrate from the olive fruit into the extracted oil is a bare 2%, as more than 90% are passed into OMW and the pomace, whereas the phenol-rich extract of olive has both hypocholesterolemic and antioxidant potentials, amongst other biological effects [8]. For decades, the use of natural products as a source of new agents in the pharmaceutical industry has been widely applied. For that, the rapid development of recovery and sustainable methods to valorize wastes and wastewaters has gained much interest to reduce the environmental hazards caused by their uncontrolled elimination [2]. Therefore, olive pomace could proffer numerous positive applications in the pharmaceutical, healthcare, and food industries, creating functional ingredients that should be of high significance in olive oil waste valorization. The present chapter is aimed at providing a general description of olive byproducts and wastes, while emphasizing olive pomace, with a special interest in its healthful compounds, especially phenolic composition, as well as potent functional and future applications.

6.2 Olive Oil Process and Its Associated By-Products

6.2.1 Olive Oil Process

The olive oil extraction may involve traditional (olive oil mechanical press), three-phase centrifugation (continuous), and two-phase centrifugation (presently in high use based on its economic and environmental significance) methods. Figure 6.1

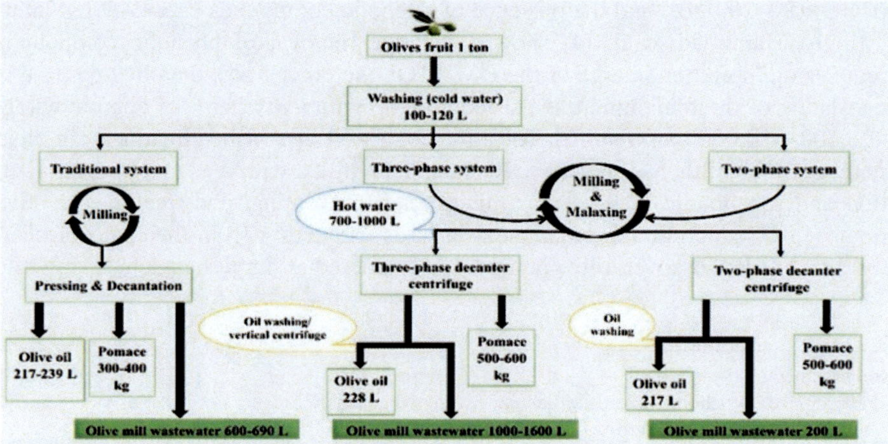

Fig. 6.1 Diagram presents steps from each olive oil processing line that generates solid and liquid wastes. (Reproduced with permission. *Springer Nature* [2])

depicts that for 1 ton of olive fruits, a large amount of freshwater ranging from 100 to 1000 L was added to generate (217–239 L) of olive oil, (300–600 kg) of pomace, and 200 and 1600 L of OMW [2]. The three-phase or continuous processing method has been used in olive-producing countries for a long time based on high-working efficiency, reduced labor, and operation costs because the centrifugal decanter allows for the separation of three flows of matter; olive oil, pomace, and wastewater [8]. The challenge soon became obvious because this system utilizes lots of fresh warm water for olive paste dilution, which in effect reduces the bioactive components of the oil produced, especially antioxidants, and of the OMW generated too. It also results in the production of 20% olive oil, 30% olive pomace, and 50% OMW, implying that the generated wastes are a quarter times more than the oil obtained [1].

6.2.2 *Associated By-Products*

As mentioned above, the olive oil process generates different byproducts that can be described as solid waste and liquid waste.

As a result, olive pomace is a solid byproduct of olive processing. It is a combined form of olive pulp, skin, stone, and water, whose characteristics may differ based on oil extraction type and time, with a production of 0.5–0.6 tons of olive pomace for every ton of olives [9]. Crude olive cake, which is obtained from crushed olive kernel shells, skin, and pulps, differing from exhausted olive cake largely based on oil and water contents upon dehydration in the oil extraction process; while a partly destoned exhaustive olive cake is an aggregate of shells and pulps [10]. Olive stones and seeds are also produced during oil extraction and are mainly utilized for electric energy, heat (wood adhesives coated with formaldehyde) or raisin production, carbon activation, metal biosorbent, and feedstock [11]. Both the stones and seeds are rich in sugars, fats, proteins, phenols, and many other bioactive compounds [12]. Generally, this byproduct is gained attention in various valorizations, which are presented in Table 6.1.

Table 6.1 Different valorization of olive pomace

Byproduct	Valorization	References
Olive leaves and small branches	Directly combusted, fed to animals, and are used up in the production of animal feed pellets, cosmetics, food	[9]
Crude olive cake (crushed olive kernel shells, skin and pulps)	Food for animals	[10]
Olive stones and seeds	Electrical energy, heat (wood adhesives coated formaldehyde) or raisins production, carbon activation, metal biosorbent, and raw materials	[13]
Olive pomace	α-, β- and γ tocopherols were recovered, except δ-tocopherol, in supercritical fluid extraction (SFE)-treated samples	[5, 14]

Therefore, OMW can be described as the main liquid effluent from olive oil extraction process, rich in high polyphenolic concentrations and some other organic matter that has escaped from olive fruits. These active compounds, including dietary fibers, proteins, sugars, and other vital organic compounds, are important in fabricating novel foods and ingredients that are capable of proffering gelling, foaming, amongst other physicochemical properties [10]. Thus, the average OMW volume generated during olive oil production is within the range of 0.3–1.2 m^3/ton of processed olives [1].

6.3 Recovery and Value-Addition

The recovery and valorization options of olive pomace, as well as its bioactivity and bioavailability, are of much more interesting and noteworthy considerations. Currently, olive pomace is mostly utilized for solvent extraction to recover the residual oil. The stone fragments can also be recovered and used as fuel to heat the kilns or to make activated carbon. However, both the two-phase decanter-derived olive pomace and the pitted olive-pomace are difficult to manage for oil extraction because pomace dehydration takes more time and energy. As a result, scientists have concentrated their efforts on finding alternative and, at the same time, sustainable uses for olive pomaces, such as the extraction of molecules of interest [15]. Thus, recovering polyphenols and other bioactive compounds at this stage from olive cake, pomace, kernel, paste, stone, and seeds can be explored for biovalorization.

Tocopherols and squalene have been reportedly recovered in olive pomace. When using two-phase olive oil production systems, it was found that α-, β- and γ tocopherols were recovered, except δ-tocopherol, in supercritical fluid extraction (SFE)-treated samples [16]. In a similar study involving the optimal conditions for recovering squalene with supercritical carbon dioxide (SC-CO_2) method in the extraction process, maximal recovery from olive pomace occurred at 17.5 MPa, 33 °C, and 5.4 g/min conditions using temperature, pressure, and CO_2 flow rate parameters, respectively [17]. Both tocopherols and squalene play multiple essential and protective roles in free radicals scavenging, skin health, and neurological healthful outcomes.

Proteins of olive kernel source also showed a very high yield in a comparison study with total phenolic compound (TPC) content, attributive to high voltage electrical discharge (HVED)-promoted suspension turbulence, thereby facilitating protein-lignocellulose degradation [18].

The same authors investigated the effects of pulsed electric fields (PEF), HVED and ultrasounds (UAE) on polyphenols recovery from the olive kernel, and they noticed a significant increase in TPC content upon energy input increment in all treatments [10, 18].

In another study, about 25 compounds were characteristically identified as phenolic alcohols or derivatives, secoiridoids, flavones, and elenolic acid derivatives, when the phenolic compounds in olive oil filter cake were recovered [19]. Other simple phenolic compounds, (caffeic and *p*-coumaric acid) and polyphenols (luteolin) were found in olive pomace where dihydroxy phenolics such hydroxytyrosol and tyrosol were the most significant [14].

These phenolics from olive pomace have immense antioxidative and general health benefits. The byproducts are important in formulating sustainable functional foods and ingredients meant for animal and human nutrition, amongst other non-edible or environmentally sustainable applications, is well-reviewed by Berbel and Posadillo [20].

Olive pomace may further be treated biotechnologically through composting, aerobic, anaerobic, biochemical, and biophysical techniques, or by evaporation and physical-chemical approaches to bioconverte into valuable products [21]. Novel techniques keep emerging to extract economically and sustainably biological compounds from olive pomace to be utilized in all ramifications of the food sector [22].

6.4 Nutraceutical and Bioavailability of Olive Pomace

6.4.1 Nutraceutical

Dietary fiber, unsaturated fatty acids, minerals, and phenolic substances are all present in significant amounts in olive pomace. Due to its ingredients, olive pomace is a good source of functional compounds, which makes it a valuable byproduct for food formulation.

In order to obtain functional food, different researchers have been examined the olive pomace or recovered functional compounds as ingredients in foods. Cedola et al. [23] reported that the antioxidant activity of olive pomace (10%) as an additive in bread; however, results showed some differences in color, smell, and taste compared with the blank. As olive pomace powders demonstrated a potential multifunctional as ingredients in food, Lin et al. [24] showed newly biscuits-enriched olive pomace, which is acceptable in texture, smell, and rich in dietary fiber. Moreover, the feasibility of using olive pomace in the ingredient mixture to obtain extruded snacks rich in fiber and polyphenols were evaluated. Results showed the stability of aromas and flavors for the obtained functional snacks with 5% of olive pomace [22]. In addition, Durante et al. [25] demonstrated that the enriched Italian snack with controlled 20% of olive pomace enriched the taralli with hydroxytyrosol, triterpenic acid, and carotenoid compared with control. Ying et al. [26] revealed that the addition of dietary fiber and polyphenols from olive pomace in two different flours (rice-oat and maize-oat) modified the physical characteristics of rice-oat flour.

While pasta is also followed for olive pomace addition due to its poorness of biologically active substances. Padalino et al. [27] enriched spaghetti with olive pomace powder (10%) and found a high amount of free phenolics, flavonoids, fiber, tocopherols, and carotenoids compared with control. However, results highlighted low elasticity and firmness in cooked samples. In the same way, Simonato et al. [28] evaluated the effect of the addition of olive pomace (5%) on the textural and sensory properties of spaghetti. The spaghetti enriched olive pomace showed lower cooking time and water absorption due to the presence of fiber. Thanks to phenolic content, the antioxidant activity improved.

Therefore, the antioxidant activity of olive pomace was reported on fish burger functionalization. Results showed that pretreating olive pomace powder with milk before producing a burger gives much better texture, smell, and taste when it is cooked [29].

Vegetable oils are well known to be linked to human health, both because they offer needed fatty acids and because they contain antioxidant compounds, particularly in virgin olive oils, which counteract oxidation. Oils have unsaturated fatty acids, which are more vulnerable to oxidation, thus antioxidants can be added to extend their shelf life. Natural antioxidants derived from olive pomace are preferred over synthetic antioxidants. Suárez et al. [30] demonstrated that extracted phenolic compounds from olive pomace enriched virgin olive oil gave higher oxidative stability. While, findings on sunflower oil enriched with polyphenols from olive pomace, showed an enhancement in oxidative activity compared to control [31]. In a similar way, the enriched olive oil with secoiridoid derivatives extracted from olive pomace demonstrated good sensory acceptability and high oxidative stability [32].

Furthermore, dairy products have also taken attention to enhance the probiotic quality. The phenolic compounds, extracted from the olive pomace, added to fermented milk revealed an increase in the antioxidant activity with no effect on the number of viable functional microorganisms [33].

On the other hand, animal foodstuffs are gained important attention due to their related part of human nutrition. The enriched chicken meat with polyphenols from olive pomace increased the functional and nutraceutical characteristics of the meat, improving oxidative stability [34]. While results on lamb meat enriched with olive pomace showed an increase in oxidative stability and level of polyunsaturated fatty acids and vitamin E without modifying the health state of lambs and the quality of meat [35]. Moreover, Sioriki et al. [36] supplied olive pomace in the diet of fish; results showed increased biological activity, which was explained by the neutralization of platelet-activating factor (PAF) enzyme responsible for lipid fraction in fish

6.4.2 Bioavailability

Further, the recovery and valorization options of olive pomace, as well as its bioactivity and bioavailability are of much more interesting and noteworthy considerations. Based on their physicochemical characteristics, olive pomace cannot be discharged directly on soils or water streams without proper treatment to avoid serious environmental issues. The availability of olive pomace to pass the intestinal barrier is important for its ultimate bioactivity. Interestingly, olive-derived phenolics such as hydroxytyrosol, ligistroside-aglycone, and tyrosol are reportedly bioavailable based on their significant levels of absorbability and excretability in animals and health human urines upon oral intake [37]. Mechanistically, a human intestinal epithelium model indicated that hydroxytyrosol transport takes place via a two-way passive diffusion, and its absorbability is largely carried out in the intestines [38].

In another study, oral intake of olive pomace in ewe feeding models showed a quick metabolic response and absorption based on the phenol's concentration levels with positive effect on the yield of milk without modifying their properties [39].

Moreover, Visioli et al. [40] reported high absorption of tyrosol/hydroxytyrosol within 24 h in a human study. In addition, a human intestinal epithelium model indicated that hydroxytyrosol transport takes place via a two-way passive diffusion, and its absorbability is largely carried out in the intestines [38].

Yet, in a more directly correlated study of verbascosides-rich olive pomace ultra-purified phenolic extract, there was rapid assimilation of verbascoside and isoverbascoside [41], indicative of the bioavailability of olive pomace and its compounds like the verbascosides.

The valorization of olive pomace is a viable strategy for addressing economic and environmental challenges while also increasing the profitability of the olive oil industry. Novel extraction techniques could be used to recover phenolic compounds, enhancing the quality of the extracts in terms of antioxidant potential and phenolic content. The application of microwave-assisted extraction gives an efficient way for the extraction of oleanolic and maslinic acids from olive pomace with high antimicrobial activity, leading to minimal processing for food application [42].

6.5 Conclusion and Future Perspectives

The current chapter highlights the nutraceutical ability of olive pomace based on its active constituents, the phenolics. The availability of the major phenols further proof that they are use in food systems. Although studies have shown the possibilities of recovering high-added value compounds, most of them are conducted on polyphenols. The phenol present in olive pomace is very beneficial and undoubtedly important in organic activities, but their preparation or incorporation into functional products requires high precautionary measures, which can affect both the bioactivity and bioavailability of phenolic compounds. Thus, other important compounds and biomolecules should be revisited for recovery studies.

Therefore, the reuse of olive pomace increases the shelf-life and/or formulating more nutritious and healthy products. Consequently, based on sensory evaluation using olive pomace as ingredients, studies concluded that the formulas and product preparation may be slightly modified to reduce the impact of some hazardous effects including the darker color and increased bitterness. They proposed that adequate labeling and marketing of enriched items could boost product adoption by generating positive expectations.

The economic importance of valorized olive by-products are enormous but the regulations guiding each territory of olive producers may be some bottlenecks that must be overcome. Meanwhile, the two-phase extraction process is still highly recommended for the millers to generate lesser olive pomace and OMW which could then be recovered and biovalorized.

References

1. Khdair A, Abu-Rumman G (2020) Sustainable environmental management and valorization options for olive mill byproducts in the Middle East and North Africa (MENA) region. Processes 8(6):671
2. Hamimed S, Landoulsi A, Chatti A (2021) The bright side of olive mill wastewater: valuables bioproducts after bioremediation. Int J Environ Sci Technol 18:4053–4074
3. Rharrabti Y, El Yamani M (2018) Olive mill wastewater: treatment and valorization technologies. In: Hussain C (ed) Handbook of environmental materials management. Springer, Cham, pp 1–28
4. Hamimed S, Jebli N, Sellami H, Landoulsi A, Chatti A (2020) Dual valorization of olive mill wastewater by bio-nanosynthesis of magnesium oxide and yarrowia lipolytica biomass production. Chem Biodivers 17(3):e1900608
5. Erses Yay AS, Birinci B, Açıkalın S, Yay K (2021) Hydrothermal carbonization of olive pomace and determining the environmental impacts of post-process products. J Clean Prod 315:128087
6. Azzaz AA, Jeguirim M, Kinigopoulou V, Doulgeris C, Goddard ML, Jellali S, Matei GC (2020) Olive mill wastewater: from a pollutant to green fuels, agricultural and water source and bio-fertilizer-hydrothermal carbonization. Sci Total Environ 733:139314. https://doi.org/10.1016/j.scitotenv.2020.139314
7. Difonzo G, Troilo M, Squeo G, Pasqualone A, Caponio F (2021) Functional compounds from olive pomace to obtain high-added value foods—a review. J Sci Food Agric 101(1):15–26. https://doi.org/10.1002/jsfa.10478
8. Zbakh H, El Abbassi A (2012) Potential use of olive mill wastewater in the preparation of functional beverages: a review. J Funct Foods 4(1):53–56
9. Romani A, Ieri F, Urciuoli S, Noce A, Marrone G, Nediani C, Bernini R (2019) Health effects of phenolic compounds found in extra-virgin olive oil, by-products, and leaf of Olea europaea L. Nutrients 11(8):1776
10. Roselló-Soto E, Barba FJ, Parniakov O, Galanakis CM, Lebovka N, Grimi N, Vorobiev E (2015) High voltage electrical discharges, pulsed electric field, and ultrasound assisted extraction of protein and phenolic compounds from olive kernel. Food Bioprocess Technol 8(4):885–894
11. Carraro L, Trocino A, Xiccato G (2010) Dietary supplementation with olive stone meal in growing rabbits. Ital J Anim Sci 4(Suppl 3):88–90
12. Ghanbari R, Anwar F, Alkharfy KM, Gilani AH, Saari N (2012) Valuable nutrients and functional bioactives in different parts of olive (Oleaeuropaea L.)—a review. Int J Mol Sci 13(3):3291–3340
13. Ordoudi S, Bakirtzi C, Tsimidou M (2018) The potential of tree fruit stone and seed wastes in Greece as sources of bioactive ingredients. Recycling 3(1):9. https://doi.org/10.3390/recycling3010009
14. Ribeiro TB, Oliveira A, Coelho M, Veiga M, Costa EM, Silva S, Pintado M (2020) Are olive pomace powders a safe source of bioactives and nutrients? J Sci Food Agric 101(5):1963–1978. https://doi.org/10.1002/jsfa.10812
15. Balli D, Cecchi L, Innocenti M, Bellumori M, Mulinacci N (2021) Food by-products valorisation: grape pomace and olive pomace (pâté) as sources of phenolic compounds and fiber for enrichment of tagliatelle pasta. Food Chem 355:129642
16. Ibanez E, Palacios J, Senorans FJ, Santa-Maria G, Tabera J, Reglero G (2000) Isolation and separation of tocopherols from olive by-products with supercritical fluids. J Am Oil Chem Soc 77(2):187–190
17. Stavroulias S, Panayiotou C (2005) Determination of optimum conditions for the extraction of squalene from olive pomace with supercritical CO2. Chem Biochem Eng Q 19(4):373–381
18. Roselló-Soto E, Koubaa M, Moubarik A, Lopes RP, Saraiva JA, Boussetta N, Barba FJ (2014) Emerging opportunities for the effective valorization of wastes and by-products generated dur-

ing olive oil production process: non-conventional methods for the recovery of high-added value compounds. Trends Food Sci Technol 45(2):296–310
19. Lozano-Sánchez J, Castro-Puyana M, Mendiola JA, Segura-Carretero A, Cifuentes A, Ibáez E (2014) Recovering bioactive compounds from olive oil filter cake by advanced extraction techniques. Int J Mol Sci 15(9):16270–16283
20. Berbel J, Posadillo A (2018) Review and analysis of alternatives for the valorisation of agro-industrial olive oil by-products. Sustainability 10(1):237
21. Uddin MA, Siddiki SYA, Ahmed SF, Rony ZI, Chowdhury MAK, Mofijur M (2021) Estimation of sustainable bioenergy production from olive mill solid waste. Energies 14:7654. https://doi.org/10.3390/en14227654
22. Cecchi L, Schuster N, Flynn D, Bechtel R, Bellumori M, Innocenti M (2019) Sensory profiling and consumer acceptance of pasta, bread and granola bar fortified with dried olive pomace (pâté): a byproduct from virgin olive oil production. J Food Sci 84(10):2995–3008
23. Cedola A, Cardinali A, D'Antuono I, Conte A, Del Nobile MA (2020) Cereal foods fortified with by-products from the olive oil industry. Food Biosci 33:100490
24. Lin S, Chi W, Hu J, Pan Q, Zheng B, Zeng S (2017) Sensory and nutritional properties of Chinese olive pomace based high fibre biscuit. Emirates J Food Agric 29:495–501
25. Durante M, Bleve G, Selvaggini R, Veneziani G, Servili M, Mita G (2019) Bioactive compounds and stability of a typical Italian bakery products "taralli" enriched with fermented olive paste. Molecules 24:3258
26. Ying DY, Hlaing MM, Lerisson J, Pitts K, Cheng L, Sanguansri L (2017) Physical properties and FTIR analysis of rice-oat flour and maize-oat flour based extruded food products containing olive pomace. Food Res Int 100:665–673
27. Padalino L, D'Antuono I, Durante M, Conte A, Cardinali A, Linsalata V (2018) Use of olive oil industrial by-product for pasta enrichment. Antioxidants 7:1–15
28. Simonato B, Trevisan S, Tolve R, Favati F, Pasini G (2019) Pasta fortification with olive pomace: effects on the technological characteristics and nutritional properties. LWT Food Sci Technol 114:108368
29. Cedola A, Cardinali A, Del Nobile MA, Conte A (2017) Fish burger enriched by olive oil industrial by-product. Food Sci Nutr 5:837–844
30. Suárez M, Romero MP, Motilva MJ (2010) Development of a phenol-enriched olive oil with phenolic compounds from olive cake. J Agric Food Chem 58:10396–10403
31. Orozco-Solano MI, Priego-Capote F, Luque De Castro MD (2011) Influence of simulated deep frying on the antioxidant fraction of vegetable oils after enrichment with extracts from olive oil pomace. J Agric Food Chem 59:9806–9814
32. Rubió L, Motilva MJ, MacIà A, Ramo T, Romero MP (2012) Development of a phenol-enriched olive oil with both its own phenolic compounds and complementary phenols from thyme. J Agric Food Chem 60:3105–3112
33. Aliakbarian B, Casale M, Paini M, Casazza AA, Lanteri S, Perego P (2015) Production of a novel fermented milk fortified with natural antioxidants and its analysis by NIR spectroscopy. LWT Food Sci Technol 62:376–383
34. Branciari R, Galarini R, Giusepponi D, Trabalza-Marinucci M, Forte C, Roila R (2017) Oxidative status and presence of bioactive compounds in meat from chickens fed polyphenols extracted from olive oil industry waste. Sustainability 9:1–13
35. Awawdeh MS, Dager HK, Obeidat BS (2019) Effects of alternative feedstuffs on growth performance, carcass characteristics, and meat quality of growing Awassi lambs. Ital J Anim Sci 18:777–785
36. Sioriki E, Smith TK, Demopoulos CA, Zabetakis I (2016) Structure and cardioprotective activities of polar lipids of olive pomace, olive pomace-enriched fish feed and olive pomace fed gilthead sea bream (Sparus aurata). Food Res Int 83:143–151
37. Chanioti S, Katsouli M, Tzia C (2021) Novel processes for the extraction of phenolic compounds from olive pomace and their protection by encapsulation. Molecules 26:1781. https://doi.org/10.3390/molecules26061781

38. Manna C, Galletti P, Maisto G, Cucciolla V, D'Angelo S, Zappia V (2000) Transport mechanism and metabolism of olive oil hydroxytyrosol in Caco-2 cells. FEBS Lett 470(3):341–344
39. Chiofalo B, Liotta L, Zumbo A, Chiofalo V (2004) Administration of olive cake for ewe feeding: effect on milk yield and composition. Small Rumin Res 55:169–176
40. Visioli F, Galli C, Grande S, Colonnelli K, Patelli C, Galli G, Caruso D (2003) Hydroxytyrosol excretion differs between rats and humans and depends on the vehicle of administration. J Nutr 133(8):2612–2615
41. Antónia Nunes M, Costa ASG, Bessada S, Santos J, Puga H, Alves RC, Oliveira MBPP (2018) Olive pomace as a valuable source of bioactive compounds: a study regarding its lipid- and water-soluble components. Sci Total Environ 644:229–236. https://doi.org/10.1016/j.scitotenv.2018.06.350
42. Biltekin Sİ, Göğüş F, Yanık DK (2022) Valorization of olive pomace: extraction of maslinic and oleanolic by using closed vessel microwave extraction system. Waste Biomass Valor 13:1599–1608. https://doi.org/10.1007/s12649-021-01620-w

Chapter 7
Orange Peel as Source of Nutraceuticals

Anum Nazir, Nizwa Itrat, Aleena Shahid, Zain Mushtaq, Surajudeen Abiola Abdulrahman, Chukwuebuka Egbuna, Babatunde Oluwafemi Adetuyi, Johra Khan, Chukwuemelie Zedech Uche, and Pere-Ebi Yabrade Toloyai

7.1 Introduction

The term "nutraceutical" is related to nutrition and pharmaceutical. Nutraceuticals are typically food or part of food that supports and enhance health status [1, 2]. It has a wide range of health benefits including well documented medical and pharmacological advantages. It promotes wellbeing by modifying and sustaining the physiological functions in the body. The rapid global market expansion of nutraceuticals is predicated on its increasing popularity and positive impact on population health

A. Nazir (✉) · N. Itrat
Department of Nutrition and Dietetics, The University of Faisalabad, Faisalabad, Pakistan

Department of Nutritional Sciences, Government College University Faisalabad, Faisalabad, Pakistan
e-mail: anum.nazir@tuf.edu.pk

A. Shahid
Department of Nutrition and Dietetics, The University of Faisalabad, Faisalabad, Pakistan

Z. Mushtaq
Institute of Soil and Environmental Sciences, University of Agriculture Faisalabad, Faisalabad, Pakistan

S. A. Abdulrahman
Health Education England, East Midlands Deanery, Leicester, UK

C. Egbuna
Chukwuemeka Odumegwu Ojukwu University, Awka, Nigeria

B. O. Adetuyi
Biochemistry Unit, Department of Biochemistry, Precious Cornerstone University, Ibadan, Nigeria

© The Author(s), under exclusive license to Springer Nature Switzerland AG 2022
C. Egbuna et al. (eds.), *Food and Agricultural Byproducts as Important Source of Valuable Nutraceuticals*, https://doi.org/10.1007/978-3-030-98760-2_7

worldwide. Nutraceuticals are becoming the new era of medicine, health, and research-oriented sector [1, 2].

Health benefits of nutraceuticals include promotion of wellbeing by increasing the nutritional value of daily diet thereby enhancing longer and healthy life, prevention of ill-health and diseases, and improving overall quality of life [2]. They are also reported to have fewer side effect on health than the pharmaceutics [2].

Distinctive biochemical constituents in nutraceuticals possess many therapeutic properties for treatment of ailments such as immunocompromised conditions, metabolic conditions, and auto-immune disorders. Nutrients, dietary and herbal supplements are key elements of nutraceuticals that are continually being explored by the global population to promote health and prevent diseases. Sources of nutraceuticals include conventional foods. Bioactive compounds present in food commodities or the food itself are nutraceuticals that possess physiological benefits and prevent the body from the numerous chronic diseases.

The macronutrients and micronutrients mandatory for healthful existence and bodily functions are embodied in nutraceuticals [2, 3].

Examples of functional food sources of nutraceuticals include fortified foods, herbals, minerals and vitamins, antioxidants, fruits, and dietary supplements [2, 3].

Nutraceuticals are obtained from vegetal and animal origin. Several food commodities and their products are well-known as nutraceuticals. With the increasing public interest in nutraceuticals, recent researches have been necessarily focused on gaining better understanding of the mechanisms of action, efficacy, and safety profile of nutraceuticals [2, 4]. Notably, the bioavailability of nutraceuticals and its activity in the body to support long-term wellbeing devoid of adverse reactions and side effects has been the major focus of researchers in recent times [5].

Nutraceuticals possess fewer side effects as compared to pharmaceuticals and this quality attracts people towards nutraceuticals. Currently, compared to medicinal drugs (pharmaceuticals), it appears that the public is increasingly in favor of nutraceuticals which provides benefits beyond their nutritional value due to their bioactive compounds. To promote sustainable ecologically friendly solutions to disease and poor health status, the interest in nutraceuticals has deservedly been gaining continuous attention among researchers and global population [6].

J. Khan
Department of Medical Laboratory Sciences, College of Applied Medical Sciences, Majmaah University, Majmaah, Saudi Arabia

Health and Basic Sciences Research Center, Majmaah University, Majmaah, Saudi Arabia

C. Z. Uche
Department of Medical Biochemistry and Molecular Biology, Faculty of Basic Medical Sciences, University of Nigeria, Nsukka, Enugu State, Nigeria

P.-E. Y. Toloyai
Department of Medical Biochemistry, Delta State University, Abraka, Delta State, Nigeria

7.2 Peel Waste and Its Potential Use

After the consumption of fruits and vegetables, it's peel and seeds are left to litre the environment, hence constituting global waste. Anything unused or not given any advantage is considered to be a waste. Rotten, over or under-ripe food commodities leads to a waste product. The chemical composition of waste peel and seeds is different, as waste peel contains high content of cellulosic and fibrous waste that can be used as phytochemicals in food sector by researchers.

Fruit and vegetables contain a wide range of potentially bioactive compounds such as antioxidant vitamin E and vitamin C, phenolic compounds, flavonoids, and carotenoids [2]. The peel and skin of these commodities is also a good source of antioxidants and phytochemicals. In some cases, the content of health-protecting anti-microbial, flavonoids, and antioxidants is much greater in waste peel and seeds. New insights are emerging regarding the utility of waste materials high in phytochemicals and nutritional value in promoting health and wellbeing.

Numerous studies have shown the potential classification of nutraceuticals in seeds and peel. Studies have shown that the peel of fruits and vegetables has major phenolic and phytochemical content as compared to the edible part. Fruits such as apples, pears, pomegranate, banana, mango, jackfruit, avocados, have a much higher quantity of bioactive compounds per 100 g of fruit as compared to the edible tissue of commodity [7]. The bioactive compound TPC in the peel of citrus fruits like oranges has 15% more phytochemicals/phenolic compounds than the pulp of the fruit [8].

The leading fruit crop globally is the citrus fruit family. Its family include fruits such as lemon, malta, lime, grapefruit, oranges, and sweet oranges, etc. The large quantity of processed citrus fruit globally leads to very high volume of by-products. Citrus waste is a rich source of pectin by-products, and oil. In the manufacturing of pectin at a commercial level, citrus fruits (orange in particular) are a significant raw material. The presence of polysaccharides in varieties of citrus fruit including orange has been the subject of recent research [9].

The quantity of natural phenolic compounds in orange peel is high. The six classes of flavonoids are present in citrus. These flavonoids provide significant health benefits and protection against diseases. Similarly, phytochemicals and liminoids which are highly beneficial to human health are present in citrus fruits. These phytochemicals possess many therapeutic effects on the body [7].

7.3 Citrus Fruit

Citrus belongs to the family Rutaceae and sub-families of *citrus, Poncirus trifoliata, and Fortunella*. Globally, numerous species of citrus are naturally planted [1]. In terms of quantity of production, citrus is the second most popular crop in the world. One hundred eight million tons of this fruit are produced every year [3].

Citrus Sinesis L. Osbeck is normally called orange, and is a major source of minerals such as potassium, calcium, folate, magnesium, niacin, and thiamine, and vitamins [10]. Orange accounts for 60% of the total citrus food production volume worldwide. A larger part of the orange fruit is used for industrial extraction yearly, contributing to the major volume of residue including the peel. Peel is the primary product, representing 50–60% of total fruit weight [11].

7.4 Nutritional Composition of Orange

To maintain health and promote wellbeing, bioactive compounds are required by the human body. The sources are naturally present in plant materials [12]. The presence of natural bioactive compounds, functional food ingredients, and phytochemicals increases the nutraceutical property of fruits and vegetables. Nutraceuticals are involved in the modulation of human body metabolism and helps to prevent degenerative and chronic diseases [13].

Oranges are the richest source of phytochemical. It contains nutritious and antioxidant properties. Nutritional profile of orange shows that it contains nutrients such as fiber, macro, and micronutrients. It has high antioxidant content which is involved in protecting the body from many chronic diseases [14].

7.5 Orange Peel

The increasing global threat of infectious diseases and malnutrition [1] has led researchers to explore the use of natural agents rich in bioactive compounds in the treatment and prevention of these infectious diseases [15].

Currently, in the household, orange peel is discarded, whereas at the commercial level orange peel is used in different ways to promote wellbeing due to its nutritional values. Regular consumption of orange peel is substantially valuable for numerous health benefits such as prevention and management of several health issues [16].

Citrus is a highly significant commercial fruit crop, grown and supplied mostly in Pakistan [17]. Citrus peel is utilized in numerous ways such as bio-ethanol manufacturing, cosmetic production, and fodder in fisheries [18]. Orange peel is used as nutraceutical due to its significant anti-microbial activity. Numerous health benefits promoted by the citrus peel include anti-fungal, anti-viral, anti-diabetic, antibacterial, anti-microbial, hypotensive mediator, and anti-mutagenic agent [19].

Peels of ***Citrus sinensis*** *are* usually discarded as waste material. Studies have shown that the orange peel is a significant source of many nutrients and bioactive compounds that provide numerous health benefits when used as a nutraceutical in daily diet.

Orange peel is a potential reserve of nutraceuticals universally, and also a source of low-cost dietary supplement due to easy accessibility and low cost of orange residue. Orange peel is a novel nutraceutical, hence, the consumption of citrus residues is a source of the inexpensive and competent bioactive compounds that promote wellbeing [20].

7.6 Nutritional Profile of Orange Peel

The beneficial constituents of orange peel include Minerals such as Potassium, Copper, Magnesium, Iron, Calcium, and Potassium; Vitamins such as B6, B5, Vit-A and Vit-C; Pigments like B-Carotene; and Fiber such as Pectin [21].

The color and texture of orange peel contributes to its flavonoids content that promotes high biological activity and wellbeing [13]. The consumption of orange peels protects against infectious agents, degenerative and metabolic diseases, free radicals and immune suppression due to presence of anti-oxidant, phytochemicals and flavonoids content. The primary residue from oranges extraction is the orange peel which contains a significant source of bioactive compounds such as phytochemicals, flavonoids associated with potential antioxidant activity in the body. Antioxidant activity from peel is due to the presence of hesperidin. Some phenol compounds present in orange peel such as **phlorin** and coniferin are involved in the potential action against the radical scavenging in the body [22].

Extensive analysis of the nutritional potential of *Citrus sinesis* L. Osbeck, peel has provided detailed information on the beneficial effects of its component parts. Proximate analysis demonstrates its high-level of carbohydrates, proteins, fats, fiber, and ash. Elemental analysis revealed the presence of nutrient intensities such as magnesium, copper, chromium, zinc and manganese. Phytochemical analysis shows the presence of bioactive composites such as saponins, flavonoids, cardiac glycosides, and reducing sugars. Researchers have also found many physicochemical properties of bioactive elements in orange peel provides great benefits for skin diseases [23].

A research conducted to examine the TPC and TFC content in orange peel extract, found numerous organic acids such as ascorbic acid, citric acid, and lactic acid, as well as flavonoids including vitexin, apigenin, catechin, and epigallocatechin. High antioxidant activity in orange peel extract is due to the presence of flavonoids and phenolics [12]. Analysis of *Citrus sinensis* L. peel indicate that it is rich in Fiber-Rich Fraction (FRFs), containing both soluble and insoluble fibers. Insoluble fiber consists of cellulose and pectin rich in polysaccharides. FRFs in orange peel possess nutraceutical properties by promoting physiological functions [24].

7.7 Orange Peel as a Nutraceutical Potential

7.7.1 Antioxidant and Anti-inflammatory Activity

Orange peel contains a complex mixture of flavonoids called chenpi. *Citrus reticulata* has been historically used as a traditional remedy in the management of various digestive disorders. The nature of bioactive compounds present in orange peel is dependent on the type of orange and from where it originates. *Xinhui* orange peel extract has great polyethoxylated flavones content. Orange peel extract has a significant level of NO, cyclooxygenase inhibitory activity. The Nobiletin and essential oils contained in the orange peel possess anti-inflammatory activity [25]. Similarly, the presence of limonene in orange peel accounts for its anti-cancerous property. The level of flavonoids in orange peel extract is considerably high, thus enhancing its antioxidant activity in the body. Such anti-oxidant activity include effects of polymethoxylated flavones, nitric oxide, and 2,2-azobis-2-methyl-propanimidami de and cyclooxygenase. Bioactive agents present in orange peel extract such as polymethoxyflavones (PMFs) possess anti-inflammatory activity in the body. PMFs are involved in the regulation of inflammation by controlling genes such as TNF-a, IL-, IL-8, and COX-2. Orange peel is known is reduce Paw edema in human. Studies showed that the anti inflammatory property of orange peel is comparatively similar to that produced by pharmaceutical drugs such as ibuprofen. Diseases related to inflammation can be managed and treated with effective orange peel extract bioactive compounds [26].

7.7.2 Anti-obesity Significance

Orange peel powder possess various anti-inflammatory health benefits such as protecting humans from disease conditions related to oxidative stress and inflammation. Pro-inflammatory response increases as fat deposit increases. These alterations lead to endothelial dysfunction and systemic inflammation [27].

Intake of citrus helps in management of endothelial dysfunction and abnormal lipid profile. Orange peels have a high antioxidants, Vitamin C, pectin, and flavonoids content which have shown great benefits in reducing endothelial dysfunction and inflammation caused by obesity [28]. Being a good source of pectin, orange peel can protect against hypercholesterolemia [29].

7.7.3 Anti-cancer Activity

The presence of PMFs in the citrus peel act as the potent anti-cancerous agent. They inhibit the carcinogens through cancer cell mobility inhibition, blockage of metastasis cascade process, enhance antiangiogenic action and induce proptosis in the body. Studies showed that the flavonoids in orange peel have high potential nutraceuticals which is used in cancer therapy. Citrus peel is a growing nutraceutical agent that act at molecular level in the prevention of cancer [30]. A molecular docking study by Egbuna et al. [31] in search of top fms-like tyrosine kinase 3 (FLT3) inhibitor against acute myeloid leukemia found natural bioactive compounds present in fruit peels and citrus as top performing. This is an indication that the peels of fruit could be a source of important nutraceuticals for the prevention and treatment of cancer.

7.7.4 Anti-diabetic Activity

Orange peel extract has a huge impact on blood sugar levels. Studies have shown that it possesses anti-diabetic property. It has been shown in a recent study to lower glucose levels by up to 19.30%. It is a rich source of phenolic compounds, pectin, and flavonoids such as narirutin, eriocitrin, hesperidin, and naringin which have significant effect against diabetes [32]. Hesperidin and naringin, found in orange peel act to decrease the action of phosphenol pyruvate and glucose-6-phosphate. The antidiabetic properties of these compounds has been reported [33–35]. They inhibits a-amylase inhibition activity which is involved in the stimulation of insulin secretion, it repairs the defects in pancreatic-beta cells, and it helps in the formation of glucose from carbohydrates [36].

7.7.5 Anti-hypercholesterolemia Activity

Flavonoids such as naringin and hesperidin act against hypercholesterolemia. Several studies reported that flavonoids are involved in lowering triglycerides and cholesterol level in the body. This happens through the inhibition of HMG-CoA reductase activity and acetyl-coenzyme A acetyltransferase activity (Wilcox). Flavonoids in citrus is known to prevent cardiovascular diseases via the reduction in the homocysteine level. The hypercholesterolemic effects of the limonene and polymethoxylated flavonoids is similar to the effect of lipid lowering drugs, albeit with minimal side effects. Flavonoids in citrus peel confers cardioprotective benefits to human health [37].

7.8 Conclusion

Orange peel is generally regarded and discarded as waste, however, recent research has shown that orange peel is beneficial to human health due to the presence of multiple bioactive nutrients. In addition to the high content of dietary fibers and phenolic compounds, phytochemical agents in orange peel promote wellbeing due to extensive anti-inflammatory, anti-oxidant, anti-obesity, anti-cancerous, anti-diabetic and anti-hypercholesterolemic properties. Given its easy accessibility, low price, nutritious value, and eco-friendly nature, orange peel is a potentially significant source of nutraceuticals that promotes general health and prevent illness.

References

1. Das L, Bhaumik E, Raychaudhuri U, Chakraborty R (2012) Role of nutraceuticals in human health. J Food Sci Technol 49(2):173–183
2. Egbuna C, Tupas GD (eds) (2020) Functional foods and nutraceuticals—bioactive components, formulations and innovations, 1st edn. Springer Nature, Cham. 646 pp
3. Food and Agriculture Organization of the United Nations (FAO) (2007) Report on functional foods, food quality and standards service (AGNS)
4. Santini A, Novellino E (2017) To nutraceuticals and back: rethinking a concept. Foods 6(9):74
5. Souto EB, Silva GF, Dias-Ferreira J, Zielinska A, Ventura F, Durazzo A, Lucarini M, Novellino E, Santini A (2020) Nanopharmaceutics: part I—clinical trials legislation and good manufacturing practices (GMP) of nanotherapeutics in the EU. Pharmaceutics 12(2):146
6. Campos JR, Severino P, Ferreira CS, Zielinska A, Santini A, Souto SB, Souto EB (2019) Linseed essential oil–source of lipids as active ingredients for pharmaceuticals and nutraceuticals. Curr Med Chem 26(24):4537–4535
7. Rudra SG, Nishad J, Jakhar N, Kaur C (2015) Food industry waste: mine of nutraceuticals. Int J Sci Environ Technol 4(1):205–229
8. Gorinstein S, Martin-Belloso O, Park YS, Haruenkit R, Lojek A, Ciz M, Caspi A, Libman I, Trakhtenberg S (2001) Comparison of some biochemical characteristics of different citrus fruits. Food Chem 74:309–315
9. Kratchanova M, Pavlovaa E, Panchev I (2004) The effect of microwave heating of fresh orange peels on the fruit tissue and quality of extracted pectin. Carbohydr Polym 56:181–185
10. Angew ON (2007) Functional foods. Trends Food Sci Technol 30:19–21
11. Mandalari G, Bennett RN, Bisignano G, Saija A, Dugo G, Lo Curto RB, Faulds CB, Waldron KW (2006) Characterization of flavonoids and pectins from bergamot (Citrus bergamia Risso) peel, a major byproduct of essential oil extraction. J Agric Food Chem 54(1):197–203
12. Di Majo D, Giammanco M, La Guardia M, Tripoli E, Giammanco S, Finotti E (2005) Flavanones in citrus fruit: structure–antioxidant activity relationships. Food Res Int 38(10):1161–1166
13. Tripoli E, La Guardia M, Giammanco S, Di Majo D, Giammanco M (2007) Citrus flavonoids: molecular structure, biological activity and nutritional properties: a review. Food Chem 104(2):466–479
14. Crowell PL (1999) Prevention and therapy of cancer by dietary monoterpenes. J Nutr 129(3):775S–778S
15. Thenmozhi M, Rajeshwari S (2010) Phytochemical analysis and anti-microbial activity of Polyalthia longifolia. Int J Pharm Bio Sci 1:1–7
16. Ramya Sree P. Nutritional and health benefits of orange peel. Available: https://www.pharmaresearchlibrary.com/nutritional-and-health-benefits-of-orange-peel/#:~:text=Orange%20

peel%20contains%20pectin%2C%20a,which%20results%20in%20better%20digestion. Accessed 03-10-2021
17. Rafiq S, Kaula R, Sofia SA, Bashira N, Nazirb F, Nayik GA (2016) Citrus peel as a source of functional ingredient: a review. J Saudi Soc Agric Sci 17(4):351–358
18. Kim JY, Oh TH, Kim BJ, Kim SS, Lee NH, Hyun CG (2008) Chemical composition and anti-inflammatory effects of essential oil from Faefugium japonicum flower. J Oleo Sci 57:623–628
19. Ghasemi K, Ghasemi Y, Ebrahimzadeh MA (2009) Antioxidant activity, phenol and flavonoid contents of 13 citrus species peels and tissues. Pak J Pharm Sci 22(3):277–281
20. Tao NG, Liu YJ, Zhang JH, Zeng HY, Tang YF, Zhang ML (2008) Chemical composition of essential oil from the peel of Satsuma mandarin. Afr J Biotechnol 7(9):1261–1264
21. Liew SS, Ho WY, Yeap SK, Sharifudin SA (2018) Phytochemical composition and in vitro antioxidant activities of Citrus sinensis peel extracts. PeerJ 6:e5331
22. Williams CA (2013) Specialized dietary supplements. In: Equine applied and clinical nutrition, Saunders, Philadelphia
23. Osarumwense PO, Okunrobo LO, Uwumarongie-Ilori EG (2013) Phytochemical screening, proximate and elemental analysis of Citrus sinensis peels (L.) Osbeck. J Appl Sci Environ Manag 17(1):47–50
24. Chau CF, Huang YL (2003) Comparison of the chemical composition and physicochemical properties of different fibers prepared from the peel of Citrus sinensis L. Cv. Liucheng. J Agric Food Chem 51(9):2615–2618
25. Chen XM, Tait AR, Kitts DD (2018) Flavonoid composition of orange peel and its association with antioxidant and anti-inflammatory activities. Food Chem 218:15–21
26. Gosslau A, YuChen K, Ho C-T, Li S (2014) Anti-inflammatory effects of characterized orange peel extracts enriched with bioactive polymethoxyflavones. Food Sci Human Wellness 3(1):26–35
27. Wang Z, Nakayama T (2010) Inflammation, a link between obesity and cardiovascular disease. Mediat Inflamm 2010:535918
28. Hashemi M, Khosravi E, Ghannadi A, Hashemipour M, Kelishadi R (2015) Effect of the peels of two citrus fruits on endothelium function in adolescents with excess weight: a triple-masked randomized trial. J Res Med Sci 20(8):721
29. Sabzghabaee AM, Khayam I, Kelishadi R, Ghannadi A, Soltani R, Badri S, Shirani S (2013) Effect of Zizyphus jujuba fruits on dyslipidemia in obese adolescents: a triple-masked randomized controlled clinical trial. Med Arch 67(3):156
30. Parmar HS, Kar A (2008) Medicinal values of fruit peels from Citrus sinensis, Punica granatum, and Musa paradisiaca with respect to alterations in tissue lipid peroxidation and serum concentration of glucose, insulin, and thyroid hormones. J Med Food 11(2):376–381
31. Egbuna C, Patrick-Iwuanyanwu KC, Onyeike EN, Khan J, Alshehri B (2021) FMS-like tyrosine kinase-3 (FLT3) inhibitors with better binding affinity and ADMET properties than sorafenib and gilteritinib against acute myeloid leukemia: in silico studies. J Biomol Struct Dyn 6:1–12. https://doi.org/10.1080/07391102.2021.1969286
32. Wilcox LJ, Borradaile NM, de Dreu LE, Huff MW (2001) Secretion of hepatocyte apoB is inhibited by the flavonoids, naringenin and hesperetin, via reduced activity and expression of ACAT2 and MTP. J Lipid Res 42(5):725–734
33. Kumar S, Egbuna C (eds) (2019) Phytochemistry: an in-silico and in-vitro update: advances in phytochemical research. Springer, Berlin
34. Tupas GD, Otero MCB, Ebhohimen EI, Egbuna C, Aslam M (2019) Anti-diabetic lead compounds and targets for drug development. In: Egbuna C, Kumar S, Ezzat SM, Ifemeje JC, Saravanan K (eds) Phytochemicals as lead compounds for new drug discovery. Elsevier, Cambridge, pp 127–138. https://doi.org/10.1016/b978-0-12-817890-4.00008-1
35. Egbuna C, Hassan S (2021) Dietary phytochemicals: a source of novel bioactive compounds for the treatment of obesity, cancer and diabetes. Springer, Cham. https://doi.org/10.1007/978-3-030-72999-8

36. Abdelbaky MS, Elmehiry HF, Ali NK (2009) Effect of some citrus peels on hypercholesterolemic rats. In: Proceedings of the 1st international and 4th Arab annual scientific conference on: academic accreditation for higher specific education institutions and programs in Egypt and Arab world reality and expectation 2009 Apr 8, pp 8–9
37. Milind P, Dev C (2012) Orange: range of benefits. Int Res J Pharm 3(7):59–63

Chapter 8
Pineapple Fruit Peels as a Source of Nutraceuticals

Obinna Chukwuemeka Uchenna Adumanya

8.1 Introduction

Fruits are known to have inherent health benefits and are used as both food and medicine in various cultures around the world. Nutraceuticals are a term that has recently been coined to describe such characteristics (both as food and medicine). The term nutraceutical is a buzzword which is a blend of two words, "nutrient" (a nourishing food component) and "pharmaceutical" (a medicinal drug) [1]. Stephen DeFelice, the founder and chairman of the Foundation for Innovation in Medicine, an American nonprofit based in Cranford, New Jersey, invented the name in 1989 [1, 2]. Nutraceutical is any substance which is a food or a part of food that provides medical or health benefits apart from providing basic nutrition [3]. Nutraceuticals are defined by the Canadian Ministry of Health as a substance separated or purified from food that is normally offered in a medical form that is not related with food and has been shown to provide physiological benefits as well as benefits against chronic diseases [4, 5].

Fruits (like pineapple) are one of the plant species having nutritional and medicinal properties which the nature has bestowed to mankind.

In terms of productivity, pineapple (*Ananas comosus* L.) (Fig. 8.1) belongs to the Bromeliaceae family. It is the third most important fruit crop in tropical regions of the world after banana and citrus fruit. Many countries, including India, Thailand, the Philippines, Brazil, China, Nigeria, Kenya, Indonesia, Mexico, and Costa Rica, grow pineapple as a commercially important fruit crop [6].

Due to its exceptional flavor and taste, it is renowned as the "Queen of Fruits" [7]. Pineapples can be eaten or served raw, cooked, juiced, or preserved. Its fruit is perishable and only available during certain seasons. Fruit production in the world

O. C. U. Adumanya (✉)
Department of Biochemistry, Faculty of Sciences and Computing, University of Agriculture and Environmental Sciences, Umuagwo, Imo State, Nigeria

© The Author(s), under exclusive license to Springer Nature Switzerland AG 2022
C. Egbuna et al. (eds.), *Food and Agricultural Byproducts as Important Source of Valuable Nutraceuticals*, https://doi.org/10.1007/978-3-030-98760-2_8

Fig. 8.1 Matured ripen pineapple fruit

is expected to reach 14.6 million tons. With an annual output of roughly 1.2 million tonnes, India is the fifth largest pineapple producer in the world [7]. Carbohydrates, sugar, organic acid, vitamins, crude fiber, and minerals are all in pineapple. Also in pineapple is bromelain enzyme, a natural detoxifier that is beneficial to the digestive system and aids in the maintenance of a healthy weight and balanced diet [6]. Its anti-inflammatory and anti-clotting capabilities are also well-known [8].

Pineapple has appreciable amounts of calcium, potassium, fiber, and vitamin C content. It has a minimal fat and cholesterol content. Vitamin B1, vitamin B6, copper, and dietary fiber are also abundant in this fruit. Bioactive substances are extra nutritional elements found in small amounts in food that have the ability to alter the health of living organisms. Phenolic compounds, ascorbic acid, β-carotene, and flavonoids are the key bioactive components of pineapple. β-glucan, proteoglycan, lectin, polysaccharides, triterpenoids, dietary fibre, lentinan, steroids, glycopeptides, terpenes, saponins, xanthones, coumarins, and alkaloid are all abundant in pineapple [9].

8.2 Pineapple Varieties

There are several varieties of pineapple, but only two are widely used and economically grown. They are the pineapple King and Queen.

8.2.1 Queen Pineapple

The queen pineapple is a perennial herb that grows to a height of 1.0–1.5 m. The stem of the plant is short and stocky, with stiff, waxy leaves. Its fruit are rich in color, yellow in color, and weigh 0.9–1.3 kg per [9]. The flesh has a deep golden yellow color, with a crisp texture and a wonderful scent and flavor. Because eyes are small and deep, removing the skin requires a thicker cut. Brownish-red, shorter, and spiky leaves.

8.2.2 King Pineapple

The King variety is a short 1–1.5 m herbaceous perennial with trough-shaped leaves 30–100 cm long and a stout stem [9]. It is a late-maturing pineapple type that is widely grown in India, Thailand, the Philippines, Brazil, China, Nigeria, Kenya, Indonesia, Mexico, and Costa Rica [6]. Fruit is large, rectangular in shape, and slightly tapering towards the crown, weighing 2–3 kg [9]. When fully mature, the fruit with large and shallow eyes turns yellow. The flesh is a pale yellow color.

8.3 Pharmaceutical and Nutraceutical

Pharmaceuticals are drugs or products used mainly in treatment or management of ailments. While nutraceuticals are food or products intended to prevent diseases or improve health.

8.3.1 Pineapple as Food

Pineapple fruits have a high moisture content, sugar content, soluble solid content, ascorbic acid content, and low crude fiber level. Joy [10] revealed the nutritional contents of matured ripe pineapple fruit (Table 8.1). As a result, pineapple can be utilized as a nutritional supplement for good personal health [11]. Pineapples are usually eaten fresh or in the form of pineapple juice. Field ripe fruits are best eaten fresh, and just the crown, rind, eyes, and core need to be removed. Pineapple can be eaten fresh, canned, or juiced, and can be found in a variety of foods, including desserts, fruit salads, jam, yogurt, ice cream, sweets, and as a meat accompaniment [8].

8.3.2 Pineapple as Medicine

Pineapple can be utilized as a nutritional supplement for better personal health. The fruits of the pineapple are high in vitamins and minerals. A single ripe pineapple fruit can provide around 16.2% of the daily vitamin C requirement [11]. Vitamin C is the body's primary water-soluble antioxidant, protecting normal cells from free radicals. Vitamin C is a potent antioxidant that aids in the synthesis of collagen in bones, blood vessels, cartilage, and muscle, as well as iron absorption. Vitamin C

Table 8.1 Nutrients in 100 grams (g) pineapple [10]

Nutrients	Amount
Energy	52 calories
Dietary fibre	1.40 g
Carbohydrate	13.7 g
Protein	0.54 g
Iron	0.28 mg
Magnesium	12 mg
Calcium	16 mg
Potassium	150 mg
Phosphorus	11 mg
Zinc	0.10 mg
Vitamin A	130 I.U
Vitamin B1	0.079 mg
Vitamin B2	0.031 mg
Vitamin B3	0.489 mg
Vitamin B6	0.110 mg
Vitamin C	24 mg

also helps to prevent urinary tract infections in pregnant women and lowers the risk of malignancies of the colon, esophagus, and stomach [8]. Malic acid accounts for 13% of the acidity in pineapple juice. Malic acid has health benefits as well. It improves immunity, promotes smooth, firm skin, aids in the maintenance of oral health, and lowers the danger of toxic metal poisoning [6].

Pineapple also contains vitamin B1, vitamin B6, copper, and dietary fiber. Pineapple is a natural anti-inflammatory and digestive help. Bromelain, found in fresh pineapples, is used to tenderize meat. Bromelain, a proteolytic enzyme found in pineapples, aids digestion by breaking down protein. Bromelain is only found in small amounts in the fruit's edible sections; all commercially accessible bromelain comes from the stem. Bromelain pills are widely used by athletes to address a variety of physical aches and ailments. Pineapple juice can assist the body hydrate and replenish its immunological system. It aids in the development of strong bones.

Pineapples are high in manganese, a trace mineral necessary for the body's bone and connective tissue formation. One cup of pineapple contains 73% of the daily manganese requirement. Pineapple benefits can help young people grow their bones and help older people strengthen their bones. The high manganese concentration of pineapple juice makes it a suitable choice for improving sperm quality and thus fertility [8]. Bromelain has been shown to reduce swelling in inflammatory disorders such as acute sinusitis, sore throat, arthritis, and gout, as well as expediting recovery following injuries and surgery. Pineapple enzymes have been successfully utilized to treat rheumatoid arthritis, as well as to expedite tissue healing following traumas, diabetic ulcers, and general surgery.

Pineapple helps eliminate plaque from artery walls and decreases blood clotting. Pineapple enzymes may help people with restricted arteries, such as those who suffer from angina. Bronchitis and throat infections can be treated with pineapples. Pineapple is a great brain tonic that helps with memory loss, depression, and melancholy. Drink pineapple juice if you're experiencing morning sickness, motion sickness, or nausea. It successfully relieves the symptoms of nausea and vomiting. Pineapple is well-known for its ability to relieve constipation and irregular bowel movements. This is due to its high fiber content, which promotes regular and smooth bowel motions. Pineapple helps to eliminate intestinal worms while also keeping the intestines and kidneys clean. It aids in the prevention of gum disease as well as the creation of plaque, keeping teeth healthy. To achieve and as a radical treatment for venereal illnesses, the flesh of very immature (toxic) fruits is purposely consumed [9].

The dried, powdered root is used to treat edema in Africa. On fractures, crushed rind is used, and on hemorrhoids, a rind decoction with rosemary is applied.

Pineapple lowers blood pressure, treats inflammation, aids weight loss, lowers death rates, and protects against diabetes and free radical damage. It restores the strength and health of damaged teeth. Additionally, it aids in the treatment of sinusitis and throat problems. Cure ailments such as asthma, obesity, swollen joints, digestive issues, and heart difficulties. Pineapples are high in manganese, which helps to build strong bones and muscles.

High antioxidant levels can also help to heal atherosclerosis and immunological illness. It does not harm the body's cells, and because it is so hot, it is used to disregard chilly weather. It is also used for a perfect, unbreakable body, as well as to avoid cancer, heart attacks, nausea, and to provide long natural hairs.

Use to treat acne, wrinkles, and aging issues, as well as to grow strong nails, soft lips, and thick hair [6].

8.4 Pineapple Fruit Peels (PFP) as Nutraceuticals

Pineapple fruit peels (Fig. 8.2) has a variety of health benefits, including reducing inflammation, boosting the immune system, enhancing libido, preventing cancer, relieving asthma, improving vision, preventing osteoarthritis, regulating blood sugar levels, promoting healthy bones, improving gum health, lowering cholesterol levels, curing swelling and irritation, preventing heart diseases, promoting healthy skin and hair, treating intestinal worms, aiding weight loss, and easing arthritis [12]. The proximate composition of pineapple fruit peel is presented in Table 8.2 [13]. However, it is frequently discarded after being cut off, despite the fact that it has health benefits when used properly [14, 15].

Vitamin B, C, folate, thiamin, pantothenic acid, enzyme 1, niacin, and fiber are all found in PFP. It's also high in minerals like magnesium, potassium, copper, manganese, calcium, iron, and other vitamins and minerals. Antioxidants with antibacterial, anti-inflammatory, anti-aging, and anticancer effects are also found in it [12, 16, 17]. These advantages are now known as nutraceutical qualities, and they include:

Fig. 8.2 Pineapple fruit peels

Table 8.2 Proximate composition of the pineapple fruit peels [13]

Composition	g/100 g peel, dry wt
Crude protein	9.13 ± 0.25
Crude lipid	1.57 ± 0.13
Total dietary fibre	42.2 ± 0.89
Ash	4.81 ± 0.03
Moisture	82.7 ± 1.91 (wet wt)
Carbohydrate	42.3

8.4.1 Antioxidant, Anti-inflammatory and Anti-cancer Benefits

PFP contains vitamin B, C, folate, thiamin, pantothenic acid, the enzyme bromelain, niacin, and fiber. Magnesium, potassium, copper, manganese, calcium, iron, and other vitamins and minerals are also abundant. It also contains antioxidants with antibacterial, anti-inflammatory, anti-aging, and anticancer properties [12, 16, 17]. These benefits are now referred to as nutraceutical characteristics, and they include the following:

Bromelain, a powerful enzyme present in high concentrations in the skin and stems of pineapples, helps the body to reduce inflammation. It works as an anti-inflammatory in the sinuses and throughout the body, and it's been suggested for reducing swelling following surgery or injury [12]. PFP's manganese, vitamin C, bromelain, and antioxidant content have been demonstrated in several studies to help fight tumors and prevent cancer [13].

8.4.2 Antimicrobial and Immunostimulant

Protecting the immune system is one of the foundations of good health. Pineapple fruit peel is a fantastic way to maintain your body healthy and hydrated, preventing colds and other ailments that necessitate the use of drugs. The high quantities of vitamin C and phytochemicals in its skin, which protect and fight infections, play these roles [15]. Bromelain and vitamin C have antibacterial, mucus-cutting, cough-suppressant, wound-healing, and overall system-boosting properties in the body [16, 18].

8.4.3 Aids Digestion and Energizes

Not only can the peels aid digestion, but they've also been shown to aid in the fight against intestinal parasites, constipation, and irritable bowel syndrome. They also help to maintain a healthy gut flora. If you need additional energy to get through your daily responsibilities, try pineapple peel tea instead of artificially sweetened juices, which will provide you with the calories you require without causing weight gain. This beverage also aids in fat burning [12].

8.4.4 Anti-hypertension and Blood Booster

Bromelain prevents blood clots from forming, and pineapple juice's copper component promotes the development of healthy red blood cells. Potassium helps keep blood vessels healthy and can help balance blood pressure by counteracting significant levels of salt.

8.4.5 Vision Protector

Beta carotene and vitamin C present in the entire plant aid in battling degenerative eye diseases like macular degeneration [19].

8.4.6 Bone and Dental Strengthener

Pineapple skins are strong in manganese, which helps to combat inflammation in the gums and tissues. Manganese promotes bone and tooth growth, strength, and repair [20, 21]. Its vitamin C and astringent characteristics maintain gums clean and healthy, which is good for dental health.

8.5 Pineapple Fruit Peels Preparation Procedure

Pineapple fruit peels can be prepared in a variety of ways to benefit your health. Here are a couple of ideas:

8.5.1 PFP Tea Making

By soaking in hot water, you can make a tea and serve it hot or cold to receive the benefits of PFP. To taste, add cinnamon sticks, cloves, and ginger to the hot water extract. Then serve immediately or chill and sweeten to taste.

8.5.2 PFP Extracts (Juice) Preparation

PFP (dried or fresh) is soaked in hot water, and the mixture is allowed to cool before being blended and used as needed.

8.5.3 Issues with Pineapple Fruit Peels Usage

PFP's high levels of vitamin C and bromelain may lead to itchy skin, rash, stomach pain, vomiting, diarrhea and breathing challenges after its ingestions or usage. Bromelain breaks down proteins, it can "eat away" at your flesh [13]. For this reason, it's also an ingredient in meat tenderizer. But as soon as you swallow the PFP, your stomach acids destroy the flesh-eating enzyme.

8.6 Conclusion

Pineapple, behind banana and citrus fruit, is the world's third most important tropical fruit crop in terms of production and consumption. Until recently, the PFP was either discarded or used infrequently by herbalists to address a variety of health issues. Its nutritional and medicinal properties have been studied and exploited extensively. PFP's nutraceutical properties include anti-inflammatory, anti-oxidant, anti-microbial, anti-viral, anti-cancer, anti-hypertensive, antidiabetic, neuroprotective, and immune system booster. These nutraceutical effects have been connected to PFP's high quantities of bioactive compounds such as bromelain, antioxidants (phenolic compounds, ferulic acid, vitamin C and vitamin A), and manganese. As a result, it's critical that one prepare and use it correctly.

References

1. Dable-Tupas G, Otero MCB, Bernolo L (2020) Functional foods and health benefits. In: Egbuna C, Dable Tupas G (eds) Functional foods and nutraceuticals. Springer, Cham. https://doi.org/10.1007/978-3-030-42319-3_1
2. Dutta F, Marak MC, Meena AK (2017) Nutraceuticals from major fruit crops. Int J Agric Innov Res 5(6):1–5
3. DeFelice SL (1995) The nutraceutical revolution: its impact on food industry. Trends Food Sci Technol 6:59–61
4. Pandey M, Verma RK, Saraf SA (2010) Nutraceuticals: new era of medicine and health. Asian J Pharm Clin Res 3:11–15
5. Egbuna C, Tupas GD (eds) (2020) Functional foods and nutraceuticals—bioactive components, formulations and innovations, 1st edn. Springer Nature, Cham. 646 pp
6. Hossain FM, Akhtar S, Anwar M (2015) Nutritional value and medicinal benefits of pineapple. Int J Nutr Food Sci 4(1):84–88
7. Baruwa OI (2013) Profitability and constraints of pineapple production in Osun state, Nigeria. J Horticult Res 21(2):59–64
8. Debnat P, Dey P, Chanda A, Bhakta T (2012) A survey on pineapple and its medicinal value. Scholars Acad J Pharm 1(1):451–457
9. Pooja S, Sushma D, Shailendra KS, Narayan L (2017) Nutraceuticals profiling of queen and king varieties of pineapple (Ananas comosus) (pineapple). Int J Chem Stud 5(3):25–31
10. Joy PP (2010) Benefits and uses of pineapple. Pineapple Research Station, Kerala Agricultural University, Kerala. http://www.kau.edu/prsvkm/Html/BenefitsofPA.htm
11. Hemalatha R, Anbuselvi S (2013) Physicochemical constituents of pineapple pulp and waste. J Chem Pharm Res 5(2):240–242
12. Wijaya GTG (2019) Research and development final project Ananas "meat tenderizer from pineapple skin and crown". Diploma thesis, Ottimmo International Mastergourmet Academy
13. Huang Y-L, Chow C-J, Fang Y-J (2011) Preparation and physicochemical properties of fiber-rich fraction from pineapple peels as a potential ingredient. J Food Drug Anal 19(3):318–323
14. Hepton A, Hodgson AS (2003) Processing. In: Bartholomew DP, Paul RE, Rohrbach KG (eds) The pineapple: botany, production and uses. CABI Publishing, Cambridge
15. Aguilera RG, Maycotte-Cervantes D, Flores CMG (2003) Pineapple peels (Ananas comosus) waste as a source of phytochemicals. In: Conference paper, Tecnologico De Monterrey
16. Saraswaty V, Risdian C, Primadona I, Andriyani R, Andayani DGS, Mozef T (2017) Pineapple peel wastes as a potential source of antioxidant compounds. Earth Environ Sci 60:012013
17. Kumari A, Pareek N, Jain N, Mishra S (2020) Nutraceutical potential of pineapple pomace and its utilization in development of value added bread. Int J Rec Sci Res 11(6):38868–38870
18. Lawal D (2013) Medicinal, pharmacological and phytochemical potentials of Annona Comosus Linn. peel—a review. Bayero J Pure Appl Sci 6(1):101–104
19. http://www.remedydaily.com (2017) Don't throw away your pineapple peels
20. Palacios C (2006) The role of nutrients in bone health, from A to Z. Crit Rev Food Sci Nutr 46(8):621–628
21. Rondanelli M, Faliva MA, Peroni G, Infantino V, Gasparri C, Iannello G, Perna S, Riva A, Petrangolini G, Tartara A (2021) Essentiality of manganese for bone health. Nat Prod Commun 16(5):1–8

Chapter 9
Jackfruit (*Artocarpus heterophyllus* Lam) Byproducts as a Source of Nutraceuticals

Pass Chidiebere Chijindu

9.1 Introduction

The search for nutraceuticals has formed an integral part of man's existence. Several cases of toxicity, insufficient and unaffordable drugs encourage constant researches into promising plants with both nutritional as well as medicinal potentials. In the course of these search and researches, several plants have been identified, studied extensively and integrated into food optimization and drug production. In recent times, recommendations have been given to promote proper diets as well as health status awareness. A lot of importance has been given to the maintenance of proper body functioning and improving the functionality of the immune system using natural products which have been known to be more effective and safe. This has birthed the concept nutraceuticals. Nutraceuticals are products derived from food sources that provide both nutritional and medicinal benefits. It is a term derived from two words—Nutrients and Pharmaceuticals. Based on the chemical composition of products, natural sources and pharmacological conditions, nutraceuticals can be further categorized to include dietary supplements, functional food, medicinal food and pharmaceuticals. In general, nutraceuticals can be used in prevention of diseases. However, they can also be applied in health maintenance, symptom control and prevention of malignant processes. Over the years, the concept of nutraceuticals has gathered a lot of attention to its nutritive, safety and therapeutic effects.

Artocarpus heterophyllus Lam (Jackfruit) is a plant which have been identified as a good source of nutritional and antioxidant compounds which hold the potential for nutraceutical development [1]. Jackfruit (*A. heterophyllus* Lam) belongs to the family Moraceae. The generic name comes from the Greek words 'artos' (bread) and 'Karpos' (fruit). The specific name '*heterophyllus*' in Latin means with leaves of different sizes and shapes and the word 'heteros' in Greek corresponds to the

---

P. C. Chijindu (✉)
Department of Biology, College of Education, Warri, Delta State, Nigeria

© The Author(s), under exclusive license to Springer Nature Switzerland AG 2022
C. Egbuna et al. (eds.), *Food and Agricultural Byproducts as Important Source of Valuable Nutraceuticals*, https://doi.org/10.1007/978-3-030-98760-2_9

word 'different' [2]. Jackfruit tree is native to India and is popular in several tropical and subtropical countries.

9.2 Origin and Distribution of Jackfruit

Jackfruit (Fig. 9.1) is considered to have originated in the rain forests of Western Ghats in South Western part of India [3]. It has been cultivated since prehistoric times and has naturalized itself in many parts of the tropics, particularly in South East Asia where it is today an important crop of India, Myanmar, China, Sri Lanka, Malaysia, Indonesia, Thailand and Philippines. It is also grown in parts of Africa, Brazil, Surinam, Caribbean, Florida and Australia. It has been introduced to many pacific islands since post European contact and is of particular importance in Fiji, where there is a large population of India descent [4]. Jackfruit tree grows in warm and moist regions [3, 5, 6]. *Artocarpus heterophyllus* was brought into Africa by Arabs [2] and according to Amadi et al. [7], Jackfruit grows abundantly in some parts of Nigeria. However, its cultivation and consumption is on a decline.

Size and Form
Jackfruit is a medium size, ever green tree that typically attains a height of 8–25 m (26–82 ft) and a stem diameter of 30–80 cm. The canopy shape is usually conical or pyramidal in young trees, spread and domed in older trees. The tree casts a very dense shade. Heavy side branching usually begins near the ground. All parts of the tree exude sticky white Latex when injured.

Flowers
This specie of plant is monoecious, having male and female inflorescence (or "spikes") on the same tree. Male and female spikes are borne separately on short, stout stems that sprout from older branches and the trunk. Male spikes are dense, fleshy, cylindrical to club shaped and up to 10 cm in length. Flowers are tiny, pale green when young turning darker with age. Female flowers are larger, elliptic or rounded with a tubular calyx. The flowers are reportedly pollinated by insects and wind, with a high percentage of cross pollination.

Fig. 9.1 Jackfruit and inner section

Leaves

The leaves are dark green, alternate, entire, simple, glossy, leathery, stiff, large (up to 16 cm in length) and elliptic to oval in form. Leaves are often deeply lobed when juvenile and on young shoots.

Fruit

Jackfruit has a compound or multiple fruit (syncarp) with a green to yellow brown exterior rind that is composed of hexagonal, bluntly conical carpel apices that cover a thick, rubbery, whitish (When ripe) banana flavoured flesh (aril) surrounding each seed. The heavy fruit is held together by a central fibrous core. Fruits are oblong cylindrical in shape, typically 30–40 cm in length.

Seeds

The seeds are light brown, rounded 2–3 cm in length by 1–1.5 cm in diameter and enclosed in a thin whitish membrane. Up to 500 seeds can be found in each fruit. Seeds are recalcitrant and can be stored up to a month in cool, humid conditions (www.Traditionaltree.org; [4]).

Varieties of Jackfruit

Jackfruits have been classified based on their phenotypic and organoleptic characters with variation in their flakes and seed colour, shape, size, odour and period of maturity [1, 4, 8–10]. However, there are two main varieties of Jackfruit; firm and soft. In the firm variety, the parianth remains firm even at full ripeness; while in the soft variety, the parianths become soft and fleshy on ripening [3, 11].

The soft variety has fruits with small fibrous, soft and spongy flakes with very sweet carpels, whereas the firm variety is crunchy with crisp carpels and not sweet as the soft variety [3]. The firm variety is considered to be of high quality [12]. Some studies have reported variations in the chemical composition of some soft and firm types [11].

9.3 Phytochemical Composition of Jackfruit

Phytochemicals in general are natural bioactive compounds found in plants that work with nutrients and fibres to act as a defense system against disease or for protection [2, 13]. Several studies have shown that Jackfruit contains many classes of phytochemicals such as carotenoids, flavonoids, volatile acid, sterols and tannins with varying concentrations depending on the variety [14–19].

More recently, more studies have revealed the presence of several phytochemicals in various parts of the Jackfruit tree. The study by Amadi et al. [7] revealed the presence of flavonoids and tannins in the leaves, pulp and seeds of Jackfruit. Another study also revealed that Jackfruit pulp is rich in carbohydrates, proteins, amino acid, polyphenol, fatty acid, vitamins and minerals; the peel extract contains phenolic acids and flavonoids while the leaves are rich in phenolic acids, flavonoids, terpenoids and stilbenoids which can be used as sources of some important nutrients [20, 21]. The wood bark of Jackfruit has also been identified as rich sources of prenylated flavonoids, stilbenoids, triterpenoids and steroids [22]. The leaves and stem show the presence of sapogenins, cycloartenone, cycloartenol, β-sitosterol and tannins [4]. The Root contains β-sitosterol, ursolic acid, Betulinic acid and cycloartenone [23]. The fruit is rich in phytocompounds such as flavonoids, stilbenoids, morin, artocarpin, dihydromorin, cyanomacurin, isoartocarpin, cycloartocarpin, artocarpesin, norartocarpetin and cycloartinone [1, 24]. The seeds of Jackfruit contains bioactive compounds such as sterols and anthraquinones [25] as well as carbohydrate, proteins, fats, phenols, flavonoids, phytosterols, coumarins and saponins [2]. The seeds also contain Jacalin a major protein [4]. A study by Adan et al. [26] on unutilized parts of Jackfruits (Peel, fibre and Core) also revealed the presence of phytochemicals such as phenolics, flavonoids and tannins as well as essential minerals including iron, (Fe), calcium (Ca), potassium (K), sodium (Na) and magnesium (Mg) in these parts.

Several phytochemicals have also been isolated and characterized in different parts of the Jackfruit tree [27]. This is summarized in Table 9.1.

Table 9.1 Isolated compounds from various parts of Jackfruit Tree

S/No	Compound	Plant part	References
1.	Tetracyclic triterpenoids • 9,19-Cyclolanost-3-one-24,25-diol and • 9,1-Cyclolanost-3-one-24; 2 S diol Along with cycloartenone and cycloartenol		[28]
2.	Flavonoids such as Artoheteroids, A-D, Morin, Artocarmin A, Albanin A, Euchrenone A, Norartocarpanone and Steppogenin	Roots	[29]
3.	A new prenylated flavonoid—3-prenyl luteolin (1)	Wood extract	[30]
4.	A 2-arylbenzofuran derivative, artocarstilbene B and benzaldehyde derivative, (E)-3-5 dihydroxy-4-(3-methyl-but-1-enyl) benzaldehyde	Leaves	[31]
5.	Seven prenylated chromones and five prenylated flavonoids including two new prenylated chromones—Artoheterophines A and B		[32]
6.	New phenolic compounds—2,3-dihydro-5,7-dihydroxy-2-(2-hydroxy-4-methoxyphenyl)-4H-benzopyran-4-one and 6-[(1S, 2S)-1,2-dihydroxy-3-methylbutyl]-2-(2,4-dihydroxyphenyl-3-(3-methyl-2-buten-1-yl)-4H-1-benzopyran 4-one	Wood extract	[33]
7.	Furanoflavone-7-(2,4-dihydroxyphenyl)-4-hydroxy-2-(2-hydroxypropan-2-yl)-2,3-dihydrofuro (3,2-g) chromen 5-one(artocarpfuranol) together with 14 known compounds including Dihydromorin, Steppogenin, Norartocarpetin, Artocarpanone, Artocarpesin, Artocarpin, Cycloartocarpin, Cycloartopesin, Artocarpetin, Brosimone 1, Cudraflavone B, Carpachromene, Isoartocarpesin and Cyanomaclurin	Wood	[34]
8.	Phenolic compounds including one isoprenylated 2-arylbenzofuran derivative, Arthoheterophyllin A and three isoprenylated flavonoids, Artoheterophyllin B, Artoheterophyllin C and Artoheterophyllin D	Twig extract	[35]
9.	Artocarheterones A-E	Ripe fruits	[36]
10.	Dihydromorin and Norartocarpetin	Heart wood	[37]

Table 9.1 (contiued)

S/ No	Compound	Plant part	References
11.	Two novel 2′,4′,6′-trioxygenated flavones, heteroflavones A and B 5-hydroxy-7,2′4′6′-tetramethoxyflavone	Root bark	[16, 38]
12.	Three phenolic compounds Artocarpesin [5,7,2′4′-tetrahydroxy-6-(3-methylbut-3-enyl) flavone], Norartocarpetin (5,7,2′,4′-tetrahydroxy flavone) and Oxyresveratrol (trans-2,4,3′5′-tetrahydroxystilbene)		[18]
13.	Carotenoids—Cartenes, β-carotene, α-carotene, β-zeacarotene, and β carotene-5,6 epoxide and a dicarboxylic carotenoid-crocetin	Ripe fruits	[15]

9.4 Nutritional Value of Jackfruit

Jackfruit is a highly nutritious plant with different parts having different concentrations of nutritional contents. When compared to other tropical fruits, Jackfruit flesh and seeds was found to contain more protein, calcium, iron and Thiamine [9, 39, 40] when compared to other tropical fruits. A study by Tiwari and Vidyarthi [41] also revealed that Jackfruit is richer than apple, apricot, avocado and Banana in some minerals and vitamins.

The caloric content of Jackfruits is considerably low where a 100 g of Jackfruit is found to contain only 94 calories. The fat content of the seeds is negligible making it a good constituent of a fat free diet [3]. It is a source of vitamins and minerals [42]. The seeds may vary in their protein and carbohydrate contents with certain varieties consisting of about 6.8% proteins [22]. The protein content of the flesh of different varieties of ripen jackfruit ranges from 0.57 to 0.97%. Rahman et al. [11] reported the presence of a high percentage of starch in Jackfruit perianth and seed. The starch and dietary fibre content of the flesh has also been found to increase with the maturity. The carbohydrate concentration of different varieties of Jackfruit seed varies from 37.4 to 42.5%.

Jackfruit is rich in Vitamin C [42] and B-complex group of vitamins [3]. It contains very good amounts of vitamin B_6 (pyridoxine), niacin, riboflavin and folic acid [43].

Jackfruit has also been found to contain essential minerals. The flakes of ripe Jackfruits have high nutritive value with 100 g of ripe flakes containing 287–323 mg potassium, 30.0–73.2 mg calcium and 11–19 g carbohydrates [44]. The ripened Jackfruit contains minerals such as calcium, magnesium, vitamins and organic acids [41]. A study by Amadi et al. [7] also revealed the presence of higher potassium, vitamin C, and Zinc in Jackfruit pulp while the leaves were significantly higher in calcium, manganese and iron.

Fibre content of Jackfruit fleshes has been found to have a slight variation in different varieties ranging from 0.50 to 0.90% [45].

The seed flour of Jackfruit has been found to be rich in proteins, carbohydrates and minerals (Sodium, potassium, magnesium and zinc) with low fat content. This makes the seed flour a good constituent in functional foods and can be consumed safely without any health risk [2].

Jackfruit seeds possess antioxidant properties which can scavenge free radicals. The seeds can be consumed as a good source of nutritional, mineral, antimicrobial, antioxidant components and it has potential for value addition and nutraceutical developments.

9.5 Useful Products of Jackfruit

Jackfruit is widely used for preparing baby foods, jams, jellies, juice, biscuits, bread, chips, nectar, deserts, beverages like squash, wine and so on [42]. Specifically, the seeds are boiled or roasted and also used as a snack. However, fresh seeds have short shelf-life and so the seeds are dried and powdered to store for a long time [1]. The fruits are used as food, vegetables, pickle, chutney, jam, jelly, paste, candies, juice, powder and confectionery. The seeds are also salted as table nuts, the seed flour is used for baking and substrate is used for solid state fermentation and for animal feed [46]. The wood is used for making furniture, doors, windows, roof construction, musical instruments, bee hives, boats and dye [9].

The latex is used in producing vanishes, glue, caulking for boats and buckets [9]. The root is used for carving and picture framing. The leaves are used as fodder for cattle and goats, for making spoon to take rice kanji and removal of methylene blue [46–48].

The peels of Jackfruit is used as absorbent for the removal of cadmium [1, 46, 49].

Several phytochemical compounds, essential elements, antioxidant and antimicrobial elements have been isolated from unutilized parts of the fruits (including the peel, fibre and the core) hence, making them important also to the food, Agriculture and Pharmaceutical industry [50]. Researchers have confirmed that there are appreciable amounts of bioactive compounds in waste fruit parts (peels, seeds and fibre) with potential benefits to human health and can also act as fungicides, bactericides and in disease control in Agriculture [26, 50, 51]. This would reduce waste, bioaccumulation and furthermore ensure a healthy environment through the recycling of unutilized parts of fruits.

Jackfruit has been considered as an underutilized fruit in commercial scale mainly due to higher percentage of inedible portions which leads to more waste generation. In addition, difficulty in peeling and separation of edible bulbs from the rind, lack of knowledge on proper post-harvest practices and inadequate processing facilities in regions where they are grown have also been identified as reasons for the under utilization of Jackfruits. Hence, adhering to correct post-harvest practices and conversion of jackfruit into minimally processed products such as having precut jackfruits, may encourage more population towards consumption of jackfruits and conversion of jackfruits waste materials into valuable products, further aiding in waste management [3]. More researches are however encouraged in discovering possible industrial applications of jackfruits and proper management of waste generated in jackfruits processing.

9.6 Jackfruit Products as Nutraceuticals

The nutraceutical/medicinal properties of Jackfruit would be discussed under ethnobotanical / traditional medicinal uses and pharmacological or biological properties.

9.6.1 Ethnobotanical/Traditional Uses of Jackfruits

Jackfruit has been found to possess several ethnobotanical and traditional applications, some of which includes: The use of the leaves in curing fever, boils and skin diseases. When heated, the leaves prove useful in curing wounds. The latex of the fruit is helpful in treating dysopia, opthalmities and pharyngitis. The latex can also be mixed with vinegar to heal abscesses, snake bite and glandular swellings [52]. The seed starch is useful in relieving biliousness while the roasted seeds are regarded as aphrodisiac. The ash produced by burning the bark is used for healing abscesses and ear problems while the ash of the jackfruit leaves burned with corn and coconut shells is used alone or mixed with coconut oil to heal ulcers also. The root of Jackfruit tree forms the remedy for skin diseases, fever and diarrhea. The heartwood of the tree is used by Buddhist forest Monastic in Southeast Asia, for dying the robes of the monks to light brown colour. In China, the pulp and seeds of jackfruit are considered as a cooling and nutritious tonic. The fruit is useful in overcoming the influences of alcohol on a person's body system. The wood of jackfruit tree has sedative properties and is useful in convulsions [53]. The wood pith causes abortion. Latex is used as anti-inflammatory agent [54].

According to Tyagi et al. [55], the extract from the seed, root and bark of Jackfruit is helpful in digestion and used also in the treatment of diarrhea and dysentery. The bark of matured tree is useful in the treatment of dysentery and in releasing the placenta after calving in cows. Its decoction and latex are used in the treatment of asthma, prevent ringworm infection and heel cracking feet. The infusion of the matured leaves and bark is said to be effective in the treatment of diabetes, gall stones and relieving asthma. The leaves are believed to possess wound healing effects, reduce pain, decrease abuses and relieve ear problems. The dried latex yield ortostenone convertible to artosterone, a compound with marked androgenic actions. However, the sedative property of Jackfruit wood with pith resulting in abortion, increase in coagulation, as well as allergy in people who are allergic to birch pollen serves as caution to its indiscriminative use [56].

9.6.2 Pharmacological or Biological Properties of Jackfruit

Jackfruit (*Artocarpus heterophyllus* Lam) is a rich source of several compounds of high value having potential beneficial physiological activities. A wide range of pharmacological investigations have been carried out based on the constituents present in Jackfruit.

These pharmacological activities are presented in Table 9.2.

In addition to the above, Jackfruit is a good source of vitamin C which protects the skin from the damage that occurs as a consequence of the natural aging process and prolonged exposure to the sun [90]. Vitamin C is also essential for the production of collagen which gives firmness and strength to the skin [91] and maintains oral health.

Niacin in Jackfruit is necessary for energy metabolism, nerve function and the synthesis of certain homones [43].

Dietary fibre present in Jackfruit also makes a good bulk laxative. This decreases the exposure time and binds to cancer causing chemical, as well as minerals and vitamins in the colon, membrane [3, 92]. High fibre content also maintains smooth bowel movements and prevents constipation [43]. The flesh and seeds of Jackfruit are considered as cooling and nutritioustonic as well [43].

Jackfruit has several mineral contents in abundance. It is rich in magnesium which is important for the absorption of calcium and helps strengthen the bones and prevents bone-related disorders such as osteoporosis. Iron in Jackfruit helps to prevent anemia and aids in proper blood circulation. Copper plays an important role in thyroid gland metabolism [3, 93].

9.7 Conclusion

The importance of Jackfruit and its product as source of Nutraceuticals cannot be over emphasized. As a dietary supplement, Jackfruit is known to contain various nutrients which can be used as immune-modulators as well as prophylactic. It has also been shown to contain nutrients which are essential in combating malnourishment. In the area of diet control and weight management, the low fat and calorie content of Jackfruit has also made it an excellent inclusion in low fat diets. Direct consumption of its products as food (beverage, snack, flour for confectionary, wine, juice, jam etc.) makes it a readily available source of nutrients for maintaining metabolic activities in the body as well as health maintenance. In addition to these, Jackfruit has been shown to elicit various physiological activities including anti-inflammatory, antioxidant, antimicrobial, antidiabetic, antimalarial, anticancer, anti-osteoporotic, tyrosinase inhibitory activity, as well as immune-modulatory effects. Due to all these numerous health benefits, the consumption of Jackfruit has increased in recent years. The presence of phytonutrients further enhances the opportunities for development of value added products and nutraceutical developments.

Table 9.2 Pharmacological activities of Jackfruit

S/No	Pharmacological activity	Mechanism of action	Phytochemicals implicated	Plant Part isolated	References
1.	Anti-inflammatory activity	Artocarpesin suppressed the LPS—Induced production of nitric oxide (NO) and prostaglandin E_2 (PGE_2) through the down regulation of inducible nitric oxide synthase (iNOS) and cyclo-oxygenase 2 (COX-2) protein expressions	(i) Artocarpesin (ii) Moracin (iii) Flavonoids	Fruits	[57–61]
2.	Antioxidant activity	(i) Inhibition of iron-induced lipid peroxidation and scavenging DPPH	Cycloheterophyllin and Artonins A and B (both Prenyl flavones), Carotenoids, Polysaccharides, Phenolic compounds, and Ascorbic acid		[62–69]
		(ii) Decrease in formation of Reactive Oxygen Species (ROS)			[1, 70]
3.	Antimicrobial activity				
	(a) Antifungal activity	(i) Inhibition of the growth of *Fusarium moniliform, Saccharomyces cerevisae*	Lectins (Jacalin)	Seeds	[71]
		(ii) Activity against *Colletotrichum gloeosporoides* and *Penicillium italicum*	Quinic acid, Catechin and Chlorogenic acid		[72]
	(b) Antibacterial activity	(i) Broad spectrum antibacterial activity of butanol fractions of root bark and fruits		Root bark and fruits	[73]
		(ii) Inhibition of the growth of *Pseudomonas aeruginosa* ATCC 27853 and *Candida albicans*		Latex	[74]
		(iii) Effective against *Streptococcus pyogenes*	Dihydromorin		[37]
		(iv) Strong antibacterial activity against *Streptococcus mutans*, *S. pyogenes*, *Bacillus subtilis, Staphylococcus aureus* and *S. epidermidis*	Isolated antibacterial compounds-Cycloartocarpin, Artocarpin, Artocarpanone and Cyanomaclurin	Heartwood	[75]
		(v) Effective against methicillin-resistant *Staphylococcus aureus, P. auriginosa* and *E. coli*	Artocarpin		[75]
		(vi) Strong antibacterial activity displayed by membranes cell	Artocarpanone		[76]
		(vii) Inhibition of *E. coli, Listeria monocytogenes, Salmonella typhimurium, Salmonella enterica, Bacillus cereus. Enterococcus faecalis,* and *Staphylococcus aureus*		Crude leaves extract	[68]
	(c) Antiviral activity	Remarkable anti-HIV-I effects with EC_{50} values ranging from 0.09 to 9.72 μm	Chromones	Fruits	[36]

(continued)

Table 9.2 (continud)

S/No	Pharmacological activity	Mechanism of action	Phytochemicals implicated	Plant Part isolated	References
4.	Antidiabetic activity	(i) Improving glucose tolerance	Total phenolics content	Crude hot water extract	[77]
		(ii) Inhibiting α-glucosidase activity		Extract of jackfruit peel, pulp, fruit, flake and seeds	[20]
		(iii) Inhibitory activity on α-amylase and α-glucosydase		Ethanolic stembark extract	[78]
5.	Antimalarial activity	Antiplasomodial activity	Flavonoids—Artonin, Artocarpone		[46]
6.	Antihelminthic activity	The shoots revealed nematicidal activity against various nematodes namely: *Rotylenchulus reniformis, Tylechorhynchus brassicae, Tylenchus filiformis* and *Meloidogyne incognita*		Shoots	[79]
7.	Antihypertensive activity	(i) Vitamin B_6 present, reduces homocysteine levels in the blood which then lowers the risk of heart disease	Vitamin B6		[77]
		(ii) Presence of potassium helps in lowering blood pressure and reversing the effects of sodium that causes rise in blood pressure which affects the heart and blood vessels	Potassium		[3]
		(iii) Antihypertensive activity	Lignans, Isoflavones, Saponins		[3]

8.	Anticancer activity	(i) Inhibiting activity against the PC-3, NCI-H46 and/or A549 cancer cell lines		Leaves	[31]
		(ii) Anticancer effects on MCF-7, H460 and SMMC-7721 human cancer cell lines	Isolated phenolic compounds		[80]
		(iii) Antiproliferative activity against Hep G_2 and MCF-7 cells	Apigenin C-glycoside identified as 2′-0-β-D-Xylosylvitexin		[81]
		(iv) Effective against cell lines like T47D, TH29 and B16F10		Seed extract	[82]
		(v) Artocarpin impaired anchorage-independent growth capability, suppressed colon cancer growth and induced GIphase cell cycle arrest which was followed by apoptic as well as autophagic cell death	Artocarpin		[83]
		(vi) Significant biological activity on both normal and cancerous cells (human colon carcinoma cells) cultured invitro	Water-soluble polysaccharides		[84]
		(vii) Cytotoxic activity against MCF-7 cell nine	Heterophyllene, Norartocarpin, Artocarpin		[85]
		(viii) Inhibition of B16 melanoma cells with less cytoxicity in vivo	3-prenyl luteolin		[30]
			Isolated prenylated flavones-based polyphenols	Wood	[14]
		(ix) Chemoprotective properties to reduce mutagenicity of aflatoxin B1 (AFBI) and proliferation of cancer cells	Compounds present	Jackfruit	[86]
		(x) Top performing compound against fms-like tyrosine kinase 3 (FLT3)	Cycloartocarpesin	Jackfruit	[87]
9.	Anti-osteoporotic activity	Flavonoids from *A. heterophyllus* was found to suppress Cathepsin-K (cat-K) which is known to play a vital role in osteoclast-mediated bone resorption evidenced as an important target for the treatment of osteoporosis.	Flavonoids		[29]

(continued)

Table 9.2 (continud)

S/No	Pharmacological activity	Mechanism of action	Phytochemicals implicated	Plant Part isolated	References
10.	Tyrosinase inhibitory activity	Potent tyrosinase inhibitor	Norartocarpetin, Artocarpesin	Twigs and woods	[34, 35, 88]
11.	Immuno-modulatory effects	Jacalin is strongly mitogenic for human $CD4^+$ T lymphocytes. Hence its usefulness as a tool for the evaluation of the immune status of patients infected with human immuno-deficiency virus (HIV)-I	Jacalin	Seeds	[89]

References

1. Sreeja Devi PS, Kumar NS, Sabu KK (2021) Phytochemical profiling and antioxidant activities of different parts of *Artocarpus heterophyllus* Lam (Moraceae): a review on current status of knowledge. Future J Pharm Sci 7(30):1–7
2. Sreeletha AS, Lini JJ, Dhanyalekshmi CS, Sabu KR, Pratapchandran R (2017) Phytochemical, proximate, antimicrobial, antioxidant, and FTIR analysis of seeds of *Artocarpus heterophyllus* Lam. Adv Biotechnol Microbiol 5(1):001–009
3. Ranasinghe RASN, Maduwanthi SDT, Marapana RAUJ (2019) Nutritional and health benefits of jackfruit (*Atocarpus heterophyllus* Lam): a review. Int J Food Sci 2019:1–2
4. Prakash O, Kumar R, Mishra A, Gupta R (2009) *Artocarpus heterophyllus* (jack fruit): an overview. Pharmacol Rev 3(6):353–358
5. Bose TK (1985) In: Mitra BK (ed) "Jackfruit" in fruits of India: tropical and subtropical. Nayaprokash, Calcutta, pp 488–498
6. De Faria AF, De Rosso VV, Marcadente AZ (2009) Carotenoid composition of jackfruit (*Artocarpus heterophyllus*) determined by HPLC-PDA-MS/MS. Plant Foods Hum Nutr 64(2):108–115
7. Amadi JAC, Ihemeje A, Afam-Anene OC (2018) Nutrient and phytochemical composition of jackfruit (Artocarpus heterophyllus). Pulp, seeds and leaves. Int J Food Nutr Sustain Agric 6(3):27–32
8. Jagadeesh SL, Reddy BS, Basavaraj N, Swamy GSK, Gorbal K, Hedge L, Raghava GSV, Kajjidoni ST (2007a) Inter tree variability for fruit quality in jackfruit selections of Western Ghats of India. Sci Hort 112(4):382–387
9. Haq N (2006) Jackfruit (*Atocarpus Heterophyllus*). In: Williams JT, Smith RW, Dunsiger Z (eds) Tropical fruit trees. Southampton Centre for underutilized crops, University of Southampton, Southampton
10. Jagadeesh SL, Reddy BS, Swamy GSK, Gorbal K, Hedge L, Ragavan GSV (2007b) Chemical composition of jackfruit (*Artocarpus heterophyllus* Lam) selections of Western Ghats of India. Food Chem 102(1):361–365
11. Rahman MA, Nahar NA, Mosihuzzman. (1999) Variation of carbohydrate composition of two forms of fruit from jack tree (*Artocarpus heterophyllus* Lam) with maturity and climatic condition. Food Chem 65:91–97
12. Shyamalamma S, Chandra SBC, Hedge M, Naryanswamy P (2008) Evaluation of genetic diversity in jackfruit (*Artocarpus heterophyllus* Lam) based on amplified fragment length polymorphism markers. Genet Mol Res 7(3):645–656
13. Edeoga HO, Okwu DE, Mbaebie BO (2005) Phytochemical constituents of some Nigerian medicinal plants. Afr J Biotechnol 4:685–688
14. Arung ET, Shimizu K, Kondo R (2007) Structured activity relationship of prenyl-substituted polyphenols from *Artocarpus heterophyllus* as inhibition of melanin biosynthesis in cultured melanoma cells. Chem Biodivers 4(9):2166–2171
15. Chandrika UG, Jansz ER, Wamaouriya ND (2004) Analysis of carotenoids in ripe jackfruit (*Artocarpus heterophyllus)* kernel and study of their bioconversion in rats. J Sci Food Agric 85(2):186–190
16. Lin CN, Lu CM (1993) Heterophylol, a phenolic compound with novel skeleton from *Artocarpus heterophyllus*. Tetrahedron Lett 34(51):8249–8250
17. Ong BT, Nazimah SAH, Osman A (2006) Chemical and flavour changes in jackfruit (*Artocarpus heterophyllus* Lam) cultivar J_3 during ripening. Postharvest Biol and Technol 40(3):279–286
18. Venkataraman K (2001) Wood of Phenolics in the chemotaxonomy of the Moraceae. Phytochemisty 11(5):1571–1586
19. Wong KC, Lim CL, Wong LL (1992) Volatile flavour constituents of Chempedak (*Artocarpus polyphema* Pers). Fruit and jackfruit (*Artocarpus heterophyllus* Lam) from Malaysia. Flavour Fragr J 7(6):307–311

20. Zhang L, Tu ZC, Xie X, Wang H et al (2017) Jackfruit (*Artocarpus heterophyllus* Lam) peel: a better source of antioxidants and α-glucosidase inhibitors than pulp, flake and seed, and phytochemical profile by HPLC-QTOF-MS/MS. Food Chem 234:303–313
21. Shafiq M, Mahmood S, Yasmia A, Khan SJ, Khan NH et al (2017) Evaluation of phytochemical, nutritional and antioxidant activity of indigenously grown jackfruit (*Artocarpus heterophyllus* Lam). J Sci Res 9:135–143
22. Baliga MS, Shivashankara AR, Haniadka R, Dsouza J, Bhat HP (2011) Phytochemistry, nutritional and pharmacological properties of *Artocarpus heterophyllus* Lam. Food Res Int 44i:1800–1811
23. Dayal R, Seshadri TR (1974) Colourless compounds of the roots of *Artocarpus heterophyllus*. Isolation of new compound artoflavone. Indian J Chem 12:895–896
24. Wetprasit N, Threesangsri W, Klamklai N, Chula-Vantol M (2000) Jackfruit lectin: properties of mitogenicity and the inhibition of herpes virus infection. Jpn J Infect Dis 53:156–161
25. Meshram RL, Umbarka SN (2011) Comparative evaluation for invitro antioxidant activity from *Artocarpus heterophyllus* Lam and *Balanites aegyptica* L. Int J Pharm Tech Res 3(4):2006–2010
26. Adan AA, Ojwang RA, Mugu EK, Muwanza BK, Nyaboga EN (2020) Phytochemical consumption and essential mineral profile, antioxidant and antimicrobial potential of unutilized parts of jackfruits. Food Res 4(4):1125–1134
27. Bhad PR, Bobde MV, Sibi G (2021) Chemical constituents and biological activities of *Artocarpus heterophyllus* Lam (jackfruit): a review. Int J Clin Microbiol Biochem Tech 4:005–009
28. Barik BR, Bhaumik T, Day AK, Kundu AB (1994) Triterpenoids from *Artocarpus heterophyllus*. Phytochemistry 35i:1001–1004
29. Yuan WJ, Yuan JB, Peng JB, Din YQ, Zhu JX et al (2017) Flavonoids from the roots of *Artocarpus heterophyllus*. Fitoterapia 117:133–137
30. Arung ET, Shimizu K, Tanaka H, Kondo R (2010) 3-Prenyl Luteolin, a new prenylated flavone with melanin biosynthesis inhibitory activity from wood of *Artocarpus heterophyllus*. Fitoterapia 81:640–643
31. Wang XL, Shan XXT, Wang SQ, Wang XN (2017) New phenolic compounds from the leaves of *Artocarpus heterophyllus*. China Chem Lett 28:37–40
32. Liu YP, Yu XM, Zhang W, Wang T, Jiang B et al (2020) Prenylated chromosomes and flavonoids from *Artocarpus heterophyllus* with their potential antiproliferative and anti-inflammatory activities. Bioorg Chem 101:104030
33. Zheng ZP, Xu Y, Qin C, Zhang S, Gu X et al (2014) Characterization of antiproliferative activity constituents from *Artocarpus heterophyllus*. J Agric Food Chem 62:5519–5527
34. Zheng ZP, Chen KW, Jk T, Li H, Wang M (2008) Isolation of tyrosinase inhibitors from *Artocarpus heterophyllus* and use of its extracts as anti-browning agents. Mol Nutr Food Res 52:1530–1538
35. Zheng ZP, Chen S, Wang S, Wang XC, Chung KW et al (2009) Chemical components and tyrosinase inhibitions from the twigs of *Artocarpus heterophyllus*. J Agric Food Chem 57:6649–6655
36. Fu YH, Gu JM, Xie YT, Yu XM, Su Q-T et al (2020) Penylated chromosomes from the fruits of *Artocarpus heterophyllus* and their potential anti HIV-1 activities. J Agric Food Chem 68:2024–2030
37. Septama AW, Jantan I, Panichayupakaranant P, Aluwi MFFM, Rahmi EP (2020) Immunosuppresive and antibacterial activities of dihydromorin and norartocarpetin isolated from *Artocarpus heterophyllus* heartwoods. Asian Pacific. J Trop Biomed 10:361–368
38. Lin C, Lu C, Huang P (1995) Flavonoids *Artocarpus heterophyllus*. Photochemistry 39(6):1447–1451
39. Bathia BS, Siddapa LG (1955) Composition and nutritive value of jackfruit. Indian J Agric Sci 25:303–306
40. Kumar AB, Singh AB, Abidi R, Upadhyay G, Singh A (1988) Proximate composition of jack-

fruit seeds. J Food Sci Tech 25:308–309
41. Tiwari AK, Vidyarthi AS (2015) Nutritional evaluation of various edible fruit parts of jackfruit (*Artocarpus heterophyllus*) at different maturity stages. Int J Chem Pharm Rev Res 1:21–26
42. Swami BB, Thakor NJ, Habdankar PM, Kalse SB (2012) Jackfruit and its many functional components as related to human health: a review. Compr Rev Food Sci Food Saf 11(6):565–576
43. Mushumbusi DG (2015) Production and characterization of jackfruit jam (master of science in food science thesis). Sokoine University of Agriculture, Morogoro
44. Sammadar HM (1985) In: Bose TK, Mitra SK (eds) "Jackfruit" in fruits of India: tropical and subtropical. Naya Prokash, Calcutta, pp 638–649
45. Goswami C, Hossain K, Kader A, Islam R (2011) Assessment of physicochemical properties of jackfruits (*Artocarpus heterophyllus* Lam) flesh. J Hortic Biotechnol 15(3):26–31
46. Vazhacharickal PJ, Sajeshkumar NK, Mathew JJ, Kuriakose AC, Abraham B, Mathew RJ, Albin AN, Thomson D, Thomas RS, Varghese N, Jose S (2015) Chemistry and medicinal properties of jackfruit (*Artocarpus heterophyllus*): a review on current status of knowledge. Int J Innov Res Rev 3(2):83–95
47. Uddin MT, Rukanuzzama M, Khan MR, Islam MA (2009a) Adsorption of methylene blue from aqueous solution by Jackfruit (*Artocarpus heterophyllus)* leaf powder. A fixed-bed column study. J Environ Manag 90(11):3443–3450
48. Uddin MT, Rukanuzzaman M, Khan MR, Islam MA (2009b) Jackfruit (*Artocarpus heterophyllus*) leaf powder: an effective absorbent for removal of methylene blue from aqueous solutions. Indian J Chem Tech 16(2):142–149
49. Hameed BH (2009) Removal of cationic dye from aqueous solution using jackfruit peel as non-conventional low cost absorbent. J Hazard Mater 162(1):344–350
50. Geraci A, Di Stefano V, Di Martino E, Schillaci D, Schicehi R (2016) Essential oil components of orange peels and antimicrobial activity. Nat Prod Res 31(6):653–659
51. Al-Zoreky NS (2009) Antimicrobial activity of pomegranate (Punica granatum L.) fruit peels. Int J Food Microbiol 134(3):244–248
52. Vaidya Gogte VM (2000) Ayurvedic pharmacology and therapeutic use of medicinal plants. Swami Prakashan and Ayurvedic Research Centre, Mumbai, pp 656–657
53. Hemborn PP (1996) Contact therapy practiced by Mundas Chotanagar (Bihar). Ethnobot 8:36–39
54. Gupta AK, Tandon N (2004) Review on Indian medicinal plants. Indian Council of Medical Research, New Delhi, pp 182–200
55. Tyagi S, Sahay S, Nanher AH, Kumar V (2015) Medicinal uses of jackfruit (*Artocarpus heterophyllus*). Compedium of XII agricultural science congress. Nutr Health:ASC2015ABS1419415762
56. Bolhaar ST, Ree R, Bruijnzeel-Koomen CA, Knulst AC, Zuidmeer L (2004) Allergy to jackfruit, a novel example of bet V I—related food allergy. Allergy 59(11):1187–1192
57. Fang SC, Hsu CL, Yen GC (2008) Anti-inflammatory effects of phenolic compounds isolated from the fruits of *Artocarpus heterophyllus*. J Agric Food Chem 56(12):160–168
58. Liu YY, Wang T, Yang RX, Tang HX, Qiang L et al (2018) Anti-inflammatory steroids from the fruits of *Artocarpus heterophyllus*. Nat Prod Res 22:1–7
59. Meera M, Ruckmani A, Saravanan R, Lakshmipathy PR (2018) Anti-inflammatory effect of ethanolic extract of spine, skin and rind of jackfruit peel. A comparative study. Nat Prod Res 32:2740–2744
60. Yao X, Wu D, Dong N, Quyang P, Pu J et al (2016) Phenolic compounds isolated from *Artocarpusheterophyllus* suppresses lipopolysaccharide-activated inflammatory responses in murine raw 264.7 macrophages. Int J Mol Sci 17:1199
61. Wu J, Wang L (2005) Development of *Artocarpusheterophyllus* Lams jam ice cream. Beverage Fast Frozen Food Industry 2:21–23
62. Ko FN, Cheng ZJ, Lin CN, Teng CM (1998) Scavenger and antioxidant properties of prenylflavones isolated from *Artocarpus heterophyllus*. Free Radiac Biol Med 25(2):160–168
63. Jagtap UB, Waghmar SR, Lokhande VH, Suprasanna P, Bapal VA (2011) Prepaation and eval-

uation of antoxidant capacity of jackfruit (*Artocarpus heterophyllus* Lam) wine and its protection role against radiation induced DNA damage. Ind Crop Prod 34:1495–1601
64. Zhu KX, Zhang Y, Nie SP, Xu F, He SZ et al (2017) Physicochemical properties and in vitro antioxidant activities of polysaccharide from *Artocarpus heterophyllus* Lam. Pulp Carbohyd Polym 155:354–361
65. Zhu K, Yao S, Zhang Y, Liu Q, Xu F et al (2019) Effects of in vitro saliva, gastric and intestinal digestion on the chemical properties, antioxidant activity of polysaccharides from *Artocarpus heterophyllus* Lam. Pulp Food Hydrocoll 87:952–959
66. Daud MN, Wibowo A, Abdullah N, Ahmad R (2018) Bioassay-guided fractionation of *Artocarpus heterophyllus* L J33 variety fruit waste extract and identification of its anti-oxidant constituents by TOFLCMS. Food Chem 266:200–214
67. Zhang X, Zhu K, Xie J, Chen Y, Tan L et al (2020) Optimization and identification of non-extractable polyphenols in the dietary fibre of jackfruit. (*Artocarpus heterophyllus* Lam) pulp released by alkaline acid and enzymatic hydrolysis: content, composition and antioxidant activities. LWT Food Sci Tech 138:110400
68. Loizzo MR, Tundis R, Chandrika UG, Abeysekera AM, Menichini F et al (2020) Antioxidant and antibacterial activities on food borne pathogens of *Artocarpus heterophyllus* Lam (Moraceae) leaves extracts. J Food Sci 75:M291–M295
69. Biworo A, Tanjung E, Khairina I, Suhartono E (2015) Antidiabetic and antioxidant activity of *Artocarpus heterophyllus* extract. J Med Bioengine 4:318–323
70. Soong YY, Barlow PJ (2004) Antioxidant activity and phenolic content of selected fruit seeds. Food Chem 88(3):411–417
71. Trindade MB, Lopes JI, Soares CA, Monteiro AL, Moreira RA, Oliva ML, Beltramini LM (2006) Structural characterization of novel chitin-binding lectins from the genus *Artocarpus* and their antifungal activity. Biochim Biophys Acta 1764(1):146–152
72. Vazquez-Gonzalez Y, Ragazzo-Sanchez JA, Calderon-Santoyo M (2020) Characterization and antifungal activity of jackfruit (*Artocarpus heterophyllus* Lam) leaf extract obtained using conventional and emerging technologies. Food Chem 330:127–211
73. Khan MR, Omoloso KM (2003) Antibacterial activity of *Artocarpus heterophyllus*. Fitoterapia 74(5):501–505
74. Siritapetawee J, Thammasirivak S, Samosomsuk W (2012) Antimicrobal activity of a 48-KDa protease (AMP48) from *Artocarpus heterophyllus* Latex. Eur Rev Med Pharmacol 16:132–137
75. Septama AW, Panichayupakaranant P (2015) Antibacterial assay-guided isolation of active compounds from *Artocarpus heterophyllus* heartwoods pharm. Biol 53:1608–1613
76. Septama AW, Pannichayupakaranant P (2017) Antibacterial activity of *Artocarpus heterophyllus* heartwoods against diarrheal pathogens and its mechanism of action on membrane permeability. J App Pharm Sci 7:64–68
77. Fernando MR, Nahinie SMD, Wickramasinghe MI, Thabrew P, Ariyanda L, Karunanayake EH (1991) Effect of *Artocarpus heterophyllus* and *Asteracanthus longifolia* on glucose tolerance in normal human subjects and in maturity—onset diabetic patients. J Ethnopharmacol 31(3):277–282
78. Ajiboye BO, Ojo OA, Adeyonu O, Imiere O, Olayide I et al (2016) Inhibitory effects on key enzymes relevant to acute type-2 diabetes and antioxidant activity of ethanolic extract of *Artocarpus heterophyllus* stem bark. J Acute Dis 5:423–429
79. Sharma W, Trivedi PC (1995) Nematocidal and nematostatic response of aqueous extract of certain plants of semi arid niche. Curr Nematol 6:45–53
80. Zhang Y, Zhu K, He S, Tan L, Kong X (2016) Characterizations of high purity starches isolated from five different jackfruit cultivars. Food Hydrocoll 52:785–794
81. Wen L, Zhao Y, Jiang Y, Yul ZX et al (2017) Identification of a flavonoid C-glycoside as potent antioxidant. Free Radic Biol Med 110:92–101
82. Burci LM, da Silva CB, Rondon JN, da Silva LM, de Andrade SF et al (2018) Acute and subacute (28 days) toxicity, hemolytic and cytotoxic effect of *Artocarpus heterophyllus* seed extracts. Toxicol Rep 61:1304–1308

83. Sun Co Z, Lee MH, Yu Y, Kang S, Dong Z et al (2017) Chemoprevention of colorectal cancer by Artocarpin, a dietary phytochemical from *Artocarpus heterophyllus*. J Agric Food Chem 65:3474–3480
84. Wiater A, Paduch R, Trojnar S, Pleszczynska M et al (2020) The effect of water soluble polysaccharide from jackruit (*Artocarpus heterophyllus* Lam) on human colon carcinoma cells cultured *in vitro*. Plant (Basel) 9:103
85. Boonyaketgoson S, Rukachaisiarikul V, Phongpaichit S, Trisuwan K (2017) Cytotoxic arylbenzofuran and stilbene derivatives from the twigs of *Artocarpus heterophyllus*. Tetrahed Lett 58:1585–1589
86. Ruiz-Montanez G, Burgos-Hernandez A, Calderono-Santoyo M et al (2014) Screening antimutagenic and antiproliferative properties of extracts isolated from jackfruit (*Artocarpus heterophyllus* Lam). Food Chem 175:409–416
87. Egbuna C, Patrick-Iwuanyanwu KC, Onyeike EN, Khan J, Alshehri B (2021) FMS-like tyrosine kinase-3 (FLT3) inhibitors with better binding affinity and ADMET properties than sorafenib and gilteritinib against acute myeloid leukemia: in silico studies. J Biomol Struct Dyn. https://doi.org/10.1080/07391102.2021.1969286
88. Nguyen NT, Nguyen MHK, Ngueyen HX, Bui NKN, Nguyen MTT (2012) Tyrosinase inhibitors from the wood of *Artocarpus heterophyllus*. J Nat Prod 75:1951–1955
89. Kabir S (1998) Jacalin, a jackfruit (*Artocarpus heterophyllus*) seed derived lectin of versatile applications in immunobiological research. J Immunol Methods 212(2):193–211
90. Lavker RM (1995) Cutaneous aging: chronologic versus photoaging. In: Gilchrest BA (ed) Photodamage. Blackwell Science, Cambridge, pp 123–135
91. Babitha S, Sandhya C, Pandey A (2004) Natural food colorants. Appl Bot Abstracts 23:258–266
92. Morton J (1987) Jackfruit. In: Fruits of warm climate. Julia F. Morton, Miami, FL
93. Singh A, Kumah S, Singh IS (1991) Functional properties of jackfruit seed flour. Lebensm Willu Tech 24:373–374

Chapter 10
Pawpaw Peels as a Source of Nutraceuticals

Odoh Uchenna Estella, Chukwuma Micheal Onyegbunam, Chukwuebuka Egbuna, Theodora Mba, and Peculiar Feenna Onyekere

10.1 Introduction

Carica papaya L. (Pawpaw), synonym *Papaya Carica*, Gaertn, is a tropical fruit, often seen in orange-red, yellow-green, and yellow-orange hues, with rich orange pulp. Along with delicious taste, whole plant parts, fruit, roots, bark, peel, seeds, and pulp have medicinal properties. It is consumed worldwide and has seen extensive research. Surprisingly, the peel has been ignored both as a consumable part and as a research interest topic. Our literature search found few reported investigations of pawpaw peel [1–5]. Dietary ripe or unripe pawpaw peel powder improves egg quality in laying hens.

Many scientific studies were carried to prove the nutritional value of papaya. These studies validated that papaya has anti-dengue, anti-cancer, antiseptic, anti-parasitic, anti-inflammatory, antidiabetic, and contraceptive features [6]. The nutraceutical properties as a food or as a quasi-drug are poorly understood or undervalued by people [7]. *C. papaya* as a natural resource can be a source of nutraceuticals or conventional medications to be used as a potential preventative or treatment option for various health issues [8, 9]. The whole parts of the papaya tree contain papain, an enzyme that helps digest proteins that are mostly found in the ripe fruit [10].

O. U. Estella (✉) · C. M. Onyegbunam · P. F. Onyekere
Faculty of Pharmaceutical Sciences, Department of Pharmacognosy and Environmental Medicines, University of Nigeria, Nsukka, Nigeria
e-mail: uchenna.odoh@unn.edu.ng; chukwuma.michael@unn.edu.ng

C. Egbuna
Chukwuemeka Odumegwu Ojukwu University, Awka, Nigeria

T. Mba
Faculty of Pharmaceutical Sciences, Department of Pharmacognosy, Enugu State University of Sciences and Technology (ESUT), Agbani, Enugu State, Nigeria
e-mail: theodora.mba@esut.edu.ng

© The Author(s), under exclusive license to Springer Nature Switzerland AG 2022

C. Egbuna et al. (eds.), *Food and Agricultural Byproducts as Important Source of Valuable Nutraceuticals*, https://doi.org/10.1007/978-3-030-98760-2_10

Papain is used in meat tenderizing, pharmaceuticals, beauty products, and cosmetics [11].

Many benefits of papaya are due to the high content of Vitamins A, B, and C, phytochemicals and also proteolytic enzymes like papain and chymopapain, which have antiviral, antifungal, and antibacterial properties [12, 13]. Extracts have antioxidant activity and antibacterial activity that may be of use for the development of new preservatives in the food and cosmetic industries [14].

The term nutraceutical was coined in 1989 by Stephen De Felice, he defined a nutraceutical "as a food, or parts ofa food, that provide medical or health benefits, including the prevention and treatment of disease". It is used to describe these medicinally or nutritionally functional foods. Nutraceuticals, have also been called medical foods, designer foods, phytochemicals, functional foods and nutritional supplements [15, 16]. Nutraceutical is derived from "nutrition" and "pharmaceutics" and is applied to products that are isolated from herbal products, dietary supplements (nutrients), specific diets, and processed foods such as cereals, soups, and beverages that other than nutrition are also used as medicine. Nutraceuticals are shown to exert a physiological benefit or provide protection against chronic disease [16]. Nutraceuticals may be used to improve health, delay the aging process, prevent chronic diseases, increase life expectancy, or support the structure or function of the body [16, 17]. A working definition of nutraceutical from a Science forum states: "a diet supplement that delivers a concentrated form of a biologically active component of food in a non-food matrix to enhance health" [18].

Description

The fruit is spherical or cylinder in form and about 15–45 cm (5.9–17.7 in.) long and 10–30 cm (3.9–11.8 in.) in diameter [19, 20]. It is ripe when it is soft and the skin has attained a yellow to orange colour and along the walls of the large central cavity are attached numerous black seeds [21]. It is a melon like fruit with with many small black seeds enclosed in skin that ranges in color from green to orange [22]. It is a large, up to 50–60 cm in diameter, weigh between 1/2 and 2 kg. It is cylindrical or pear-shaped and the central cavity is surrounded by hundreds of small seeds, sometimes there are seedless varieties of the fruit (Fig. 10.1) [10, 23].

10.2 Nutritional Content of *Carica papaya*

Pawpaw contains high level of vitamin C, Niacin, Calcium and potassium than apples, oranges and bananas [24]. Pawpaw has all the essential amino acids and loaded with antioxidants. It contains Carotenoids (β-carotene, crytoxanthin), energy about 163 kJ, carbohydrates, sugars, vitamin A, dietary fibre, and minerals such as Calcium, potassium and sodium [12]. It also helps in prevention of various chronic diseases, including cardiovascular problems, parasitic and bacterial disorders, diabetes, and cancer [25, 26]. The composition of leaves are: ascorbic acid (140 mg), vitamin E (136 mg) per 100 g,β-carotene equivalent (11,565 μg), thiamine (0.09 mg),

Fig. 10.1 Pawpaw (*C. papaya*) fruit (**a**) Ripe pawpaw fruit (**b**) Pawpaw fruit with tree (**c**) Unripe pawpaw fruit

riboflavin (0.48 mg), niacin (2.1 mg) calories (74 g), water (77.5 g), protein (7 g), fat (2 g), total carbohydrates (11.3), fiber (1.8 g), ash (2.2 g), calcium (344 mg), phosphorous (142 mg), iron (0.8 mg), sodium (16 mg) and potassium (652 mg). The presence of iron signifies that the leaves can be used against anemia, tuberculosis, and growth disorders [27]. As a source of phytochemicals, they can be used as a cleanser in herbal remedies [28]. The essential vitamins and minerals, such as Ca, Mg, Mn, Fe, Zn, K, P, and vitamins like A, C, E, and B [29]. The vitamins A and C is good for better eyesight, prevention of early age blindness in children, and also good for those who frequently suffer from cold, cough, or flu, because intake of papaya boosts their immune system.

Unripe fruits of papaya are used to stimulate reproductive organs there by boosting male fertility and has an enzyme called arginine which is known to boost blood flow around the penis. It boosts nitric acid in the body and helps to relax the muscles surrounding the blood vessels that supply the man-hood with blood causing the blood vessels to expand and allow blood flow. A more concentrated form of arginine is used to treat erectile dysfunction. The papain and chymopapain can help lower inflammation and improve healing from burns. Papaya contains cancer fighting lycopene which is a key intermediate in the biosynthesis of many important carotenoids, such as beta-carotene and xanthophylls and thus another useful compound not readily found in the plant kingdom but found in pawpaw is Fibrin. Fibrin helps to reduce the occurrence of blood clots and also enhance the quality of blood cells making blood to flow freely through the circulatory system and also important in stroke prevention [12]. *C. papaya* L. is known and consumed all around the world. Due to its recognition as a plant with nutritive and medicinal potentials it has been investigated by many researchers. Surprisingly the peel has been ignored both as food and research topic. Available literature contains very few mention of the consumption of the peel or published research works on the peels. It is the aim of this chapter to bring together available information on *C. papaya* L. peelings, highlight its potentials and stimulate research interest in it, especially its value as a nutraceutical. Compared to the seeds papaya peel flour stood out for the following parameters: fiber, ash, phenolic compounds, vitamin C, soluble solids, titratable acidity, and the minerals copper, sulfur, and potassium. On the other hand, papaya seed flour stood out for protein, lipids, and the minerals calcium, iron, magnesium, and zinc [30].

10.3 Chemistry of *C. papaya* and Its Peels

The preliminary phytochemical analysis of *C. papaya* fruits showed the presence of phytochemicals, including kaempferol, quercetin, and caffeic acid [2]. An array of active compounds with versatile action are present in the various plant tissues, i.e., linalool in fruit pulp [31], dehydrocarpaine I and II, alkaloids, carpaine, and pseudocarpaine in leaves [32], in the latex, glutaminyl cyclase; cysteine endopeptidases and class-II and III chitinase [33], quercetin, and kaempferol in shoots [34]. Total phenolic content was analyzed using the Folin-Ciocalteau method and the total phenolic content (TPC) was expressed as gallic acid equivalent (GAE). The highest TPC, i.e., 1735.1 mg/L GAE, was obtained at a temperature of 120 °C and a time of 5 h in a solid-solvent ratio of 1:40 g/mL, while the lowest TPC of 616.57 mg/L GAE was obtained at a temperature of 90 °C and a time of 3 h at a solid-solvent ratio of 1:20 g/mL. With such a high phenolic content, Sekaki papaya (*C. papaya*) peel can be used as a natural antioxidant and protect the human body from various free-radical-associated diseases [35]. Papaya is one of the fruits with essentiual vitamins (vitamin A, C, B1, and B2, thiamine, folate, riboflavin, niacin) and minerals (calcium, potassium, iron, and fiber) The energy value of papaya is 200 kJ/100 g. Glucose (29.8 g/100 g), fructose (21.9 g/100 g), and sucrose (48.3 g/100 g) are the main sugars in papaya. About 100 g of fresh fruit contains 108 mg of ascorbic acid, higher than oranges (67 mg/100 g of fresh fruit) [36]. Linalool is the papaya's highly abundant volatile [37] and 94% in Solo papaya varieties. On the other hand, the oxide cis-linalool is abundant; linalool ranks second highest in concentration [38]. It contains aromatic compounds, such as 3-methyl butanol, butanol, terpineol, and benzyl alcohol, become abundant at the ripe stage [39]. Among 103 esters, methyl butyrate is in high concentration [40]. Various compounds other than linalool, are also present in fruit, i.e., benzyl isothiocyanate [38, 41], and terpene hydrocarbons. Butanol, 3-methyl butanol, benzyl alcohol, and α-terpineol show maximum concentration in the third maturation stage connected with fruit ripeness [42].

The following compounds: Ferulic acid (277.49–186.63 mg/100 g), p-coumaric acid (229.59–135.64 mg/100 g), and caffeic acid (175.51–112.89 mg/100 g) have been identified from the fruit of *C. papaya*. Lycopene (0.36–3.40 mg/100 g), β-cryptoxanthin (0.28–1.06 mg/100 g), β-carotene (0.23–0.50 mg/100 g), and vitamin C (25.07–58.59 mg/100 g) are also present. The total flavonoid (191.06 mg QAE/100 g) of the fruit than that of papaya peel [Total phenol = 126.75 mg GAE/100 g; Total flavonoid = 166.11 mg QAE/100 g]. The antioxidant results revealed that the extract of unripe pawpaw peel had a higher ferric reducing antioxidant property (112.35 mg AAE/100 g) compared to the unripe papaya seed extract (102.78 mg AAE/100 g). Saponin and steroids showed positive result while cardiac glycosides and anthraquinones were absent in the phytochemical analysis of the extracts of both unripe papaya peel and seed. The use of unripe papaya peel and seed as herbal materials could be of therapeutic use in the management/treatment of some oxidative stress-induced human ailments due to their antioxidant and phytochemical potencies [1].

Proteases from papaya peels showed maximum casein hydrolysis in buffer pH 8 and at 75 °C [43]. The papaya latex is well known for being a rich source of the four cysteine endopeptidases, namely papain, chymopapain, glycyl endopeptidase, and caricain [44]. Apart from papain and chymopapain, *C. papaya* contains many biologically active compounds. *C. papaya* lipase, or CPL, a hydrolase, is tightly bonded to the water-insoluble fraction of crude papain and is considered as a "naturally immobilized" biocatalyst [45]. There are several applications of CPL which include (1) fats and oils modification, derived from the sn-3 selectivity of CPL as well as from its preference for short-chain fatty acids; (2) esterification and inter-esterification reactions in organic media, and (3) the asymmetric resolution of different non-steroidal anti-inflammatory drugs (NSAIDs), 2-(chlorophenoxy) propionic acids, and non-natural amino acids [46].

Thus another beneficial compound not readily found in the plant kingdom but found in pawpaw is Fibrin. Fibrin improves the quality of blood cells, blood flow through the circulatory system essential in stroke prevention and thereby reduces the risk of blood clots [12].

The main compounds produced by the fruit are esters and alcohols. The most abundant esters are ethyl acetate, and ethyl butanoate, methyl butanoate, and butyl acetate comprising 88% of the volatiles in fully ripe fruit. Butanol is the most abundant alcohol. Among the volatiles produced, ethyl butanoate, ethyl acetate, ethyl hexanoate, and ethyl 2-methyl butanoate are reported to be the most potent odour compounds [46].

Lycopene, β-cryptoxanthin, and β-carotene are the main carotenoids that have been identified in papaya [47]. Papain, caricain, chymopapain, and glycine endopeptidase can survive acidic pH conditions and pepsin degradation. However, at low pH, a conformational transition occurs in papain, caricain, chymopapain, and glycine endopeptidase that instantaneously converts their native forms into molten globules that are quite unstable and rapidly degraded by pepsin. So for them to be effective orally against gastrointestinal nematodes, they have to be protected against both acid denaturation and proteolysis.

10.4 Pharmacological Activities of *C. papaya*

C. papaya is renowned for its medicinal and pharmacological properties [6]. *C. papaya* fruit showed good antidiarrheal activity [48], antitrichomonal activity [49], anxiolytic [50], antiinflamatory and wound healing properties [51, 52], antioxidant and antimicrobial activities [14], antibacterial and wound healing effect of *C. papaya* [53] Antimicrobial activity of *C. papaya* [54, 55], antifungal activity [56, 57], hypoglycemic and antihyperglycemic effects effects [58, 59], antihyperglycemic effects [60], hepatoprotective effect [61, 62], antifertility activity [63, 64], antisickling activity [65–67], antihelminthic and antiamoebic activity of *C. papaya* seeds [68, 69], cytoprotective and antimotility. The cytoprotective and antimotility properties of the extracts may account for the anti-ulcer property of the unripe fruit [70], anti-inflammatory and immunomodulatory activities [71], in vitro cytoprotective and Cell apoptosis inhibition. Aqueous extract of papaya seeds at 1 mg/mL showed cytoprotective against H_2O_2 induced cell toxicity. Aqueous extract of papaya seeds at a 1 mg/mL concentration inhibited H_2O_2 induced apoptosis by approximately 30%. Seed extract at 1 mg/mL inhibited oxidative stress-induced cell apoptosis, reduced mitochondrial dysfunction, and impeded release of cytochrome C [72], anticancer activities. The anticancer activities of papaya has also been studied [73, 74], Anticancer and immunomodulatory activities [75, 76], antineoplastic activity [77], antioxidant [78, 79], antiplasmodial activity [80, 81]. Treatment of dengue fever [82]. *C. papaya* leaves aqueous extract exhibited potential activity against Dengue fever, antisickling, and reversal of sickling activities [83, 84]. The consumption of unripe and semi-ripe papaya fruits must be done with caution. A study with rats at different stages of gestation showed that it could be unsafe during pregnancy [85].

10.5 Conclusion

Among all the parts of the pawpaw tree, the peelings have been the most neglected as food or for research purpose. Even the less brightly colored though not less attractive seeds have attracted more interest. This has manifested as a lack of information on research on *C. papaya* peelings. Most of the available reports are not up-to-date, and the current ones are nanotech-focused. The number of research on the fruit generally should motivate interest in the peelings since it does not exist separate from the fruit. *C. papaya* peelings should not be seen as a waste but as a research item or industrial raw materials, including as natural sources of nutraceuticals. Extracts have antioxidant activity and antibacterial activity that may be useful for developing new preservatives in the food and cosmetic industries.

References

1. Dada FA, Nzewuji FO, Esan AM, Oyeleye SI, Adegbola VB (2016) Phytochemical and antioxidant analysis of aqueous extracts of unripe pawpaw (*Carica papaya* Linn.) fruit's peel and seed. Int J Res Rev Appl Sci 27(3):68
2. dos Santos CM, de Abreu CMP, Freire JM, de Rezende Queiroz E, Mendonça MM (2014) Chemical characterization of the flour of peel and seed from two papaya cultivars. Food Sci Technol 34(2):353–357. https://doi.org/10.1590/fst.2014.0048
3. Lydia E, Riyazudin M, John S, Thiyagarajan S (2016) Investigation on the phytochemicals present in the fruit peel of *Carica papaya* and evaluation of its antioxidant properties. Int J Health Allied Sci 5:247–252. https://doi.org/10.4103/2278-344X.194127
4. Espin N, Islam MN (1998) Stabilization of papain from papaya peels. Food Sci Technol Int 4:179–187. https://doi.org/10.1177/108201329800400304
5. Diarra SS, Waqa T, Tonga K, Gilu M, Manu'a OP, Andrew S (2020) Dietary ripe or unripe papaya (*Carica papaya*) peel powder improves egg quality in laying hens. Livestock Res Rural Dev 32:Article #143. Retrieved August 11, 2021, from http://www.lrrd.org/lrrd32/9/siaka32143.html
6. Egbuna C, Ifemeje JC (2015) Comparative studies on the phytochemical properties of five Nigerian medicinal plants. J Adv Med Pharm Sci 6(2):1–12
7. Karunamoorthi K, Kim H-M, Jegajeevanram K, Xavier J, Vijayalakshmi J (2014) Papaya: a gifted nutraceutical plant—a critical review of recent human health research. Int J Genuine Tradit Med 4(1/e2):1–17. https://doi.org/10.5667/tang.2013.0028
8. Kong YR, Jong YX, Balakrishnan M, Bok ZK, Weng JKK, Tay KC, Goh BH, Ong YS, Chan KG, Lee LH et al (2021) Beneficial role of Carica papaya extracts and phytochemicals on oxidative stress and related diseases: a mini review. Biology 10(4):287. https://doi.org/10.3390/biology10040287
9. Auria JD, Chen F, Pichersky E (2002) Characterization of an acyl trans feras e capable of synthesizing benzylbenzoate and other volatile esters in flowers and damaged leaves of Clarkia breweri. Plant Physiol 130:466–476
10. EDI (2013) EDIS document FE913, a publication of the Food and Resource Economics Department, Florida Cooperative Extension Service, Institute of Food and Agricultural Sciences, University of Florida, Gainesville, FL. Available at: http://edis.ifas.ufl.edu. Accessed 22 Mar 2013
11. Aravind G et al (2013) Traditional and medicinal uses of *Carica papaya*. J Med Plants Stud 1(1):07–15, 20
12. Maisarah AM et al (2013) Antioxidant analysis of different parts of *Carica papaya*. Int Food Res J 20(3):1043–1048
13. Pati TA, Jain L (2020) Natural preservatives from fruit peels. Acta Sci Microbiol 3(6):152–157. issn:2581-3226
14. Bull E, Rapport L, Lockwood B (2000) What is a nutraceutical? Pharm J 13:57. https://doi.org/10.1211/PJ.2021.1.71102
15. Nwosu OK, Ubaoji KI (2020) Nutraceuticals: history, classification and market demand. In: Egbuna C, Dable TG (eds) Functional foods and nutraceuticals. Springer, Cham. https://doi.org/10.1007/978-3-030-42319-3_2
16. Egbuna C, Tupas GD (eds) (2020) Functional foods and nutraceuticals—bioactive components, formulations and innovations, 1st edn. Springer Nature, Cham. 646 pp
17. Zhao J (2007) Nutraceuticals, nutritional therapy, phytonutrients, and phytotherapy for improvement of human health: a perspective on plant biotechnology application. Bentham Science Publishers [Last accessed on 2012 Mar 24]. Available from: http://www.bentham-science.com/biot/samples/biot1-1/Zhao.pdf
18. Zeisel SH (1999) Regulation of "nutraceuticals". Science 285:1853–1855
19. Varzakas T, Zakynthinos G, Verpoort F (2016) Plant food residues as a source of nutraceuticals and functional foods. Foods 5:88. https://doi.org/10.3390/foods5040088. https://www.itis.gov/

20. Heywood VH, Brummitt RK, Culham A, Seberg O (2007). Flowering plant families of the world. Firefly Books, Buffalo. 424 pp
21. Attar SC, Sadia GA, Mysore NR, Ramesh YA, Mysore NR, Ramesh SR (2004) Optimization of enzymatic liquefaction of papaya (*Carica papaya* L.) and jackfruit (Artocarpus heterophyllus Lam.) pulp using response surface methodology. Food Agric Environ 2(2):108–113
22. Auria JD, Chen F, Pichersky E (2002) Characterization of an acyl trans feras e capable of synthes izing benzylbenzoate and other volatile esters in flowers and damaged leaves of Clarkia breweri. Plant Physiol 130:466–476
23. Teixeira da Silva JA, Rashid Z, Nhut DT, Sivakumar D, Gera A, Souza MT Jr, Tennant PF (2007) Papaya (*Carica papaya* L.) biology and biotechnology. In: Tree and forestry science and biotechnology. Global Science Books, pp 48–66
24. Snake J, Desmond L (1997) Cooking with papaya. Kentucky State University Cooperative. Extension Program Guide. 129atwood Research, Frankfort, KY, pp 40601–42355
25. Aravind G, Debjit B, Duraivel S, Harish G (2013) Traditional and medicinal uses of *Carica papaya*. J Med Plants Stud 1(1):7–15
26. Murakami A, Ohigashi H, Koshimizu K (1994) Possible anti-tumor promoting properties of traditional Thai foods and some of their active constituents. Asia Pac J Clin Nutr 3:185–191
27. Sherman PW, Billing J (1999) Darwinian gastronomy: why we use spices. Biosciences 49:453–463
28. Claude B, Paule S (1979) The manual of natural living, 1st edn. Thorsons, London, pp 98–101
29. Atta KB (1999) The power of garlic. Cardiovascular Disease Prevention Association, Buea, p 72
30. Seigler DS, Pauli GF, Nahrstedt A, Leen R (2002) Cyanogenic allosides and glucosides from Passiflora edulis and *Carica papaya*. Phytochemistry 60:873–882
31. Miranda-Osorio PH, Castell-Rodríguez AE, Vargas-Mancilla J, Tovilla-Zárate CA, Ble-Castillo JL, Aguilar-Domínguez DE, Juárez-Rojop IE, Díaz-Zagoya JC (2016) Protective action of *Carica papaya* on cells in Streptozotocin-induced diabetic rats. Int J Environ Res Public Health 13:446
32. Winterhalter P, Katzenberger D, Schreier P (1986) 6,7-Epoxy-linalool and related oxygenated terpenoids from *Carica papaya* fruit. Phytochemistry 25:1347–1350
33. Khuzhaev VU, Aripova SF (2000) Pseudocarpaine from *Carica papaya*. Chem Nat Compd 36:418
34. Azarkan M, Clantin B, Bompard C, Belrhali H, Baeyens-Volant D, Looze Y, Villeret V, Wintjens R (2005) Crystallization and preliminary X-ray diffraction studies of the glutaminyl cyclase from *Carica papaya*latex. Acta Crystallogr Sect F Struct Biol Cryst Commun 61:1
35. Miean KU, Mohamed S (2001) Flavonoid (myricetin, quercetin, kaempferol, luteolin, and apigenin) content of edible tropical plants. J Agric Food Chem 49:3106–3112
36. Jamal P, Akbar I, Jaswir I, Zuhanis Y (2017) Quantification of total phenolic compounds in papaya fruit peel Pertanika. J Trop Agric Sci 40(1):87–98. http://irep.iium.edu.my/60113/1/Pertanika%20journal%20paper%20published.pdf
37. Lim Y, Lim T, Tee J (2007) Antioxidant properties of several tropical fruits: a comparative study. Food Chem 103:1003–1008
38. Flath RA, Forrey RR (1977) Volatile components of papaya (*Carica papaya* L., Solo variety). J Agric Food Chem 25:103–109
39. Franco MRB, Rodrigues-Amaya DB (1993) Volatile components of two papaya cultivars. Arquivos Biol Technol 36:613–632
40. Almora K, Pino JA, Hernandez M, Duarte C, Gonzalez J, Roncal E (2004) Evaluation of volatiles from ripening papaya (*Carica papaya* L., var. Maradol roja). Food Chem 86:127–130
41. MacLeod AJ, Pieris NM (1983) Volatile components of papaya (*Carica papaya* L.) with particular reference to glucosinolate products. J Agric Food Chem 31:1005–1008
42. Flath RA, Light DM, Jang EB, Mon TR, John JO (1990) Headspace examination of volatile emissions form ripening papaya (*Carica papaya* L., Solo variety). J Agric Food Chem 38:1060–1063

43. Karina A, Jorge AP, Mercedes H, Cira D, Juan G, Elda R (2004) Evaluation of volatiles from ripening papaya (*Carica papaya* L., var. *Maradol roja*). Food Chem 86:127–130
44. Chaiwut P, Nitsawang S, Shank L, Kanasawud P (2007) A comparative study on properties and proteolytic components of papaya Peel and latex proteases. Chiang Mai J Sci 34(1):109–118. www.science.cmu.ac.th/journal-science/josci.html
45. Azarkan M, El Moussaoui A, van Wuytswinkel D, Dehon G, Looze Y (2003) Fractionation and purification of the enzymes stored in the latex of *Carica papaya*. J Chromatogr B Analyt Technol Biomed Life Sci 790:229–238
46. Domínguez de María P, Sinisterra JV, Tsai S-W, Alcántara AR (2006) *Carica papaya* lipase (CPL): an emerging and versatile biocatalyst. Biotechnol Adv 24:493–499
47. Li D, Xu Y, Xu G, Gu L, Li D, Shu H (2006) Molecular cloning and expression of a gene encoding alcohol acyltransferase (MdAAT2) from apple (cv. Golden Delicious). Phytochemistry 67:658–667
48. Prabhu AK, Devadas SM, Lobo R, Udupa P, Chawla K, Ballal M (2017) Antidiarrheal activity and phytochemical analysis of *Carica papaya* fruit extract. J Pharm Sci Res 9(7):1151–1155. www.jpsr.pharminfo.in
49. Calzada F, Yépez-Mulia L, Tapia-Contreras A (2007) Effect of Mexican medicinal plant used to treat trichomoniasis on Trichomonasvaginalistrophozoites. J Ethnopharmacol 113:248–251
50. Kebebew Z, Shibeshi W (2013) Evaluation of anxiolytic and sedative effects of 80% ethanolic-*Carica papaya* L. (Caricaceae) pulp extract in mice. J Ethnopharmacol 150:665–671
51. Rahman A (2013) Health benefits, chemistry and mechanism of *Carica papaya* a crowning glory advances in natural. Science 6(3):26–37
52. Nafiu AB, Rahman MT (2015) Anti-inflammatory and antioxidant properties of unripe papaya extract in an excision wound model. Pharm Biol 53:662–671
53. Anibijuwon II, Udeze AO (2009) Antimicrobial activity of *Carica papaya* on some pathogenic organisms of clinical origin from South-Western Nigeria. Ethnobot Leaf 4:850–864
54. Baskaran C, Bai VR, Velu S, Kumaran K (2012) The efficacy of *Carica papaya* leaf extract on some bacterial and fungal strain by well diffusion method. Asian Pac J Trop Dis 2012:347–349
55. Doughari JH, Elmahmood AM, Manzara S (2007) Studies on the antibacterial activity of root extracts of *Carica papaya* L. Afr J Microbiol Res 1:37–41
56. Baskaran C, Velu S, Kumaran K (2012) The efficacy of *Carica papaya* leaf extract on some bacterial and a fungal strain by well diffusion method. Asian Pac J Trop Dis 2:S658–S662
57. Chávez-Quintal P, González-Flores T, Rodríguez-Buenfil I, Gallegos- TS (2011) Antifungal activity in ethanolic extracts of *Carica papaya* L. Cv. Maradol leaves and seeds. Indian J Microbiol 51:54–60
58. Juárez-Rojop IE, Díaz-Zagoya JC, Ble-Castillo JL, Miranda-Osorio PH, Castell-Rodríguez AE, Tovilla-Zárate CA, Rodríguez-Hernández A, Aguilar-Mariscal H, Ramón-Frías T, Bermúdez-Ocaña DY (2012) Hypoglycemic effect of *Carica papaya* leaves in streptozotocin-induced diabetic rats. BMC Complement Altern Med 12:236. http://www.biomedcentral.com/1472-6882/12/236
59. Oboh G, Olabiyi AA, Akinyemi AJ, Ademiluyi AO (2014) Inhibition of key enzymes linked to type 2 diabetes and sodium nitroprusside-induced lipid peroxidation in rat pancreas by water-extractable phytochemicals from unripe pawpaw fruit (*Carica papaya*). J Basic Clin Physiol Pharmacol 25:21–34
60. Sasidharan S, Sumathi V, Jegathambigai NR, Latha LY (2011) Antihyperglycaemic effects of ethanol extracts of *Carica papaya* and Pandanus amaryfollius leaf in streptozotocin-induced diabetic mice. Nat Prod Res 25:1982–1987
61. Adeneye AA, Olagunju JA, Banjo AF, Abdul SF, Sanusi OA, Sanni OO et al (2009) The aqueous seed extract of *Carica papaya* Linn. Prevents carbon tetrachloride induced hepatotoxicity in rats. Int J Appl Res Nat Prod 2:19–32
62. Sadeque MZ, Begum ZA, Umar BU, Ferdous AH, Sultana S, Uddin MK (2012) Comparative efficancy of dried fruits of *Carica papaya* Linn. And vitamin E on preventing hepatotoxicity in rats. Faridpur Med College J 7:29–32

63. Lohiya NK, Goyal RB, Jayaprakash D, Ansari AS, Sharma S (1994) Antifertility effects of aqueous extract of *Carica papaya* seeds in male rats. Planta Med 60:400–404
64. Poharkar RD, Saraswat RK, Kotkar S (2010) Survey of plants having antifertility activity from Western Ghat area of Maharashtra state. J Herb Med Toxicol 4:71–75
65. Mojisola OC, Adebolu EA, Alani DM (2008) Antisickling properties of *Carica papaya* Linn. J Nat Prod 1:56–66
66. Imaga NO, Gbenle GO, Okochi VI, Akanbi SO, Edeoghon SO, Oigbochie V et al (2009) Antisickling property of *Carica papaya* leaf extract. Afr J Biochem Res 3:102–106
67. Karunamoorthi K, Kim HM, Jegajeevanram K, Xavier J, Vijayalakshmi J (2014) Papaya: a gifted nutraceutical plant a critical review of recent human health research. TANG 4:21
68. Kanthal LK, Mondal P, Somnath DE, Jana S, Aneela S, Satyavathi K (2012) Anthelmintic activity of *Carica papaya* latex using *Pheritima* Posthuma. Int J Life Sci Pharm Res 2:10–12
69. Okeniyi JA, Ogunlesi TA, Oyelami OA, Adeyemi LA (2007) Effectiveness of dried *Carica papaya* seeds against human intestinal parasitosis: a pilot study. J Med Food 10:194–196
70. Ezike AC, Akah PA, Okoli CO, Ezeuchenne NA, Ezeugwu S (2009) *Carica papaya* (paw-paw) unripe fruit may be beneficial in ulcer. J Med Food 12(6):1268–1273. https://doi.org/10.1089/jmf.2008.0197
71. Mohr T, Desser L (2013) Plant proteolytic enzyme papain abrogates angiogenic activation of human umbilical vein endothelial cells (HUVEC) in vitro. BMC Complement Altern Med 13:231
72. Panzarini E, Dwikat M, Mariano S, Vergallo C, Dini L (2014) Administration dependent antioxidant effect of *Carica papaya* seeds water extract. Evid Based Complement Alternat Med 2014:281508
73. Farvid MS, Cho E, Chen WY, Eliassen AH, Willett WC (2014) Dietary protein sources in early adulthood and breast cancer incidence: prospective cohort study. BMJ 348:3437. https://doi.org/10.1136/bmj.g3437
74. Pandey S, Walpole C, Cabot PJ, Shaw PN, Batra J, Hewavitharana AK (2017) Selective antiproliferative activities of *Carica papaya* leaf juice extracts against prostate cancer. Biomed Pharmacother 89:515–523. https://doi.org/10.1016/j.biopha.2017.02.050
75. Fauziya S, Krishnamurthy R (2013) Papaya source material for anticancer. CIB Tech J Pharm Sci 2:25–34
76. Nguyen TT, Shaw PN, Parat MO, Hewavitharana AK (2013) Anticancer activity of *Carica papaya*: a review. Mol Nutr Food Res 57:153–164
77. Sianipar MP, Suwarso E, Rosidah R (2018) Antioxidant and anticancer activities of hexane fraction from *Carica papaya* L. male flower. Asian J Pharm Clin Res 11:81–83
78. Vuong QV, Hirun S, Roach PD, Bowyer MC, Phillips PA, Scarlett CJ (2013) Effect of extraction conditions on total phenolic compounds and antioxidant activities of *Carica papaya* leaf aqueous extracts. J Herb Med 3:104–111
79. Maisarah AM, Amira BN, Asmah R, Fauziah O (2013) Antioxidant analysis of different parts of *Carica papaya*. Int Food Res J 20:1043–1048
80. Karunamoorthi K, Tsehaye E (2012) Ethnomedicinal knowledge, belief and self-reported practice of local inhabitants on traditional antimalarial plants and phytotherapy. J Ethnopharmacol 141:143–150
81. Julianti T, Oufir M, Hamburger M (2014 Aug) Quantification of the antiplasmodial alkaloid carpaine in papaya (*Carica papaya*) leaves. Planta Med 80(13):1138–1142. https://doi.org/10.1055/s-0034-1382948
82. Ahmad N, Fazal H, Ayaz M, Abbasi BH, Mohammad I, Fazal L (2011) Dengue fever treatment with *Carica papaya* leaves extracts. Asian Pac J Trop Biomed 1(4):330–333. https://doi.org/10.1016/S2221-1691(11)60055-5
83. Oduola T, Adeniyi FAA, Ogunyemi EO, Bello IS, Idowu TO (2006) Antisickling agent in an extract of unripe pawpaw (*Carica papaya*): is it real? African Journal of Biotechnology 5(20):1947–1949

84. Mojisola OC, Adebolu EA, Alani DM (2008) Antisickling properties of *Carica papaya* Linn. J Nat Prod 1(56–66):45
85. Adebiyi A, Adaikan PG, Prasad RN (2002) Papaya (*Carica papaya*) consumption is unsafe in pregnancy: fact or fable? Scientific evaluation of a common belief in some parts of Asia using a rat model. Br J Nutr 88:199–203

Chapter 11
Nutritional and Nutraceutical Potentials of Residual Cakes from Seeds of Moringa (*Moringa oleifera* L.), Sacha Inchi (*Plukenetia volubilis* L.) and Hibiscus Flower (*Hibiscus sabdariffa* L.) After Oil Extraction

Dairon Iglesias Guevara, Claudia Chávez Hernández, Sirley González Laime, Ariel Martínez García, and Juan Abreu Payrol

11.1 Introduction

In the food industry, many by-products derived from the treatment of raw materials are generated for their transformation into processed products, which are still rich in certain recoverable substances for their use for food purposes. The residual oilseed pastes or cakes that are generated during the edible oil production process are rich in protein. Thus, a by-product of the food industry can be used as a raw material to obtain other food products [1].

The main objective of obtaining protein concentrates from defatted oilseed meal is to eliminate as completely and selectively as possible the soluble non-protein compounds present. As a result, a product rich in insoluble sugars and proteins will

D. I. Guevara
Departamento de Alimentos, Instituto de Farmacia y Alimentos, Universidad de La Habana, La Habana, Cuba

C. C. Hernández · S. G. Laime
ECTI Sierra Maestra, La Habana, Cuba
e-mail: cchavez@bionaturasm.cu; sirley@bionaturasm.cu

A. M. García
ECTI Sierra Maestra, La Habana, Cuba

Instituto de Ciencia y Tecnología de Materiales, Universidad de La Habana, La Habana, Cuba
e-mail: arielmg@imre.uh.cu

J. A. Payrol (✉)
ECTI Sierra Maestra, La Habana, Cuba

Escuela Latinoamericana de Medicina, La Habana, Cuba
e-mail: jabreu@bionaturasm.cu; jabreu@elacm.sld.cu

© The Author(s), under exclusive license to Springer Nature Switzerland AG 2022

C. Egbuna et al. (eds.), *Food and Agricultural Byproducts as Important Source of Valuable Nutraceuticals*, https://doi.org/10.1007/978-3-030-98760-2_11

be obtained. Vegetable protein concentrates result from an enrichment of the material in its protein content, through a gradual separation of its non-protein components (lipids, fiber, carbohydrates, minerals), in such a way that their nutritional properties are not lost [2].

The Food Codex defines plant protein products (VPP) as protein isolates and concentrates that can be used in the food industry. The following is taken into account: for a protein isolate the percentage of protein is greater than 90%; for a protein concentrate the protein percentage is between 65 to 90% and a protein flour when the protein percentage is between 50 to 65% [3].

Although protein concentrate has better chemical characteristics than flour, such as higher protein richness, it still has high content of other unwanted components in the final product. These compounds include fiber, reducing sugars, phenols, lipids and glucosinolates, as in the case of rapeseed (*Brassica napus*). Fiber (cellulose, hemicellulose, lignin, pectic substances, etc.) is the majority component among the undesirable and of great importance, due to its abundance in some protein sources such as defatted sunflower cake (*Helianthus annuus*) [4].

11.2 Applications of Protein Concentrates from Defatted Seed Meal

The main reason for the use of protein concentrates and isolates in foods is the improvement in nutrition. The obtaining and elaboration of vegetable protein extracts is carried out in order to take advantage of their nutritional characteristics and techno-functional properties. Table 11.1 shows some of the main applications of isolated concentrates and protein hydrolysates [2]. The most relevant properties in food processing are: water absorption, oil absorption, emulsion, foaming, gelling and solubility [5].

Protein hydrolysis is applied in various processes such as digestion, mediated by the action of proteases such as pepsin, trypsin or chymotrypsin. In this case, the

Table 11.1 Application of protein concentrates, isolates and hydrolyzates in the food industry

Food	Uses
Pastry	Improve texture and color
Sausages	Improve texture due to their water retention capacity
Lean meats	Reduce cholesterol levels
Children's formulas	They have a high nutritional value and are hypoallergenic
Nutritional drinks	Provide formulation flexibility and hypolipidemic effects
Boneless poultry meat	Improve color and texture
Soups, sauces and dressings	They improve texture due to their properties of viscosity, emulsification, water retention and hypolipidemic effects
Food coating films	Improve cohesion and adhesion properties

original raw material, usually a protein isolate or concentrate, is transformed into another product, a protein hydrolyzate, by the action of external proteases that do not come from endogenous microorganisms [6].

Studies carried out on the protein extraction of defatted sunflower flour (*Helianthus annuus* L.) by adding the protease alcalase during the alkaline extraction showed that the addition of 0.1% (v/v) of the enzyme improves the yields of protein extraction from 57.5 to 87.4%, giving an extract with a 22% degree of hydrolysis. In addition, an increase of up to 4.5 times in protein solubility is obtained at low pHs, which is correlated with the degree of hydrolysis [7].

Protein concentrates from broad bean (*Vicia faba*) have been used in the preparation of films to coat food, using films prepared with fresh egg albumin as a comparison standard. Concentrates in suspensions of 5 g of protein per 100 g of water, adding glycerol as a plasticizer, can produce films whose characteristics of permeability to water vapor, solubility and water content depend on the drying conditions of the material [8].

Proteins from tomato processing industries (*Solanum lycopersicum*) by *Bacillus subtilis* A were fermented for 14 h, to produce hydrolysates with antioxidant and antibacterial activities. The effects of different initial pH levels, incubation temperature, fermentation time, protein concentration and inoculum size on proteolytic activity, amino acid and peptide release, antioxidant and antibacterial activities of hydrolysates were evaluated and optimized by using the response surface methodology (RSM). The hydrolysates showed a remarkable activity of elimination of 2,2′-diphenyl-1-picrylhydrazyl (DPPH) (up to 70%), reducing power of ferric ions (FRAP) and inhibitory activity against *B. cereus* (up to 69.8%) and *E. coli* (up to a maximum of 70%). In general, a good correlation was observed between the concentration of amino acids and peptides and the antioxidant and antibacterial activities (in particular for the inhibitory activity of *B. cereus*) [9].

Defatted soybean meal was used to obtain protein concentrates and isolates. The determination of the best extraction conditions allowed to obtain an isolate with 84.79% of crude protein. The hydrolysis of the isolate obtained by sequential hydrolysis (endopeptidases, papain and exopeptidases) presented a degree of hydrolysis of 80%. With this, a powder supplement (isolate and soluble starch) was formulated with 70% protein and an energy intake of 340 kcal [10].

Soy proteins were modified by alkali treatment at pH 10, followed by hydrolysis with papain. Solubility, water holding capacity (WHC), surface hydrophobicity, foaming and emulsifying properties of unmodified, alkali-treated and papain-modified soy protein were compared. PMSP (papain modified soy protein) showed higher solubility (100% at pH 7.0), WHC (3.13) and hydrophobicity (40.8), while unmodified soy protein had 68.5% solubility, WHC 0.21 and hydrophobicity 8.1. PMSP had a foaming capacity (22.0 mL) similar to egg white (21.2 mL) at pH 7 and higher foam stability (36.4) compared to the unmodified control (32.9) [11].

Camellia oleifera seed cake protein (SCP) was hydrolyzed by five commercial proteases (Flavorzyme, Trypsin, Neutrase, Papain, Alcalase). The amino acid composition, molecular weight distribution, antioxidant activity and functional property of seed cake protein hydrolysates (SCPH) were investigated. Enzymatic hydrolysis improved the solubility of proteins significantly, but affected the foaming and emulsifying property. The hydrolyzate generated by alcalase had the highest degree of hydrolysis (DH) and antioxidant activity and showed excellent protein solubility in a wide pH range, while the hydrolyzate prepared with Flavorzyme showed better copper chelating capacity and emulsifying stability with a distribution low molecular weight. The study showed that the degree of hydrolysis could improve or reduce functional properties according to real applications [12].

A process has been designed to obtain protein isolates from defatted rapeseed flour (*Brassica napus*). The method includes basic extraction of soluble proteins followed by acid precipitation at the isoelectric point. The precipitate is washed with water (pH 4.5), ethanol and acetone, obtaining a protein isolate with 86% protein and reducing the content of polyphenols and soluble sugars by more than 90% compared to defatted flour [13].

In the search for new proteins and peptides with biological activities, bioinformatics is an important tool. In fact, bioinformatics allows the prediction of structure-function relationships of proteins, the identification of protein domains, and the computer simulation of proteolytic processes. All this information can be extracted from the large number of biologically active peptides that have already been isolated. The structural motifs in the active peptides serve as a source of information to be used in the search for new bioactive molecules. All this information is arranged in databases such as PepBank, EROP-Noscow, BioPD, PeptideDB, APD, BIOPEP, etc. The BIOPEP database contains protein information from 54 plants hydrolyzed with 21 endopeptidases. These data revealed that wheat gliadins were the plant proteins most susceptible to the release of bioactive peptides [14–16].

Correctly documented BP (bioactive peptides) can be predicted from a known protein amino acid sequence. Computer-aided databases are available to predict BP located within a parent protein. Other databases predict the precursor protein of a BP from a known amino acid sequence [17].

11.3 Moringa (*Moringa oleifera* L.)

M. oleifera seeds have been reported to contain 19–47% oil, 10–52% protein, and 2.5–20% glucosinolates. Moringa seed proteins, which have antimicrobial potential, can be made into food, drink, and feed [18].

On the other hand, the aqueous extract of the *M. oleifera* seed grain was found to contain positively charged peptides with low molecular mass. The *Moringa oleifera* coagulant protein (MOCP) comprised several small proteins with a molecular mass of approximately 6–7 kDa, as determined by SDS-PAGE under reducing conditions, with an isoelectric point greater than 10.0 [19].

Moringa seed flour (MSF) is used as a natural coagulant and plays an important role in the water purification industry. MSF contains a considerable amount of high-quality protein (~52%), with all essential amino acids, and can serve as a potential source of functional protein. A moringa seed protein concentrate can be used as an alternative to other proteins for human use in food due to its balanced amino acid profile [20].

The proteins in *M. oleifera* seeds are soluble in water and can be efficiently extracted with water and aqueous salts [21, 22]. However, these studies are primarily for analytical purposes and are for a single group of substances, practical methods for utilizing *M. oleifera* seed resources are rarely reported [18].

Recently, three Mo-CBP (*Moringa oleifera* Chitin Binding Proteins), called Mo-CBP2, Mo-CBP3 and Mo-CBP4, were isolated from the mature seeds of *M. oleifera*. They are highly thermostable basic glycoproteins, with pIs ranging from 10.5 to 10.9. They consist of a small polypeptide chain (3.8–4.6 kDa) and a large polypeptide chain (7.9–8.4 kDa) linked by disulfide bonds [23–25]. Furthermore, they have a potent antifungal action in vitro but lack chitinase, β-1,3-glucanase, and hemagglutinating activity [26]. Mo-CBP3, low molecular weight (14 kDa) cationic and basic peptide from crude extract of *M. oleifera* seeds, which has been the subject of several studies, as it is widely used for water purification and also possesses antifungal and antibacterial activities [27].

Mo-CBP3 belongs to the albumin 2S protein family, which are synthesized as precursors, which are then proteolytically cleaved to form the mature protein. Four isoforms of this precursor have been isolated from *M. oleifera* seeds, namely Mo-CBP3-1, Mo-CBP3-2, Mo-CBP3-3 and Mo-CBP3-4, which differ from each other only by a few amino acid residues [26–28]. The Mo-CBP3-1 precursor (pMo-CBP3-1) consists of 163 amino acids including the N-terminal signal peptide and linker peptides. The structural solutions (NMR) of several 2S albumin precursors have been deposited in the Protein Data Bank (http://www.rcsb.org) [26].

2S albumins, the storage proteins of seeds, are the main sources of carbon and nitrogen and are involved in plant defense. The mature mMo-CBP3-1 form of *Moringa oleifera*, an isoform chitin-binding protein 3-1, is a thermostable, antifungal, antibacterial and flocculating 2S albumin, widely used for water treatment and potentially interesting for development of antifungal drugs and transgenic crops. The crystal structure of mMo-CBP3-1 determined at a resolution of 1.7 Å showed that it is composed of two proteolytically processed helical chains, stabilized by four disulfide bridges that are stable, resistant to changes in pH and have a melting temperature (Tf) of about 98 °C. Surface arginines and the polyglutamine motif are the key structural factors for the flocculating, antibacterial and antifungal activities observed. This represented the first crystal structure of a 2S albumin and the proprotein model indicates the structural changes that occur after mMo-CBP3-1 formation and determines the structural motif and charge distribution patterns for the various observed activities [26].

11.4 Sacha Inchi (*Plukenetia volubilis* L.)

Sacha Inchi (*Plukenetia volubilis* L.), member of *Euphorbiaceae* family, is native to the rainforests of South America. Also known as Inca peanut, wild peanut, Sacha peanut or mountain peanut, it has been cultivated for centuries by the indigenous population. Until now, most of the studies on Sacha Inchi have dealt with plant development and physiology, characterization of seed oil, in vitro regeneration systems and potential applications of the oil in biofuel production and in the cosmetic, pharmaceutical and food industries [29].

Plukenetia volubilis L. has an unusual nutritional composition, since its edible seeds contain very high amounts of polyunsaturated essential fatty acids, α-linolenic acid and linoleic acid (ω-3 and ω-6; 35.2–50.8% and 33.4–41.0% of the total lipid fraction, respectively). Furthermore, the seeds are high in protein (approximately 22–30%) and are a source of antioxidants [30, 31]. Due to the composition of the lipid fraction of *P. volubilis*, commercial interest in the oil is increasing and, consequently, possible applications of the press cake remaining after oil extraction are being explored. The largest fraction of the press cake is protein; defatted seeds or defatted seed meal of *P. volubilis* contain approx. 53–59% protein [30–33]. The protein is predominantly soluble and more than 90% of the proteins in seed meal could be solubilized by aqueous solvents [32].

It should also be noted that fresh and unprocessed (raw) seeds and leaves of *P. volubilis* contain saponins, alkaloids, and lectins that may be slightly cytotoxic [34]. However, heat processing effectively reduces these potential phytotoxins and therefore roasting is recommended prior to ingestion of *P. volubilis* seeds and leaves [35, 36].

The protein content of the seed may vary with, among other things, the extraction method, cultivar and protein assay used. In a comparison study, enzyme-assisted extraction of defatted Sacha Inchi seeds [54.2 °C, 5.6% enzyme (alcalase), 50:1 (v/m) solvent-to-sample ratio, pH 9.0, 40.4 min] gave a 44.7% protein yield. In contrast, the protein yield of an alkaline extraction of defatted seeds [54.2 °C, 42:1 (v/m) solvent to sample ratio, NaCl 1.65 M, pH 9.5, 30 min] was only 29.7% [37].

Proteins isolated from Inca peanut seeds are soluble in aqueous buffers and have a molecular weight (MW) range of 35–63 kDa. They are composed primarily of albumin, globulin, prolamine, and glutelin [37]. Albumin is a 3S glycoprotein, dimeric, basic, nutritionally complete, for plant storage, with high levels of cysteine, tyrosine, threonine, and tryptophan [38, 39].

In flour made from seeds defatted with hexane, albumin (43.7%) is the predominant aqueous soluble protein, followed by glutelin (31.9%), globulin (27.3%), and prolamine (3.0%). Soluble seed meal proteins are composed primarily of monomeric polypeptides of 32–35 kDa and ~60–62 kDa, most in the range of 20–40 kDa. These seed proteins have disulfide-linked polypeptides. Heat denatured Sacha Inchi seed albumin was shown to be highly digestible by tosyl phenylalanyl chloromethyl ketone (TPCK)-trypsin, tosyl-L-lysyl-chloromethane hydrochloride (TLCK) -chymotrypsin and pepsin using in vitro assays. Therefore, Sacha Inchi seed proteins are of good nutritional quality and can be developed for human nutrition, especially in the growing market for gluten-free foods [40].

A water soluble storage albumin from Inca peanuts (Inca Peanut Albumin, IPA) represented ~25% (m/m) of the defatted seed meal weight, which represents 31% of the total seed protein. It is a 3S storage protein composed of two glycosylated polypeptides, with estimated molecular weights (MW) of 32.8 and 34.8 kDa, respectively. IPA has an estimated sugar content of 4.8%. IPA is a basic protein (pI ~9.4) and contains all essential amino acids in adequate amounts compared to the standard recommended by FAO/WHO for an adult human [38].

The structural and functional properties and the immunomodulatory activity of an albumin fraction isolated from Inca peanut seed (Albumin Fraction—Inca Peanut Seed, AF-IPS) were determined. Structural characterization revealed that AF-IPS contains two polypeptides with molecular weight ranges of 25–45 and 10–15 kDa. AF-IPS is primarily composed of α-helix, β-sheet, and β-turn secondary structures. The Differential Scanning Calorimetry analysis indicated that the denaturation temperature of AF-IPS at room temperature is 101.93 °C. AF-IPS had excellent solubility (63.0%), water retention capacity (1.59 g/g), foam (350.2%) and emulsification capacity (13.0 mL/g). Heat treatment improved protein solubility, oil retention capacity and foam and emulsion capacity. AF-IPS exhibited immunomodulatory activity by stimulating proliferation and enhancing TNF-α splenic lymphocyte secretion and by increasing cellular lysosomal enzyme and pinocytic activities and by moderately promoting NO and H_2O_2 production from RAW 264.7 cells. AF-IPS has potential applications in functional foods and pharmaceuticals [39].

11.5 Roselle (*Hibiscus sabdariffa* L.)

Roselle (*Hibiscus sabdariffa* L.) is an attractive alternative source of oil, it belongs to *Malvaceae* family. The yield of seeds was registered in 3039.4–6103.5 kg/ha, they are used in the cultivation at a rate of 1 kg per hectare. The quantities of roselle seeds produced that are not used for the recultivation process are of no economic value [41, 42]. The seeds contain 9.2–11.6% moisture, 30.1–31.0% protein, 5.8–6.9% total ash, 1.2–4.1% crude fiber, 36.3–38.1% total carbohydrates and over 20% total lipids [43, 44].

Several studies report that roselle seed oil and seed extract have high antioxidant activities [45] and may be a potential source of natural antioxidants [46]. Roselle seed oil has been shown to reduce hyperlipidemia and hypercholesterolemia in rats fed with a diet rich in saturated fatty acids [44]. On the other hand, seeds are an important source of protein (33–34%), fats (15–22%), total dietary fiber (18–23%) and carbohydrates (≈13%), they also constitute an important source of Vitamin E, a fat-soluble compound that participates in many functions of the body, ranging from keeping the immune system strong against viruses and bacteria, is involved in the formation of red blood cells and helps with the use of vitamin K. It also helps to dilate blood vessels, preventing blood from clotting within them. It favors the interaction of cells with each other and protects tissues from damage caused by free radicals (antioxidant effect). The lipid and vitamin E composition of these seeds are of special attention, so they could be used as an alternative source of obtaining fats in order to improve health in general [47].

After removing the calyces, the velvety capsules containing the *H. sabdariffa* seeds are removed as a by-product. Furthermore, it is wasteful if the seeds are left intact, with no effort to exploit their usefulness and benefits. Research on the usefulness of seeds has been very limited, they are rarely studied in comparison with calyces or with studies on other seeds, only one review article on this topic has been published. However, roselle seeds are among the seeds that contain the most protein compared to other seeds [47].

In Africa, the cake remaining from pressed oil extraction is cooked seasoned with *kambo*, a local condiment. In Nigeria, the seeds are fermented in a seasoning known as *mungza ntusa*. The seeds are roasted and ground into a powder and then consumed in oily soups and sauces as a meal. Roselle seeds are used for medical purposes, for example in Myanmar to treat weakness and in Taiwan as a diuretic, laxative and tonic [47]. Roasted seeds have been used as a substitute for coffee, are said to have aphrodisiac properties, and are considered an excellent chicken feed [47–49]. Lambs fed diets containing Jamaica flower seeds had higher average daily weight gain, feed conversion efficiency, and final body weight than the control group [50].

Roselle seeds are a potential source of protein with high digestibility [42, 43, 51, 52]. Globulin was the main protein fraction of the seeds rather than albumins. High levels of albumin raise the content of sulfur-containing amino acids (cysteine and methionine). Its lysine content was similar to that of the FAO reference protein. The limiting amino acids were valine, isoleucine and tryptophan, while those containing total sulfur were not limiting. Jamaica flower seeds could be used as a protein source and as a complementary food mix for low lysine sources [43]. Furthermore, the amino acid profiles of roselle seed protein concentrate and its protein isolate are similar to those of defatted seed meal. The extraction method to obtain proteins did not show adverse effects on the amino acid profile [43, 53].

H. sabdariffa seeds have shown no toxicity [42, 54]. Diets with the inclusion of Jamaica flower seeds did not interfere with food intake in rats. All rats remained healthy and active during the experiments [53].

The defatted meal extract of *H. sabdariffa* seed has a chymotrypsin inhibitory activity almost half that of trypsin [42]. It has proven to be a valuable food resource as it is an excellent source of dietary fiber. Biscuits incorporated with roselle seed powder have exhibited improved antioxidant properties. Therefore, the seed powder can be used as a source of dietary fiber and developed as a functional ingredient in food products [55].

11.6 Conclusion

In the residual cake of the oil expression of the seeds of *Moringa oleifera*, *Plukenetia volubilis* and *Hibiscus sabdariffa* there is a high content of proteins, valuable not only from the nutritional point of view. Among its constituents are lectins, which may have intrinsic antiviral action, but also have the potential to generate bioactive peptides by hydrolysis with proteases, taking advantage of the conserved motifs linked to certain biological activities, thus becoming important supplements with nutraceutical value, useful for strengthen the human organism, in particular in conditions of threat to health. This value is complemented by the low toxicity that has been found.

As the residual cakes of the oil expression of the seeds of these material species are scarcely studied in this direction and constitute a growing source of residual waste from oil extraction, the study of these potentialities becomes an interesting and useful topic for research, allowing to obtain proteins and peptides of potentially high added value. Therefore, a study of the residual cake from the extraction of oils from the seeds of these plants could yield valuable results applicable in obtaining protein concentrates, nutraceutical and medicinal products.

References

1. Ulloa JA, Rosas Ulloa P, Ramírez Ramírez JC, Ulloa Rangel BE (2012) Producción de aislados proteicos a partir de subproductos industriales. Revista Fuente Nueva Época 4(11):7
2. Herrera Saenz PDR (2016) Caracterización Tecno-Funcional del Extracto Proteico del Salvado de Arroz. ESPOL, Guayaquil
3. Vallejos Ibarra YE (2018) Obtención de concentrados proteicos de la harina de arveja (Pisum sativum) y determinación de su actividad antioxidante por el método del ácido tiobarbitúrico (TBA). Universidad Técnica de Ambato, Facultad de Ciencia e Ingeniería en Alimentos, Ambato
4. Pszczola D (2004) Ingredients. Food Technol 58:56–69
5. Saavedra JP (2010) Caracterización del proceso de obtención de aislados de proteína de Lupinus silvestre del Estado de Hidalgo. Instituto Politécnico Nacional, Mexico City
6. Vioque J, Clemente A, Pedroche J, del Mar Yust M, Millán F (2001) Obtention and uses of protein hydrolysates. Grasas Aceites 52(2):132–136
7. Yust MM, Pedroche J, Megías C, Girón-Calle J, Alaiz M, Millán F et al (2003) Improvement of protein extraction from sunflower meal by hydrolysis with alcalase. Grasas Aceites 54(4):419–423
8. Rozo AP, Bermúdez AS. Preparación de películas para alimentos a partir de concentrados proteicos de haba (*Vicia faba*). Rev. Colomb. Quim. [Internet]. 1 de enero de 2003 [citado 12 de abril de 2022];32(1):55-64. Disponible en: https://revistas.unal.edu.co/index.php/rcolquim/article/view/746
9. Moayedi A, Hashemi M, Safari M (2016) Valorization of tomato waste proteins through production of antioxidant and antibacterial hydrolysates by proteolytic Bacillus subtilis: optimization of fermentation conditions. J Food Sci Technol 53(1):391–400
10. Munive Ledesma PA (2009) Elaboración de un suplemento alimenticio en polvo para consumo humano a partir de una mezcla de hidrolizado de soya y almidón de maíz. Escuela Politécnica Nacional, Quito
11. Were L, Hettiarachchy NS, Kalapathy U (1997) Modified Soy Proteins with Improved Foaming and Water Hydration Properties. J Food Sci 62(4):821–824
12. Li X, Deng J, Shen S, Li T, Yuan M, Yang R et al (2015) Antioxidant activities and functional properties of enzymatic protein hydrolysates from defatted Camellia oleifera seed cake. J Food Sci Technol 52(9):5681–5690
13. Gonçalves N, Vioque J, Clemente A, Sánchez Vioque R, Bautista-Gallego J, Millán F (1997) Obtención y caracterización de aislados proteicos de colza. Grasas Aceites 48(5):282–289
14. García MC, Puchalska P, Esteve C, Marina ML (2013) Vegetable foods: a cheap source of proteins and peptides with antihypertensive, antioxidant, and other less occurrence bioactivities. Talanta 106:328–349

15. Dziuba J, Niklewicz M, Iwaniak A, Darewicz M, Minkiewicz P (2004) Bioinformatic-aided prediction for release possibilities of bioactive peptides from plant proteins. Acta Aliment 33(3):227–235
16. Dziuba M, Darewicz M (2007) Food proteins as precursors of bioactive peptides—classification into families. Food Sci Technol Int 13(6):393–404
17. López-Barrios L, Gutiérrez-Uribe JA, Serna-Saldívar SO (2014) Bioactive peptides and hydrolysates from pulses and their potential use as functional ingredients. J Food Sci 79(3):R273–R283
18. Chen R, Wang X-J, Zhang Y-Y, Xing Y, Yang L, Ni H et al (2019) Simultaneous extraction and separation of oil, proteins, and glucosinolates from Moringa oleifera seeds. Food Chem 300:125162
19. Gassenschmidt U, Jany KD, Bernhard T, Niebergall H (1995) Isolation and characterization of a flocculating protein from Moringa oleifera Lam. Biochim Biophys Acta Gen Subj 1243(3):477–481
20. Jain A, Subramanian R, Manohar B, Radha C (2019) Preparation, characterization and functional properties of Moringa oleifera seed protein isolate. J Food Sci Technol 56(4):2093–2104
21. Baptista ATA, Silva MO, Gomes RG, Bergamasco R, Vieira MF, Vieira AMS (2017) Protein fractionation of seeds of Moringa oleifera lam and its application in superficial water treatment. Sep Purif Technol 180:114–124
22. Dezfooli SM, Uversky VN, Saleem M, Baharudin FS, Hitam SMS, Bachmann RT (2016) A simplified method for the purification of an intrinsically disordered coagulant protein from defatted Moringa oleifera seeds. Process Biochem 51(8):1085–1091
23. Leite Pereira M, David de Oliveira H, Tadeu Abreu de Oliveira J, Menezes Gifoni J, de Oliveira Rocha R, de Oliveira Bezerra de Sousa D et al (2011) Purification of a Chitin-Binding Protein from Moringa oleifera Seeds with Potential to Relieve Pain and Inflammation. Protein Pept Lett 18(11):1078–1085
24. Batista AB, Oliveira JTA, Gifoni JM, Pereira ML, Almeida MGG, Gomes VM et al (2014) New insights into the structure and mode of action of Mo-CBP3, an antifungal chitin-binding protein of Moringa oleifera seeds. PLoS One 9(10):e111427
25. Neto JXS, Pereira ML, Oliveira JTA, Rocha-Bezerra LCB, Lopes TDP, Costa HPS et al (2017) A chitin-binding protein purified from Moringa oleifera seeds presents anticandidal activity by increasing cell membrane permeability and reactive oxygen species production. Front Microbiol 8:980
26. Ullah A, Mariutti RB, Masood R, Caruso IP, Gravatim Costa GH, Millena de Freita C et al (2015) Crystal structure of mature 2S albumin from Moringa oleifera seeds. Biochem Biophys Res Commun 468(1):365–371
27. Freire JEC, Vasconcelos IM, Moreno FBMB, Batista AB, Lobo MDP, Pereira ML et al (2015) Mo-CBP3, an antifungal chitin-binding protein from Moringa oleifera seeds, is a member of the 2S albumin family. PLoS One 10(3):e0119871
28. Garcia TB, Soares AA, Costa JH, Costa HPS, Neto JXS, Rocha-Bezerra LCB et al (2019) Gene expression and spatiotemporal localization of antifungal chitin-binding proteins during Moringa oleifera seed development and germination. Planta 249(5):1503–1519
29. Hu X-D, Pan B-Z, Fu Q, Niu L, Chen M-S, Xu Z-F (2018) De novo transcriptome assembly of the eight major organs of Sacha Inchi (Plukenetia volubilis) and the identification of genes involved in α-linolenic acid metabolism. BMC Genomics 19(1):380
30. Dash P, Ghosh G (2017) Amino acid composition, antioxidant and functional properties of protein hydrolysates from Cucurbitaceae seeds. J Food Sci Technol 54(13):4162–4172
31. Ruiz C, Díaz C, Anaya J, Rojas R (2013) Análisis proximal, antinutrientes, perfil de ácidos grasos y de aminoácidos de semillas y tortas de 2 especies de Sacha Inchi (Plukenetia volubilis y Plukenetia huayllabambana). Rev Soc Quím Perú 79(1):29–36
32. Sathe SK, Kshirsagar HH, Sharma GM (2012) Solubilization, fractionation, and electrophoretic characterization of Inca peanut (Plukenetia volubilis L.) proteins. Plant Foods for Hum Nutr 67(3):247–255

33. Chirinos R, Aquino M, Pedreschi R, Campos D (2017) Optimized methodology for alkaline and enzyme-assisted extraction of protein from Sacha Inchi (Plukenetia volubilis) kernel cake. J Food Process Eng 40(2):e12412
34. Srichamnong W, Ting P, Pitchakarn P, Nuchuchua O, Temviriyanukul P (2018) Safety assessment of Plukenetia volubilis (Inca peanut) seeds, leaves, and their products. Food Sci Nutr 6(4):962–969
35. Bueno-Borges LB, Sartim MA, Gil CC, Sampaio SV, Rodrigues PHV, Regitano-d'Arce MAB (2018) Sacha inchi seeds from sub-tropical cultivation: effects of roasting on antinutrients, antioxidant capacity and oxidative stability. J Food Sci Technol 55(10):4159–4166
36. Kodahl N (2020) Sacha inchi (Plukenetia volubilis L.)—from lost crop of the Incas to part of the solution to global challenges? Planta 251(4):80
37. Chirinos R, Zorrilla D, Aguilar-Galvez A, Pedreschi R, Campos D (2016) Impact of roasting on fatty acids, tocopherols, phytosterols, and phenolic compounds present in Plukenetia huayllabambana seed. J Chem 2016:1–11
38. Sathe SK, Hamaker BR, Sze-Tao KW, Venkatachalam M (2002) Isolation, purification, and biochemical characterization of a novel water soluble protein from Inca peanut (Plukenetia volubilis L.). J Agric Food Chem 50(17):4906–4908
39. Li P, Wen J, Ma X, Lin F, Jiang Z, Du B (2018) Structural, functional properties and immunomodulatory activity of isolated Inca peanut (Plukenetia volubilis L.) seed albumin fraction. Int J Biol Macromol 118(Pt B):1931–1941
40. Wang S, Zhu F, Kakuda Y (2018) Sacha inchi (Plukenetia volubilis L.): nutritional composition, biological activity, and uses. Food Chem 265:316–328
41. Samy MS (1980) Chemical and nutritional studies on roselle seeds (Hibiscus sabdariffa L.). Z Ernahrungswiss 19(1):47–49
42. Abu-Tarboush HM, Ahmed SAB (1996) Studies on karkade (Hibiscus sabdariffa): protease inhibitors, phytate, in vitro protein digestibility and gossypol content. Food Chem 56(1):15–19
43. El-Adawy TA, Khalil AH (1994) Characteristics of roselle seeds as a new source of protein and lipid. J Agric Food Chem 42(9):1896–1900
44. Ali RF, El-Anany AM (2017) Hypolipidemic and hypocholesterolemic effect of roselle (Hibiscus sabdariffa L.) seeds oil in experimental male rats. J Oleo Sci 66(1):41–49
45. Mohd-Esa N, Hern FS, Ismail A, Yee CL (2010) Antioxidant activity in different parts of roselle (Hibiscus sabdariffa L.) extracts and potential exploitation of the seeds. Food Chem 122(4):1055–1060
46. Nyam KL, Teh YN, Tan CP, Kamariah L (2012) In vitro antioxidant activities of extract and oil from roselle (Hibiscus sabdariffa L.) seed against sunflower oil autoxidation. Malays J Nutr 18(2):265–274
47. Ismail A, Ikram EHK, Nazri HSM (2008) Roselle (Hibiscus sabdariffa L.) seeds nutritional composition protein quality and health benefits. Food 2(1):1–16
48. Omobuwajo T, Sanni L, Balami Y (2000) Physical properties of sorrel (Hibiscus sabdariffa) seeds. J Food Eng 45(1):37–41
49. Al-Wandawi H, Al-Shaikhly K, Abdul-Rahman M (1984) Roselle seeds: a new protein source. J Agric Food Chem 32(3):510–512
50. Beshir AA, Yagoub YM, Babiker SA (2009) Performance of Sudanese desert lambs fed graded levels of water melon (Citrullus lanatus) seed cake instead of groundnut cake. Pak J Nutr 8(5):525–529
51. Abu-Tarboush HM, Ahmed SAB, Al Kahtani HA (1997) Some nutritional and functional properties of karkade (Hibiscus sabdariffa) seed products. Cereal Chem 74(3):352–355
52. Yagoub AE-GA, Mohamed BE, Ahmed AHR, El Tinay AH (2004) Study on furundu, a traditional Sudanese fermented roselle (Hibiscus sabdariffa L.) seed: effect on in vitro protein digestibility, chemical composition, and functional properties of the total proteins. J Agric Food Chem 52(20):6143–6150

53. Hainida E, Ismail A, Hashim N, Mohd.-Esa N, Zakiah A (2008) Effects of defatted dried roselle (Hibiscus sabdariffa L.) seed powder on lipid profiles of hypercholesterolemia rats. J Sci Food Agric 88(6):1043–1050
54. Fagbenro OA (2005) Soybean meal substitution with roselle (Hibiscus sabdariffa L.) seed meal in dry practical diets for fingerlings of the African catfish, Clarias gariepinus (Burchell 1822). J Anim Vet Adv 4(4):5
55. Nyam KL, Leao SY, Tan CP, Long K (2014) Functional properties of roselle (Hibiscus sabdariffa L.) seed and its application as bakery product. J Food Sci Technol 51(12):3830–3837

Chapter 12
Whey Protein from Milk as a Source of Nutraceuticals

Chinaza Godswill Awuchi

12.1 Introduction

Whey protein comprise of protein mixture obtained from whey, the liquid milk component that separates during the production of cheese. During the production of cheese, there is coagulation of the fats in the milk, resulting in the separation of whey as by-product. Whey can also be considered as the liquid part of milk remaining after the curdling and straining of the milk. As a byproduct of cheese or casein production, whey along with whey proteins, has many commercial and nutraceutical applications. Sweet whey results from the production of rennet hard cheese, such as Swiss cheese, cheddar, etc. Sour whey (acid whey) is obtained during the production of dairy products' acid types, including strained yogurt, cottage cheese, etc. Whey proteins contain β-lactoglobulin (BLG), α-lactalbumin (LALBA), proteose peptones, immunoglobulins, lactoferrin, and serum albumin [1–3]. Whey protein is the globular proteins' collection from whey. Cow milk protein contains 80% casein protein and 20% whey protein, while human milk protein is 30% casein protein and 70% whey protein. The whey protein fraction constitutes around 10% of the whey's total dry solids. The protein is usually a mixture of immunoglobulins, bovine serum albumin (6%), alpha-lactalbumin (13–19%), beta-lactoglobulin (~48–58%), etc. [1–3]. They are soluble in nature, in spite of the pH. The common analytical techniques, including hyphenated techniques, used in biosciences can be used to quantify these whey protein components [4, 5]. Figure 12.1 shows the common proteins found in whey proteins.

C. G. Awuchi (✉)
Department of Biochemistry, Kampala International University, Bushenyi, Uganda

School of Natural and Applied Sciences, Kampala International University, Kampala, Uganda

Department of Healthcare Management, Unicaf University, Kampala, Uganda
e-mail: awuchi.chinaza@kiu.ac.ug

© The Author(s), under exclusive license to Springer Nature Switzerland AG 2022
C. Egbuna et al. (eds.), *Food and Agricultural Byproducts as Important Source of Valuable Nutraceuticals*, https://doi.org/10.1007/978-3-030-98760-2_12

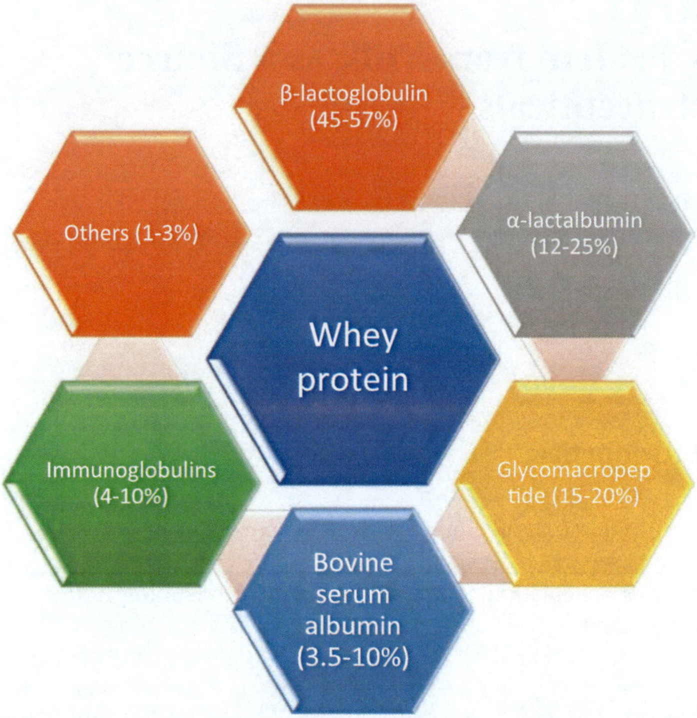

Fig. 12.1 % Protein contents of whey proteins

Cysteine, an amino acid, in whey protein is required for glutathione synthesis in the body; glutathione is a powerful cellular antioxidant in the body [6]. Whey protein may decrease cancer risk in animals. Whey protein has been recognized as potential source for formulating novel nutraceuticals due to its several health and therapeutic benefits. One of the underlying mechanisms through which whey protein components show their biological effects is possibly through intracellular translation of cysteine to glutathione, an important intracellular antioxidant [7]. Studies have shown that whey protein has great therapeutic properties for the treatment of cardiovascular diseases, AIDS, cancer, etc.

This chapter focuses on the therapeutic and health benefits of whey protein for the formulation of nutraceuticals. The chapter also explored the biological, biochemical, and physiological properties of whey protein and its components. The use of other food and agricultural byproducts in the formulation of nutraceuticals have been extensively discussed in other chapters. Whey protein is usually available as a supplement from dietary source, with several health and therapeutic claims attributed to it and its components [8, 9]. Whey is a major component of several protein-based powders primarily used by bodybuilders and athletes every day to obtain the required daily intake of protein. Whey protein contains high leucine level, one of the three amino acids with branched chain, which makes it suitable for the repair and growth of muscles.

12.2 Types of Whey Protein

There are several popular types of whey protein. Their main difference is in the way they have been processed. The three major types of whey protein include hydrolysate, concentrate, and isolate. Whey protein hydrolysates are partially hydrolyzed and predigested for easier metabolizing, although they are generally costly. Whey protein that is highly hydrolyzed might be less allergenic compared to other types of whey protein. Native whey proteins are obtained from skim milk directly without being byproduct of the production of cheese, and is made as whey protein isolate or whey protein concentrate. Whey protein concentrate contains 70–80% protein, with some lactose and fat; its flavor is the best [10–12]. Whey protein isolate contains at least 90% protein, less lactose and fat; it does not have many beneficial nutrients commonly seen in whey protein concentrates. Whey protein hydrolysate, also called hydrolysed whey, is pre-digested for faster absorption, and results in 28–43% more spike in the levels of insulin than whey protein isolate. Overall, the best option appears to be whey protein concentrate, since it is more cost effective and retains many beneficial nutrients naturally found in whey [12]. Additionally, many individuals prefer the taste, mainly because of the fat and lactose. For those who have challenges tolerating concentrate partly due to being lactose intolerant, or those who lay emphasis on protein while maintaining low carbohydrates and fat, whey protein isolate or whey protein hydrolysate can be the preferred option [12]. These are the main types of whey protein; isolate, hydrolysate, and concentrate. They vary in their protein content, digestibility, taste, and price.

12.3 Health and Nutritional Benefits of Whey Protein

Whey protein has very high protein efficiency ratio (PER) and very high biological value (BV). Biological value is a measure of amount of absorbed protein from a food which becomes incorporated into the body proteins of an organism. This incorporated protein is used for functions such as tissue formation in the body. Biological value is calculated as the percentage of protein efficiently utilized and PER measures the gain in weight per gram of consumed protein. Proteins with high biological value are very vital and essential to humans [13–16]. Whey protein contains all essential amino acids (EAAs) in sufficient amounts compared with protein from other sources, including corn, wheat gluten, soy, etc. The whey protein's amino acids are absorbed faster in comparison with free amino acid (AA) and proteins from other source [17]. As stated earlier, whey proteins are rich source of EAAs, including the branched chain AAs, such as isoleucine, valine, and leucine. The amino acid profile of whey protein is similar to the proteins of the muscle, with almost all the AAs in similar magnitudes [18]. The whey proteins branched-chain AAs play an important role in the repair and growth of tissue. Leucine is an important amino acid in protein metabolism during protein synthesis. In addition, whey protein contains sulfur-containing amino acids, which is why whey protein amino acids such as methionine and cysteine boost the functions of the immune system by conversion to glutathione intracellular. Fluid whey from cattle milk has around 50% and 20% of the nutrients and milk protein contents, respectively, corresponding to an average protein content of 4–7 g/L. While the major whey protein components include α-lactalbumin, β-lactoglobulin, BSA, and the fraction of proteose-peptone, the minor whey protein components include immunoglobulins, lactoferrins, ceruloplasmins, as well as certain enzymes such as lipase, xanthine oxidase, and lysozyme (Table 12.1) [18, 19].

Whey proteins are found in milk and dairy products, with high calcium bioavailability linked to vitamin D activities, a major factor that increase calcium absorption in the intestine. Calcium is essential to bones [23]. The whey protein bioactive peptides contain approximately 3–20 AAs and might exhibit antihypertensive, citomodulatory, immunomodulatory, antithrombotic, antimicrobial, and opioid activities; they can modulate our mood and intestinal microbiota, and show action against allergies, infections, and atopic dermatitis in children and infants that consume infant formula [24–26]. Whey proteins derived bioactive peptides can also be generated during food processing. Enzymatic hydrolysis is a common method for the production of peptide fragments used in hypoallergenic infant formulas [24, 26].

12.4 Whey Protein Applications as Nutraceutical

The high digestibility of whey proteins is mainly because of improved plasma amino acids that initiates protein synthesis, making whey proteins quite attractive in clinical nutrition and therapeutical applications [21]. Whey proteins are administered in the patients' postoperative care and strongly recommended for growth and repair of body cells [17]. As excellent source of EAAs, if carefully and professionally administered, whey proteins help in developing nutraceuticals and functional formula for clinical applications [20]. The common nutraceutical and health benefits of whey proteins are shown in Table 12.2.

12.4.1 Muscle Synthesis and Resistance Exercise

Resistance exercise, concentric (shortening), isometric (non-lengthening), and eccentric (muscle lengthening) contractions result in damage to the skeletal muscle and also generate inflammatory markers [38]. Anabolic interventions using amino acid supplements and protein hydrolysates expedite repair. The ingestion of leucine and its metabolic derivative, β-hydroxy-β-methylbutyrate, is beneficial to soreness recovery. Resistance exercise such as weight-lifting raises products of oxidation in plasma and disturbs leukocyte functionality and redistribution [39]. Nanoparticles of whey protein isolate were formulated using ethanol desolvation, and their $ZnCl_2$

Table 12.1 Molecular weight and percent of whey protein fraction

Whey protein	% Whey protein component	Nutraceutical functions	Number of AA residues	Molecular weight (Da)	Concentration (g/L)	References
β-Lactoglobulin	45 to 57	BLG contains high levels of branched chain AAs (25.1%), and binds hydrophobic molecules, contributing in reducing intestinal lipid absorption	162	1.8×10^4	1.3	[12]
α-Lactalbumin	12 to 25	α-Lactalbumin has the highest content of tryptophan (6%) of all the proteins from dietary sources. It is a rich source of cysteine, threonine, leucine, and lysine. It can bind to minerals, including calcium and zinc, and improve their absorption	123	1.42×10^4	1.2	[20]
Bovine serum albumin (BSA)	3.5 to 10	BSA has good lipid binding properties and good amino acid profile	582	6.9×10^4	0.4	[21]
Immunoglobulins	4 to 10	Four classes of immunoglobulins present in whey include IgE, IgM, IgA, and IgG. Their functions include improved immunity and antioxidant protection	–	1.5×10^5 to 9×10^5	0.7	[12]
Glycomacropeptide	15 to 20	Glycomacropeptide is formed from κ-casein digestion during cheese production, and has high levels of EAAs that improve the absorption of minerals	64	7×10^2	1.2	[21]
Lactoferrin	3	Lactoferrin inhibits proinflammatory cytokines production and protects against hepatitis.	700	7.7×10^4	0.1	[20]
Lactoperoxidase	0.4	It has strong antimicrobial properties and forms part of the innate immune system	612	7.8×10^4	0.03	[22]
Lysozyme	1	It has antimicrobial properties and forms part of the innate immune system	129	1.4×10^4		[22]

Table 12.2 Nutraceutical and health benefits of whey protein

Nutraceutical and health benefits	Study evidence	References
Anticarcinogenic properties	A study reported that when whey protein hydrolysate was fed to rat models with induced colon cancer, they significantly developed less microscopic and macroscopic tumours in comparison with the group that consumed untreated whey protein	[27]
Anti-diabetic properties	Whey protein reduces blood glucose level in healthy people, reduces ghrelin (hunger hormone) secretion, boosts satiety hormones release (glucagon like-peptide 1 [GLP-1], leptin, and cholecystokinin), and maintains muscle mass. A study showed that that the amino acid cysteine, which is abundant in whey protein, can be applied as auxiliary therapeutic measure in the control of vascular and glycaemia inflammation in diabetic individuals	[28, 29]
Antihypertensive effects	Bovine whey proteins' beta-lactoglobulin peptide sequence has high inhibitory actions against Angiotensin I converting enzyme (ACE), which catalyzes Angiotensin I translation into Angiotensin II, a potential vasoconstrictor. It reduces the risks of hypertension	[7, 21]
Antioxidant properties	Antioxidant properties of whey proteins have been reported, with potential benefits against oxidative stress, commonly reported in chronic non-communicable diseases (CNCDs) and in preterm neonates	[20]
Cardioprotective and hypotensive effects	Whey protein can reduce the risk of cardiovascular diseases. The risks of cardiovascular diseases can be ameliorated with quick-absorbable whey protein extracts	[30]
Immune improvement in HIV patients	Whey protein isolates are used in the treatment HIV patients and are widely recommended by health and nutrition experts. Whey protein consumption can significantly increase glutathione status. A study concluded that supplementation with whey protein increased CD4 cell counts in HIV patients	[31]
Immunomodulatory properties	Whey protein improves innate mucosal immunity in early stages of life and has a protective effect in certain immune conditions	[32]
Infants' immunity	As a result of its high beneficial immune properties, whey proteins are among the most suitable protein choices for mothers to feed their infants. The antimicrobial and immunomodulatory properties of whey proteins are associated with the stimulation of immune cells and protective microflora proliferation in the GI tract of humans and animals, indicating the possible whey protein role in strengthening immune system in infants and children	[26]

(continued)

Table 12.2 (continued)

Nutraceutical and health benefits	Study evidence	References
Muscle synthesis and resistance exercise	Whey protein hydrolysates were studied for their capability of translocating and accumulating GLUT 4 in membrane, consequently augmenting skeletal muscle glucose trapping. L-leucyl-L-isoleucine (a peptide) and L-isoleucine (an amino acid) in the whey protein hydrolysate made the most contribution	[33]
Phenylketonuria treatment	A study reported that glycomacropeptide of whey protein with low content of phenylalanine is suitable as the major source of protein in the diet of PKU patients. Glycomacropeptide in such diet can significantly ameliorate the effects associated with phenylketonuria	[34]
Postoperative care	Whey proteins are administered in the patients' postoperative care and strongly recommended for growth and repair of body cells	[17]
Prebiotic and gut functions	Fortification with whey protein may ameliorate tolerance to enteric nutrition and have influence on inflammation. Yeast and lactic acid bacteria (LAB) require to remain viable to exercise their therapeutic properties. Whey protein gels protect the microorganisms such as probiotics from adverse conditions	[35]
Skin protective effects	A group of researchers studied the effects of whey peptides consumption at 200 and 400 mg/kg two times a day on skin alterations (wrinkle formation, suppleness, and thickness) induced by chronic UV-B radiation in mice. The whey peptides ameliorated dermatoheliosis through preventing increase in melanin granule formation, wrinkling, and dermal stiffness. The whey peptides reduced matrix metalloproteinase (pro-MMP-9 and MMP-2) expression and vascular endothelial growth factor (VEGF) expression	[36]
Treatment of obesity and overweight	A study done for 12 weeks indicated that preloads of whey protein concentrate done 30 min before the main meal of ad libitum has more health benefits than the one of soy protein isolates on anthropometry, appetite, body composition, and calorie intake of obese men	[37]

incorporation capacity analyzed. The level of zinc loaded into the suspensions of the nanoparticles was within zinc requirements per day for a healthy adult. The nanoparticles were stable after storage at 22°C for one month [40]. Glucose transporter 4 (GLUT 4) of cell surface is the main glucose transporter isoform which is expressed in the skeletal muscles that determine the muscle glucose transport rates in cell membranes, in responding to muscle contraction and insulin. Whey protein hydrolysates were studied for their capability of translocating and accumulating GLUT 4 in membrane, consequently augmenting skeletal muscle glucose trapping. L-leucyl-L-isoleucine (a peptide) and L-isoleucine (an amino acid) in the whey protein hydrolysate made the most contribution [33]. Whey protein supplementation effects compared with casein diet effects on the muscle functional recoveries, including excitability, elasticity, extensibility, and contractility, have been studied using rat models. The whey protein supplementation caused a quicker recovery from sustained injury due to concentric and isometric exercise compared with the casein diet [41]. Churchward-Venne et al. [42] studied the effects of beverage supplementation with varied leucine doses or mixture of branched chain AAs on myofibrillar protein synthesis following resistance exercise. Results indicated that beverage with low protein content (6.25 g) could be equally effective as high protein counterpart (25 g) at promoting the rates of myofibrillar protein synthesis when supplemented with high levels of leucine (5 g) [42]. Leucine constitutes 10% of the amino acids in whey protein and is vital in improving muscle hypertrophy. Lollo et al. [43] compared the effects of health parameters, body composition, and performance produced after intake of whey protein hydrolysate for 12 weeks in players. The whey protein hydrolysate intervention caused significant decrease in lactate dehydrogenase and creatine kinase (muscle damage markers) [43]. In a randomized study, Volek et al. [44] compared the effects of daily consumption of whey protein and soy protein on the growth of muscle mass using subjects under resistance exercise. The gains in lean body mass were significantly higher for whey protein consumption group than the group that consumed soy protein, and the considerable response was associated with increased leucine levels and quicker absorption [44].

12.4.2 Management of Phenylketonuria

Phenylketonuria (PKU) is an inborn metabolic error and genetic disorder characterized by mutation in phenylalanine hydroxylase gene and reduced metabolism of phenylalanine (an amino acid), resulting in excess phenylalanine accumulation [45]. Untreated PKU or failure in phenylalanine usage results in several health complications, including seizures, osteopenia, mental disorders, intellectual disability, and behavioral problems. The common strategy for managing PKU is by consuming diet low in phenylalanine. Van Calcar and Ney [34] reported that glycomacropeptide of whey protein with low content of phenylalanine is suitable as the major source of protein in the diet of PKU patients. Glycomacropeptide in such diet can significantly ameliorate the effects associated with PKU. Phenylalanine plasma level, systemic inflammation markers, food intake, and energy expenditure can be efficiently lowered in comparison with casein diet, which has high level of phenylalanine. Solverson et al. [46] reported significantly lower respiratory exchange ratio and the total fat mass in PKU mice that consumed glycomacropeptide. Osteopenia characterized by skeletal fragility because of decreased minerals in bones is among the common complications associated with PKU [47]. Demirdas et al. [47] and Ney et al. [48] reported that osteopenia was ameliorated after consumption of glycomacropeptide diet. There was improvement in bone health and a higher radial bone growth [48]. Strisciuglio and Concolino [49] concluded that glycomacropeptide diet is healthier alternative over regular synthetic AAs, as it enhances satiety and taste of diets resulting in improved patient compliance.

12.4.3 Anticancer Effects

Many studies suggest that whey protein can have therapeutic effects on patients with cancer. Whey protein hydrolysis may enhance the efficacy of the anticancer activities. Attaallah et al. [27] reported that when whey protein hydrolysate was fed to rat models with induced colon cancer, they significantly developed less microscopic and macroscopic tumours in comparison with the group that consumed untreated whey protein. Castro et al. [50] studied whey protein anticancer effect using model of melanoma B16F10 cells, and reported significant increase in caspase-3 expression in the media that contained whey protein isolate. The Caspase-3 role in mediating cell death via apoptosis is common [51]. A Caucasian female aged 48 years with recurring cervical cancer was given whey protein (10 g three times in a day) and a weekly testosterone enanthate intramuscular injection prior to and during standard-of-care chemotherapy. Improvements in physical activity, lean body mass, and general quality of life were observed due to the combination therapy [52]. Zhang et al. [53] studied whey protein hydrolysate's protective effect against oxidative damage on pheochromocytoma PC12 cells of rat. At 100–400 μg/ml dose of hydrolysate, there was 20–30% increase in the viable cells in comparison with those incubated in hydrogen peroxide, suggesting whey protein hydrolysate may have antioxidant potentials [53].

12.4.4 Skin Protection

Dermatoheliosis, also known as photoaging, is the characteristic changes to skin induced by chronic exposure to UV radiations. Kimura et al. [36] studied the effects of whey peptides consumption at 200 and 400 mg/kg two times a day on skin alterations (wrinkle formation, suppleness, and thickness) induced by chronic UV-B radiation in mice. The whey peptides ameliorated dermatoheliosis through preventing increase in melanin granule formation, wrinkling, and dermal stiffness. The whey peptides reduced matrix metalloproteinase (pro-MMP-9 and MMP-2) expression and vascular endothelial growth factor (VEGF) expression. In addition, they prevented increase in number of apoptotic, 8-hydroxy-2′-deoxyguanosine (8-OHdG)-positive and Ki-67-positive cells caused by chronic UV-B radiation [36]. The whey peptides inhibited type IV collagen degradation, proliferation, DNA damage, and angiogenesis induced by the radiation [36].

12.4.5 Infant Foods/Formula

Bovine milk is a rich source of protein, with whey protein constitute 20% of the total protein. Formulating physiologically suitable foods/formula for infants is required to meet up with their nutritional requirements [20, 21]. Lactose performs many biological functions in infants, and are found in whey protein concentrate, isolate, and hydrolysate; although in low levels, but beneficial to infants [20]. The monosaccharide units in lactose are glucose and galactose, which perform many metabolically related developments in babies and children, including energy provision and protein sparing actions [54]. As a result of its high beneficial immune properties, whey proteins are among the most suitable protein choices for mothers to feed their infants.

12.4.6 Gut Prebiotics and Functions

Prebiotics are food compounds that induce the activities and growth of probiotics (beneficial microorganisms), including fungi and bacteria [55]. Gut dysfunctions, such as compromised intestinal barrier, abnormal patterns of motility, and delay in gastric emptying, are severe conditions in patients that are critically ill. Fortification with whey protein may ameliorate tolerance to enteric nutrition and have influence on inflammation [35]. Yeast and lactic acid bacteria (LAB) require to remain viable to exercise their therapeutic properties. Gels of whey protein protect the microorganisms from adverse conditions. Gerez et al. [56] studied the encapsulation efficacy of pectin- and whey protein-loaded *Lactobacillus rhamnosus* CRL 1505 for improved rate of survivability in bile milieu and low pH. The results showed that whey protein layer with pectin beads can be applied as carrier for probiotics in acidic functional food. Zhao et al. [57] isolated the dipeptide glycine-L-tyrosine (Gly-Tyr) that has specific binding capacity to calcium from whey protein hydrolysates. The peptide chelating mechanism was studied and the main binding sites were the amino/imino group's nitrogen and the carbonyl group's oxygen atom [57]. The study concluded that the peptide may boost the absorption of calcium in GI tract; the development of the calcium-peptide chelate supplement for improved bioavailability of calcium has promising potentials. Walsh et al. [58] studied the "effects of carbonation on probiotic survivability, physicochemical, and sensory properties of milk-based symbiotic beverages". *Bifidobacterium* and *Lactobacillus acidophilus* remained viable in yogurt stabilized with whey protein concentrate and high-methoxyl pectin [58]. Microparticles of alginate and whey protein isolate showed suitability for *Saccharomyces boulardii* (probiotic yeast) oral delivery systems. A study done with the simulation of intestinal and gastric fluid showed 40% survival rate for encapsulated yeast in comparison with 10% survival rate for free yeast [59].

12.4.7 Management of Obesity and Overweight

Obesity is a health condition characterized by excessive accumulation of body fat to an extent that may have negative effects on health, usually defined by body mass index (BMI) of 30 and above; on the other hand, overweight is characterized by more body fat than is optimal for health, usually defined by BMI within 25–29.9; normal weight and underweight are defined by BMI of 18–24.9 and below 18 respectively. Change of diet helps to combat obesity and overweight; whey protein is beneficial for this purpose [54]. In as study, ad libitum was prepared using high-fat diets that contain adequate whey protein levels or leucine supplement and fed to mice for 1 week. The study concluded that the diet rich in protein has ameliorating effects on metabolic disorders that are surely because of liver lipogenesis attenuation-mediated satiety modulation [60]. A study done for 12 weeks indicated that preloads of whey protein concentrate done 30 min before the main meal of ad libitum has more health benefits than the one of soy protein isolates on anthropometry, appetite, body composition, and calorie intake of obese men [37].

12.4.8 Anti-Diabetic Effects

Diabetes mellitus (DM), also called diabetes, is various metabolic disorders that manifest as a high level of blood sugar for a long period of time [61, 62]. Different types of DM include type 1 DM (T1DM), type 2 DM (T2DM), gestational diabetes, and other minor types [63, 64]. T2DM constitute about 90–95% of all cases of DM [63, 64]. DM is one of the major public health issues that has several complications, including angiopathy, loss of vision, decreased blood flow resulting in tissue hypoxia, diabetic ketoacidosis, foot ulcers, nerve damage, difficult wound healing, etc. [65]. T2DM can be treated using hypoglycemic drugs, exercise, and diet control/restrictions [66]. Whey protein reduces blood glucose level in healthy people, reduces ghrelin (hunger hormone) secretion, boosts satiety hormones release (glucagon like-peptide 1 [GLP-1], leptin, and cholecystokinin), and maintains muscle mass [28]. Jain [29] showed that that the amino acid cysteine, which is abundant in whey protein, can be applied as auxiliary therapeutic measure in the control of vascular and glycaemia inflammation in diabetic individuals. Badr et al. [67] studied whey protein effects on diabetic wounds recuperation on a mouse model induced with T2DM. In comparison to untreated T2DM mice, whey protein supplementation significantly enhanced diabetic lesions closure through restricting the inflammatory cytokines access via the maintenance of normal levels of IL-6, IL-1β, TNF-α, and IL-10. The supplementation with whey protein modulated the chemokines TGF-β, CX3CL1, KC, MIP-2, and MIP-1α expression in wound tissues in comparison to mice with untreated T2DM. Salehi et al. [68] studied the whey protein's insulin secreting effects and reported that increased serum levels of threonine, lysine, valine, isoleucine, and leucine was the major mechanism of action for this effect. Mortensen et al. [69] studied the effects of various whey protein fractions on postprandial hormone and lipid responses in T2DM. The whey protein hydrolysate and whey protein isolate addition to fat-rich diet reduced the responses of postprandial triglyceride in subjects with T2DM. The whey protein fractions induced very high insulin responses [69]. Toedebusch et al. [70] carried out a study titled "postprandial leucine and insulin responses and toxicological effects of a novel whey protein hydrolysate-based supplement in rats". They formulated and fed supplement of whey protein hydrolysate to rats in a period of 30 days, and reported a high level of leucine followed by an increase in the level of insulin [70]. Whey proteins are metabolized as amino acids and peptides in the gut which stimulate hormones in the gut (peptide tyrosine tyrosine (PYY), cholecystokinin) and incretin hormones (GLP-1 and gastric inhibitory peptide) that additionally induce the secretion of insulin from pancreatic β-cells. The peptides may inhibit dipeptidyl peptidase-4 (DPP-4) (incretins inhibitor) in proximal gut, and also inhibit the degradation of incretin [71]. Akhavan et al. [72] studied the whey protein action mechanism on reducing postprandial spikes in levels of glucose in a randomized study over 230 min. Whey protein caused reduced plasma insulin, C-peptide, and glucose level, and increased PYY and GLP-1 levels than preloads of glucose. The authors suggested that pre-meal whey protein consumption reduces post-meal glycaemia through insulin-independent and insulin-dependent action mechanisms [72]. Supplementation with whey protein hydrolysate and its leucine supplements can enhance insulin resistance [73].

12.4.9 Immunomodulation Properties

Whey protein improves innate mucosal immunity in early stages of life and has a protective effect in certain immune conditions [32]. Occurrence of a chronic skin disease known as atopic dermatitis characterized by scaly, itchy, and swollen rashes is growing worldwide, with infants mostly affected. Alexander et al. [74] conducted a systematic review and reported that atopic dermatitis incidence was significantly reduced in infants partially fed with whey protein hydrolysate formula in comparison with the infants fed with bovine milk. The study concluded that whey protein-based infant formula has protective effects against atopic dermatitis [74]. Badr et al. [75] studied the whey protein concentrate effects on plasma cytokine profiles, blood parameters, and proliferation/migration of immune cells in mice. The TNF-α, IL-10, IL-1β, and IL-1α plasma levels, as well as the cholesterol levels and ROS levels were significantly lesser in the group that received whey protein treatment in comparison with the control [75]. In the group that received whey protein, the IL-8, glutathione, IL-7, IL-4, and IL-2 levels improved significantly. In addition, the proliferation capability of monocytes, macrophages, and lymphocytes in response to different antigens stimulation increased [75]. The cytokines CXCL12 (CXC chemokine ligand-12) and CCL21 (CC chemokine ligand-21) attract and tether immune cells to their direction [75]. Badr et al. [75] reported that in vitro migration of T cells, dendritic cells, and B cells towards CXCL12 and CCL21 increased significantly in the patients that received whey protein in comparison with the control group. In a different study, it was concluded that consuming 20 g of whey protein isolate per day enhanced and ameliorated the conditions of psoriasis patients [76].

12.4.10 Hypotensive and Cardioprotective Properties

Whey protein consumption can reduce the risk of cardiovascular diseases (such as ischemic stroke), although the exact role of whey protein peptides in regulating vascular endothelial functions still need more studies to sufficiently be conclusive. Ballard et al. [30] carried out a study and reported that the ingestion of whey protein-derived extract (NOP-47) results in improved endothelial functions and can increase the AAs level of postprandial plasma. Arterial dilation improvement was independent of the vasoactive compounds' circulation, including prostacyclin, nitric oxide, and hyperpolarizing factor derived from endothelium [30]. The study concluded that the risks of cardiovascular diseases can be ameliorated with quick-absorbable whey protein extracts [30]. Sheikholeslami and Ahmadi [77] reported whey protein supplementation effects and the effects of resistance training on risk factors of cardiovascular diseases and antioxidant status in young men with overweight BMI. The results suggest that whey protein consumption and resistance training have synergistic effects, which were demonstrated by very high total antioxidant property, high density lipoprotein (HDL) level, and glutathione level [77].

12.4.11 Immune-Boosting for HIV Patients

Whey protein isolates are used in the treatment HIV patients and are widely recommended by health and nutrition experts. Glutathione deficiency is one of the major problems associated with HIV. People infected with HIV have low concentrations of glutathione in the lymphocytes of their blood. Introducing whey protein to their diets elevates glutathione level. As a strong antioxidant, glutathione helps maintain muscular tissue functional and structural integrity [7, 21]. Low cell levels of glutathione encourage the growth of HIV, while high glutathione greatly decreases the replication capacity of the virus. Whey protein consumption can significantly increase glutathione status. Sattler et al. [31] concluded that supplementation with whey protein did not increase lean body mass or weight in HIV patients who ate adequately, but it increased their CD4 cell counts. This shows that whey protein can boost immune cells in HIV patients, while helping them maintain normal weight.

12.4.12 Food and Nutraceutical Additives

Whey protein is widely applied in food production mainly due to its functional properties such as emulsification properties, foaming capacity, thermal stability, gelation capacity, etc., and its nutritional and nutraceutical properties. Whey protein improves food quality, including nutritional value, texture enhancement, and improvement in sensory properties. Many studies have shown the nutritional and structural effects of whey protein in foods, including functional foods, bakery preparation, energy bars, pasta, beverage, yoghurt, infant formula, etc. [78, 79]. Krzeminski et al. [78] in their study showed the capability of non-heated complex of high methoxyl pectin and whey protein as texturing agent and fat replacer in a reduced-fat yoghurt. Formulations of the skim milk with adjusted composition conferred a texture similar as that of whole-fat yoghurt. Kuhn and Cunha [80] carried out a study on the effects of high-pressure homogenization on emulsions of whey protein isolate and flaxseed oil and reported good emulsion stability induced by whey protein. Merging of the droplets led to the generation of protein aggregates with high molecular weight, which resulted to reduction in the emulsification capacity (EC) [80]. High-pressure homogenization (20–100 MPa) and higher passes number led to emulsions of more stability [80]. Akalın et al. [81] reported milk/whey protein-based ingredients' effects on probiotic yogurt microstructure in refrigeration period of 28 days. A 2% sodium calcium caseinate fortification improved yoghurt viscosity, adhesiveness, and firmness [81], while 2% whey protein concentrate improved water binding capacity more than sodium calcium caseinate. Yoghurt fortified with whey protein concentrate had better texture and lesser whey collection at the yoghurt surface. Nadeem et al. [82] formulated a whey protein-fortified date bar and a plant protein for school children. The nutrients were optimized using response surface methodology, with whey protein concentrate (6.05%) shown ideal for this application [82]. In another study, Yadav et al. [83] used pearl millet, whey protein concentrate, water, barley flour, and carboxy methyl cellulose to formulate pasta, and reported promising results.

12.4.13 Encapsulation, Delivery Systems, Active Packaging, Edible Coating

Several foods, essential oils, vitamins, and additives can be packaged or encapsulated in whey protein gels for stability and reduction of rancidity. Sustaining iron accessibility in foods fortified with iron is usually challenging. Whey protein isolate cold-set gelling capability was evaluated to address this [84]. A study entrapped iron in the ascorbate presence with cold-set gelation, and studied TNO Intestinal Model (TIM) for the optimization of the iron-ascorbate ratio [84]. The whey protein's cold-set gelation done for ascorbate and iron showed suitable in improving the iron in vitro bioaccessibility and recovery, which, from 10%, improved to 80%. The ascorbate gel strengthening effect was associated with better controlled release and encapsulation efficiency of iron. Mehyar et al. [85] evaluated the possibility of whey protein isolate-loaded cardamom essential oil microencapsulation. The microcapsules of whey protein isolate at 30% concentration optimally entrapped the essential oil during storage at known temperature conditions. The microcapsules of whey protein isolate had spherical shapes, regular and even contours, and smooth texture [85]. Pérez-Masiá et al. [86] achieved significant folic acid encapsulation with matrix of whey protein concentrate, which was associated with the interactions between folic acid and whey protein. The result indicated that whey protein can be better for this purpose than resistant starch commercially used for the stability of folic acid [86]. Janjarasskul et al. [87] studied edible ascorbic acid-impregnated film of whey protein isolate and its oxygen-scavenging potentials. The whey protein isolate film had proper tensile strength and decreased the permeability of oxygen, showing suitability for food products with oxygen sensitivity [87]. Gülseren et al. [40] made a comparison between encapsulation competency of nanoparticles of whey protein isolate desolvated suspension with methoxyl pectin and another without methoxyl pectin. The methoxyl pectin-containing suspension showed resistance to homogenization, with improved stability. The complex also showed better interfacial pressures under pH 3 storage, in comparison with the suspension of whey protein isolate nanoparticles without methoxyl pectin; this suggests their potential use as surfactants.

12.4.14 Antihypertensive Properties

Whey proteins have been studied for use as dietary supplements for the reduction of blood pressure. Bovine whey proteins' beta-lactoglobulin peptide sequence has high inhibitory actions against Angiotensin I converting enzyme (ACE), which catalyzes Angiotensin I translation into Angiotensin II, a potential vasoconstrictor [7, 21]. It reduces the risks of hypertension.

12.5 Applications of Whey Protein for Functional Foods and Nutraceutical Development

Functional foods are foods that have additional function usually associated with disease prevention and/or health promotion by the addition of new ingredients or more of the existing ingredients; they improve overall health and well-being, as well as reduce disease risks [20, 88]. Whey protein plays immunomodulatory roles due to its high concentrations of cysteine and IgA/IgG immunoglobulins. Cysteine is an amino acid that increase glutathione production. Glutathione is the main antioxidant system in the body, and can prevent cellular aging and oxidative stress. Consequently, whey protein has immune properties via the stimulation of antibodies production by the lymphocyte [88]. The antimicrobial and immunomodulatory properties of whey proteins are associated with the stimulation of immune cells and protective microflora proliferation in the GI tract of humans and animals, indicating the possible whey protein role in strengthening immune system in children and adults [26]. Bioactive peptides of whey protein have mineral binding capacity, and can go along with the fact that intestinal microbiota regulation improves absorption of mineral in the digestive and gastrointestinal tracts [26]. Peptides counteracts calcium inhibitory effects on the absorption of iron and also act in synergy with ascorbic acid (vitamin C) to promote iron absorption and bioavailability [89]; it is fairly scientific to state that whey protein peptides improve iron absorption and bioavailability. Opioid peptides are small molecules synthesized in vivo that act as hormones and neurotransmitters in the nervous system. Consuming dairy products, with whey proteins and low in fat can inhibit insulin resistance. Studies have also associated such products with blood pressure control and healthy weight because of high stimulus of satiety, consequently decreasing the occurrence of dental caries and coronary diseases [88]. Table 12.3 shows various studies that investigated different applications of whey protein fractions as nutraceuticals and other food applications.

Raikos and Dassios [26] reported that infant formulas formulated with whey protein hydrolysates have higher ACE inhibitory effect compared with infant formulas formulated without whey protein hydrolysates. Peptides with low molecular weight, usually less than 3000 Da, are mostly responsible for this action [20]. Consequently, the hydrolysed fragment sizes can be crucial for antihypertensive properties, since tripeptides and dipeptides can be absorbed readily from intestine and quickly transported to the target sites, including the blood [26, 94, 95]. Other foods have been formulated, studied, or proposed without whey protein [96–100]. However, the antioxidant properties of whey proteins have been reported, with potential benefits against oxidative stress, commonly reported in chronic noncommunicable diseases (CNCDs) and in preterm neonates [26, 96]. This property is as a result of the high levels of sulfur-containing AAs, including methionine and cysteine that are integral part of enzymes and organisms' antioxidants systems [88].

12.6 Conclusion

Whey protein is the globular proteins' collection from whey, the liquid milk component remaining after milk curdling and straining. While the major whey protein components include α-lactalbumin, β-lactoglobulin, BSA, and the fraction of proteose-peptone, the minor whey protein components include immunoglobulins, lactoferrins, ceruloplasmins, as well as certain enzymes such as lipase, xanthine oxidase, and lysozyme. Three types of whey proteins include whey protein concentrate, whey protein isolate, and whey protein hydrolysate, with various compositions. Whey protein contains sulfur-containing amino acids, e.g., methionine and cysteine that boost immune functions of glutathione. Whey proteins are suitable for

Table 12.3 Studies that highlight the applications of whey proteins in nutraceuticals and functional foods

Title of study	Aim	Type of study	Conclusion	References
"Assessing Whey Protein Sources, Dispersion Preparation Method and Enrichment of Thermomechanically Stabilized Whey Protein Pectin Complexes for Technical Scale Production"	The study aimed at optimizing the production of whey protein pectin complexes with thermal stability using lab scale scraped-surface heat exchanger before technical scale upscaling	Research	Dispersions of whey protein powder-based complexes had bigger particles that have lower particle volume within submicron ranges, compared with the complexes formulated with whey protein concentrates. The authors concluded that various sources of whey proteins are suitable for producing pectin–whey protein complexes for applications in food and nutraceutical industries	[90]
"Direct ethanolic extraction of polar lipids and fractional crystallization from whey protein phospholipid concentrate"	Extraction of dairy lipid from whey protein phospholipid concentrate using exposure directly to ethanol, food-grade solvent	Research	The study proved that fractional crystallization and ethanol extraction are suited for whey protein phospholipid extraction and sphingomyelin enrichment	[91]
"Emerging trends in nutraceutical applications of whey protein and its derivatives"	This study aimed at providing invaluable insights into functional foods fortified with whey protein, and their relative technological challenges and improvement opportunities.	Review	The study acknowledged whey protein as a significant source of important nutrients. The authors concluded that several food formulas enriched with whey protein can be formulated for target human populations, including infants, athletes, phenylketonuria patients, diabetics, geriatrics, cardiac-risk group, etc.	[17]

(continued)

Table 12.3 (continued)

Title of study	Aim	Type of study	Conclusion	References
"Preparation and Characterization of Whey Protein-Based Polymers Produced from Residual Dairy Streams"	The study investigated alternative ways to formulate polymers from milk proteins, a renewable source, aimed at polyethylene replacement	Research	The study carried out protein-based thermoset elastomers synthesis with whey protein concentrate/isolate. The whey protein concentrates and isolate showed significant potentials for application as polymers via copolymer free-radical polymerization. Such polymers can be used in functional food applications	[92]
"Whey and protein derivatives: Applications in food products development, technological properties and functional effects on child health"	The study aim was to demonstrate use of whey and whey protein in foods and their influence on infants and children health	Review	This study concluded that whey protein has several nutritional benefits and food technological applications for improvement in early stages of life; although further studies are required	[20]
"Whey Proteins and Its Derivatives: Bioactivity, Functionality, and Current Applications"	The study aimed at showing the current uses, functionality, bioactivity, and of whey protein and its derivatives	Review	It highlights the functional characteristics, bioactive properties, applications, and processing limitations of various fractions of whey protein and its derivatives in food formulations, packaging, and encapsulation	[22]
"Whey proteins characterization, free amino acids profile and antimicrobial study of some dairy drinks" based on Milk Serum	To characterize certain dairy drinks using milk serum in regards with free amino acids and whey proteins using reversed phase HPLC	Preprint (Research)	The results obtained proved that dairy drinks formulated from milk serum are important sources of bioactive components for human health	[93]

(continued)

Table 12.3 (continued)

Title of study	Aim	Type of study	Conclusion	References
"Whey Proteins—A Potential Nutraceutical"	This study aimed at exploring the potentials of whey proteins as nutraceuticals	Review	Many whey protein enriched food formulas can be formulated for target groups, including infants, athletes, phenylketonuria patients, diabetics, geriatrics, cardiac-risk group, etc.	[21]

nutraceuticals and functional food formulations, and have been shown to have cardioprotective and hypotensive effects, immune improvement in HIV patients, immunomodulatory properties, anticarcinogenic properties, anti-diabetic properties, antihypertensive effects, antioxidant properties, infants' immunity, muscle synthesis and resistance exercise, phenylketonuria treatment, postoperative care, prebiotic and gut functions, skin protective effects, and treatment of obesity and overweight. Whey proteins are suitable as the basis of various nutraceutical formulations; however, more control studies are required to explore their applications for other targeted therapies.

References

1. Wiley AS (2014) Cultures of milk: the biology and meaning of dairy products in the United States and India. Harvard University Press, Cambridge, MA, p 11. isbn:978-0-674-72905-6
2. Karimi Alavijeh M, Meyer AS, Gras SL, Kentish SE (2020) Simulation and economic assessment of large-scale enzymatic N-acetyllactosamine manufacture. Biochem Eng J 154:107459. https://doi.org/10.1016/j.bej.2019.107459
3. Ryan MP, Walsh G (2016) The biotechnological potential of whey. Rev Environ Sci Biotechnol 15(3):479–498. https://doi.org/10.1007/s11157-016-9402-1
4. Awuchi CG, Twinomuhwezi H, Awuchi CG (2021) Hyphenated techniques. In: Egbuna C, Patrick-Iwuanyanwu K, Shah MA, Ifemeje JC, Rasul A (eds) Analytical techniques in biosciences: from basics to applications. Elsevier, Amsterdam, Netherlands. https://doi.org/10.1016/B978-0-12-822654-4.00015-4
5. Awuchi CG, Twinomuhwezi H, Igwe VS, Sarvarian M, Amagwula IO (2021) Protein functions and structure prediction with I-TASSER, a novel technology. J Chem Biol Phys Sci 11(4):555–567. https://doi.org/10.24214/jcbps.B.11.4.55567
6. Awuchi CG, Igwe VS, Amagwula IO, Echeta CK (2020) Health benefits of micronutrients (vitamins and minerals) and their associated deficiency diseases: a systematic review. Int J Food Sci 3(1):1–32
7. Chauhan B, Kumar G, Kalam N et al (2013) Current concepts and prospects of herbal nutraceutical: a review. J Adv Pharm Technol Res 4(1):4–8
8. EFSA Panel on Dietetic Products, Nutrition and Allergies (2010) Scientific opinion on the substantiation of health claims related to whey protein. EFSA J 8(10):1818. https://doi.

org/10.2903/j.efsa.2010.1818
9. Willis B, Lopez G, Patel K, Frank K (2018) Whey protein. Examine.com
10. Melnik BC (2012) Excessive leucine-mTORC1-signalling of cow milk-based infant formula: the missing link to understand early childhood obesity. J Obes 2012:14. https://doi.org/10.1155/2012/354721
11. Melnik BC, John SM, Schmitz G (2015) Milk consumption during pregnancy increases birth weight, a risk factor for the development of diseases of civilization. J Transl Med 13:13. https://doi.org/10.1186/s12967-015-0541-x
12. Gunnars, K (2018) Whey protein 101: the ultimate beginner's guide. Retrieved from https://www.healthline.com/nutrition/whey-protein-101#muscle-mass-and-strength. Accessed 1 Sept 2021
13. Awuchi CG, Igwe VS, Amagwula IO (2020) Nutritional diseases and nutrient toxicities: a systematic review of the diets and nutrition for prevention and treatment. Int J Adv Acad Res 6(1):1–46
14. Awuchi CG, Igwe VS, Amagwula IO (2020) Ready-to-use therapeutic foods (RUTFs) for remedying malnutrition and preventable nutritional diseases. Int J Adv Acad Res 6(1):47–81
15. Awuchi CG, Igwe VS, Echeta CK (2019) The functional properties of foods and flours. Int J Adv Acad Res 5(11):139–160
16. Awuchi CG (2019) Proximate composition and functional properties of different grain flour composites for industrial applications. Int J Adv Acad Res 2(1):43–64
17. Patel S (2015) Emerging trends in nutraceutical applications of whey protein and its derivatives. J Food Sci Technol 52(11):6847–6858
18. Muza SF, Peres AP, Degáspari CH (2014) Desenvolvimento de iogurte enriquecido com proteína do soro do leite. Cadernos da Escola de Saúde 1:79–89. ISSN 1984-7041
19. Alves MP, Moreira RO, Júnior PHR, Martins MCF, Perrone IT, Carvalho AF (2014) Soro de leite: Tecnologias Para o processamento de coprodutos. Revista do Instituto de Laticínios Cândido Tostes 69:212–226. https://doi.org/10.14295/2238-6416.v69i3.341
20. Batista MA, Natália CAC, Marialice PCS (2018) Whey and protein derivatives: applications in food products development, technological properties and functional effects on child health. Cogent Food Agriculture 4(1):1509687. https://doi.org/10.1080/23311932.2018.1509687
21. Arora S, Singh TP, Rekha RS (2019) Whey proteins—a potential nutraceutical. Res Rev J Dairy Sci Technol 8(2):1–5
22. Minj S, Anand S (2020) Whey proteins and its derivatives: bioactivity, functionality, and current applications. Dairy 1:233–258. https://doi.org/10.3390/dairy1030016
23. Awuchi CG, Nwankwere ET (2018) Residual calcium content of sweet potato slices after osmotic pre-treatment with salt (NaCl) solution. Am J Food Nutr Health 3(1):8–15
24. Chung CS, Yamini S, Trumbom PR (2012) FDA's health claim review: whey-protein partially hydrolyzed infant formula and atopic dermatitis. Pediatrics 130:e408–e414. https://doi.org/10.1542/peds.2012-0333
25. Mezzomo TR, Nadal J (2014) A segurança alimentar e nutricional do público infanto-juvenil: O leite como componente. Demetra 9:503–513. https://doi.org/10.12957/demetra.2014.9485
26. Raikos V, Dassios T (2014) Health-promoting properties of bioactive peptides derived from milk proteins in infant food: a review. Dairy Sci Technol 94:91–101. https://doi.org/10.1007/s13594-013-0152-3
27. Attaallah W, Yilmaz AM, Erdoğan N et al (2012) Whey protein versus whey protein hydrolyzate for the protection of azoxymethane and dextran sodium sulfate induced colonic tumors in rats. Pathol Oncol Res 18:817–822. https://doi.org/10.1007/s12253-012-9509-9
28. Sousa GTD, Lira FS, Rosa JC, Oliveira EP, Oyama LM, Santos RV, Pimentel GD (2012) Dietary whey protein lessens several risk factors for metabolic diseases: a review. Lipids Health Dis 11:1–9. https://doi.org/10.1186/1476-511X-11-1
29. Jain SK (2012) L-cysteine supplementation as an adjuvant therapy for type-2 diabetes. Can J Physiol Pharmacol 90:1061–1064. https://doi.org/10.1139/y2012-087
30. Ballard KD, Kupchak BR, Volk BM et al (2013) Acute effects of ingestion of a novel whey-

derived extract on vascular endothelial function in overweight, middle-aged men and women. Br J Nutr 109:882–893. https://doi.org/10.1017/S0007114512002061

31. Sattler FR, Natasa R, Mulligan K, Kevin EY, Susan LK, Andrew Z, Beverly AS, Robert Z, Bruce B (2008) Evaluation of high-protein supplementation in weight-stable HIV-positive subjects with a history of weight loss: a randomized, double-blind, multicenter trial. Am J Clin Nutr 88(5):1313–1321
32. Pérez-Cano FJ, Marín-Gallén S, Castell M et al (2007) Bovine whey protein concentrate supplementation modulates maturation of immune system in suckling rats. Br J Nutr 98(Suppl. 1):S80–S84. https://doi.org/10.1017/S0007114507838074
33. Morato PN, Lollo PC, Moura CS, Batista TM, Carneiro EM, Amaya- Farfan J (2013) A dipeptide and an amino acid; present in whey protein hydrolysate increase translocation of GLUT-4 to the plasma membrane in Wistar rats. Food Chem 139:853–859. https://doi.org/10.1016/j.foodchem.2012.12.062
34. Van Calcar SC, Ney DM (2012) Food products made with glycomacropeptide, a low-phenylalanine whey protein, provide a new alternative to amino acid-based medical foods for nutrition management of phenylketonuria. J Acad Nutr Diet 112:1201–1210. https://doi.org/10.1016/j.jand.2012.05.004
35. Abrahão V (2012) Nourishing the dysfunctional gut and whey protein. Curr Opin Clin Nutr Metab Care 15:480–484. https://doi.org/10.1097/MCO.0b013e328356b71e
36. Kimura Y, Sumiyoshi M, Kobayashi T (2014) Whey peptides prevent chronic ultraviolet B radiation-induced skin aging in melanin-possessing male hairless mice. J Nutr 144:27–32. https://doi.org/10.3945/jn.113.180406
37. Tahavorgar A, Vafa M, Shidfar F et al (2014) Whey protein preloads are more beneficial than soy protein preloads in regulating appetite, calorie intake, anthropometry, and body composition of overweight and obese men. Nutr Res. https://doi.org/10.1016/j.nutres.2014.08.015
38. Morton JP, Kayani AC, McArdle A, Drust B (2009) The exercise-induced stress response of skeletal muscle, with specific emphasis on humans. Sports Med 39:643–662. https://doi.org/10.2165/00007256-200939080-00003
39. Freidenreich DJ, Volek JS (2012) Immune responses to resistance exercise. Exerc Immunol Rev 18:8–41
40. Gülseren I, Fang Y, Corredig M (2012) Complexation of high methoxyl pectin with ethanol desolvated whey protein nanoparticles: physico-chemical properties and encapsulation behaviour. Food Funct 3:859–866. https://doi.org/10.1039/c2fo10235h
41. Martin V, Ratel S, Siracusa J et al (2013) Whey proteins are more efficient than casein in the recovery of muscle functional properties following a casting induced muscle atrophy. PLoS One 8:e75408. https://doi.org/10.1371/journal.pone.0075408
42. Churchward-Venne TA, Breen L, Di Donato DM et al (2014) Leucine supplementation of a low-protein mixed macronutrient beverage enhances myofibrillar protein synthesis in young men: a double-blind, randomized trial. Am J Clin Nutr 99:276–286. https://doi.org/10.3945/ajcn.113.068775
43. Lollo PCB, Amaya-Farfan J, Faria IC et al (2014) Hydrolysed whey protein reduces muscle damage markers in Brazilian elite soccer players compared with whey protein and maltodextrin. A twelve-week in-championship intervention. Int Dairy J 34:19–24. https://doi.org/10.1016/j.idairyj.2013.07.001
44. Volek JS, Volk BM, Gómez AL et al (2013) Whey protein supplementation during resistance training augments lean body mass. J Am Coll Nutr 32:122–135. https://doi.org/10.1080/07315724.2013.793580
45. Williams RA, Mamotte CDS, Burnett JR (2008) Phenylketonuria: an inborn error of phenylalanine metabolism. Clin Biochem Rev 29:31–41
46. Solverson P, Murali SG, Brinkman AS et al (2012) Glycomacropeptide, a low-phenylalanine protein isolated from cheese whey, supports growth and attenuates metabolic stress in the murine model of phenylketonuria. Am J Physiol Endocrinol Metab 302:E885–E895. https://doi.org/10.1152/ajpendo.00647.2011

47. Demirdas S, Coakley KE, Bisschop PH et al (2015) Bone health in phenylketonuria: a systematic review and meta-analysis. Orphanet J Rare Dis 10:17. https://doi.org/10.1186/s13023-015-0232-y
48. Ney DM, Blank RD, Hansen KE (2014) Advances in the nutritional and pharmacological management of phenylketonuria. Curr Opin Clin Nutr Metab Care 17:61–68. https://doi.org/10.1097/MCO.0000000000000002
49. Strisciuglio P, Concolino D (2014) New strategies for the treatment of phenylketonuria (PKU). Meta 4:1007–1017. https://doi.org/10.3390/metabo4041007
50. Castro GA, Maria DA, Bouhallab S, Sgarbieri VC (2009) In vitro impact of a whey protein isolate (WPI) and collagen hydrolysates (CHs) on B16F10 melanoma cells proliferation. J Dermatol Sci 56:51–57. https://doi.org/10.1016/j.jdermsci.2009.06.016
51. Takata T, Tanaka F, Yamada T et al (2001) Clinical significance of caspase-3 expression in pathologic-stage I, nonsmall-cell lung cancer. Int J Cancer 96(Suppl):54–60. https://doi.org/10.1002/ijc.10347
52. Dillon EL, Basra G, Horstman AM et al (2012) Cancer cachexia and anabolic interventions: a case report. J Cachex Sarcopenia Muscle 3:253–263. https://doi.org/10.1007/s13539-012-0066-6
53. Zhang Q-X, Ling Y-F, Sun Z et al (2012) Protective effect of whey protein hydrolysates against hydrogen peroxide-induced oxidative stress on PC12 cells. Biotechnol Lett 34:2001–2006. https://doi.org/10.1007/s10529-012-1017-1
54. Awuchi CG, Amagwula IO (2021) Biochemistry and nutrition of carbohydrates. Global J Res Agric Life Sci 1(1):4–12
55. Awuchi CG, Twinomuhwezi H, Awuchi CG, Amagwula IO (2022) Immune foods for fighting coronavirus disease 2019. In: Rudrapal M, Egbuna C (eds) Medicinal plants, phytomedicine and traditional herbal remedies for drug discovery and development against COVID-19. Bentham Books, Sharjah. In Press
56. Gerez CL, Font de Valdez G, Gigante ML, Grosso CRF (2012) Whey protein coating bead improves the survival of the probiotic *Lactobacillus rhamnosus* CRL 1505 to low pH. Lett Appl Microbiol 54:552–556. https://doi.org/10.1111/j.1472-765X.2012.03247.x
57. Zhao L, Huang Q, Huang S et al (2014) Novel peptide with a specific calcium-binding capacity from whey protein hydrolysate and the possible chelating mode. J Agric Food Chem. https://doi.org/10.1021/jf502412f
58. Walsh H, Cheng J, Guo M (2014) Effects of carbonation on probiotic survivability, physicochemical, and sensory properties of milk-based symbiotic beverages. J Food Sci 79:M604–M613. https://doi.org/10.1111/1750-3841.12381
59. Hébrard G, Hoffart V, Beyssac E et al (2010) Coated whey protein/alginate microparticles as oral controlled delivery systems for pro-biotic yeast. J Microencapsul 27:292–302. https://doi.org/10.3109/02652040903134529
60. Freudenberg A, Petzke KJ, Klaus S (2013) Dietary L-leucine and L-alanine supplementation have similar acute effects in the prevention of high-fat diet-induced obesity. Amino Acids 44:519–528. https://doi.org/10.1007/s00726-012-1363-2
61. Tufail T, Ijaz A, Noreen S, Arshad MU, Gilani SA, Bashir S, Din A, Shahid MZ, Khan AA, Khalil AA, Awuchi CG (2021) Pathophysiology of obesity and diabetes. In: Egbuna C, Hassan S (eds) Dietary phytochemicals. Springer, Cham, pp 29–42. https://doi.org/10.1007/978-3-030-72999-8_2
62. Egbuna C, Awuchi CG, Kushwaha G, Rudrapal M, Patrick-Iwuanyanwu KC, Singh O, Odoh UE, Khan J, Jeevanandam J, Kumarasamy S, Narayanan M, Chukwube VO, Palai S, Găman M-A, Uche CZ, Ogaji DS, Ezeofor NJ, Mtewa AG, Patrick-Iwuanyanwu CC, Kesh SS, Shivamallu C, Saravanan K, Tijjani H, Akram M, Ifemeje JC, Olisah MC, Chikwendu CJ (2021) Bioactive compounds effective against type 2 diabetes mellitus: a systematic review. Curr Topics Med Chem 21:1. https://doi.org/10.2174/1568026621666210509161059
63. Yasmin I, Khan WA, Naz S, Iqbal MW, Awuchi CG, Egbuna C, Hassan S, Patrick-Iwuanyanwu KC, Uche CZ (2021) Etiology of obesity, cancer, and diabetes. In: Egbuna

C, Hassan S (eds) Dietary phytochemicals. Springer, Cham, pp 1–27. https://doi.org/10.1007/978-3-030-72999-8_1
64. Awuchi CG (2021) Medicinal plants, bioactive compounds, and dietary therapies for treating type 1 and type 2 diabetes mellitus [online first]. IntechOpen. https://doi.org/10.5772/intechopen.96470. Available from: https://www.intechopen.com/online-first/medicinal-plants-bioactive-compounds-and-dietary-therapies-for-treating-type-1-and-type-2-diabetes-m
65. Casqueiro J, Casqueiro J, Alves C (2012) Infections in patients with diabetes mellitus: a review of pathogenesis. Indian J Endocrinol Metab 16(Suppl. 1):S27–S36. https://doi.org/10.4103/2230-8210.94253
66. Awuchi CG, Echeta CK, Igwe VS (2020) Diabetes and the nutrition and diets for its prevention and treatment: a systematic review and dietetic perspective. Health Sci Res 6(1):5–19
67. Badr G, Badr BM, Mahmoud MH et al (2012) Treatment of diabetic mice with undenatured whey protein accelerates the wound healing process by enhancing the expression of MIP-1α, MIP-2, KC, CX3CL1 and TGF-β in wounded tissue. BMC Immunol 13:32. https://doi.org/10.1186/1471-2172-13-32
68. Salehi A, Gunnerud U, Muhammed SJ et al (2012) The insulinogenic effect of whey protein is partially mediated by a direct effect of amino acids and GIP on β-cells. Nutr Metab (Lond) 9:48. https://doi.org/10.1186/1743-7075-9-48
69. Mortensen LS, Holmer-Jensen J, Hartvigsen ML et al (2012) Effects of different fractions of whey protein on postprandial lipid and hormone responses in type 2 diabetes. Eur J Clin Nutr 66:799–805. https://doi.org/10.1038/ejcn.2012.48
70. Toedebusch RG, Childs TE, Hamilton SR et al (2012) Postprandial leucine and insulin responses and toxicological effects of a novel whey protein hydrolysate-based supplement in rats. J Int Soc Sports Nutr 9:24. https://doi.org/10.1186/1550-2783-9-24
71. Jakubowicz D, Froy O (2013) Biochemical and metabolic mechanisms by which dietary whey protein may combat obesity and type 2 diabetes. J Nutr Biochem 24:1–5. https://doi.org/10.1016/j.jnutbio.2012.07.008
72. Akhavan T, Luhovyy BL, Panahi S et al (2014) Mechanism of action of pre-meal consumption of whey protein on glycemic control in young adults. J Nutr Biochem 25:36–43. https://doi.org/10.1016/j.jnutbio.2013.08.012
73. Tong X, Li W, Xu J-Y et al (2014) Effects of whey protein and leucine supplementation on insulin resistance in non-obese insulin-resistant model rats. Nutrition 30:1076–1080. https://doi.org/10.1016/j.nut.2014.01.013
74. Alexander DD, Schmitt DF, Tran NL et al (2010) Partially hydrolyzed 100% whey protein infant formula and atopic dermatitis risk reduction: a systematic review of the literature. Nutr Rev 68:232–245. https://doi.org/10.1111/j.1753-4887.2010.00281.x
75. Badr G, Ebaid H, Mohany M, Abuelsaad AS (2012) Modulation of immune cell proliferation and chemotaxis towards CC chemokine ligand (CCL)-21 and CXC chemokine ligand (CXCL)-12 in undenatured whey protein-treated mice. J Nutr Biochem 23:1640–1646. https://doi.org/10.1016/j.jnutbio.2011.11.006
76. Prussick R, Prussick L, Gutman J (2013) Psoriasis improvement in patients using glutathione-enhancing, nondenatured whey protein isolate: a pilot study. J Clin Aesthet Dermatol 6:23–26
77. Sheikholeslami VD, Ahmadi KGF (2012) Changes in antioxidant status and cardiovascular risk factors of overweight young men after six weeks supplementation of whey protein isolate and resistance training. Appetite 59:673–678. https://doi.org/10.1016/j.appet.2012.08.005
78. Krzeminski A, Prell KA, Busch-Stockfisch M et al (2014) Whey protein–pectin complexes as new texturising elements in fat-reduced yoghurt systems. Int Dairy J 36:118–127. https://doi.org/10.1016/j.idairyj.2014.01.018
79. Godswill AC, Twinomuhwezi H, Igwe VS, Amagwula IO (2020) Food additives and food preservatives for domestic and industrial food applications. J Anim Health 2(1):1–16
80. Kuhn KR, Cunha RL (2012) Flaxseed oil – whey protein isolate emulsions: effect of high pressure homogenization. J Food Eng 111:449–457. https://doi.org/10.1016/j.jfoodeng.2012.01.016

81. Akalın AS, Unal G, Dinkci N, Hayaloglu AA (2012) Microstructural, textural, and sensory characteristics of probiotic yogurts fortified with sodium calcium caseinate or whey protein concentrate. J Dairy Sci 95:3617–3628. https://doi.org/10.3168/jds.2011-5297
82. Nadeem M, Salim-ur-Rehman MA et al (2012) Development, characterization, and optimization of protein level in date bars using response surface methodology. Sci World J 2012:518702. https://doi.org/10.1100/2012/518702
83. Yadav DN, Balasubramanian S, Kaur J et al (2014) Non-wheat pasta based on pearl millet flour containing barley and whey protein concentrate. J Food Sci Technol 51:2592–2599. https://doi.org/10.1007/s13197-012-0772-2
84. Martin AH, de Jong GAH (2012) Enhancing the in vitro Fe(2+) bio-accessibility using ascorbate and cold-set whey protein gel particles. Dairy Sci Technol 92:133–149. https://doi.org/10.1007/s13594-011-0055-0
85. Mehyar GF, Al-Isamil KM, Al-Ghizzawi HM, Holley RA (2014) Stability of cardamom (*Elettaria cardamomum*) essential oil in microcapsules made of whey protein isolate, guar gum, and carrageenan. J Food Sci. https://doi.org/10.1111/1750-3841.12652
86. Pérez-Masiá R, López-Nicolás R, Periago MJ et al (2015) Encapsulation of folic acid in food hydrocolloids through nanospray drying and electrospraying for nutraceutical applications. Food Chem 168:124–133. https://doi.org/10.1016/j.foodchem.2014.07.051
87. Janjarasskul T, Tananuwong K, Krochta JM (2011) Whey protein film with oxygen scavenging function by incorporation of ascorbic acid. J Food Sci 76:E561–E568. https://doi.org/10.1111/j.1750-3841.2011.02409.x
88. Shah NH (2014) Effect of health on nutrition/dairy foods and human nutrition. Int J Indian Psychol 2:60–64. ISSN 2348-5396
89. Walczyk T, Muthayya S, Wegmüller R, Thankachan P, Sierksma A, Frenken LGJ, Hurrell RF (2014) Inhibition of iron absorption by calcium is modest in an iron-fortified, casein- and whey-based drink in Indian children and is easily compensated for by addition of ascorbic acid. J Nutr 144:1703–1709. https://doi.org/10.3945/jn.114.193417
90. Filla JM, Stadler M, Heck A, Hinrichs J (2021) Assessing whey protein sources. Dispersion preparation method and enrichment of thermomechanically stabilized whey protein pectin complexes for technical scale production. Foods 2021(10):715. https://doi.org/10.3390/foods10040715
91. Ferraris Q, Qian MC (2021) Direct ethanolic extraction of polar lipids and fractional crystallization from whey protein phospholipid concentrate. JDS Commun 2:177–181. https://doi.org/10.3168/jdsc.2021-0076
92. Chalermthai B, Wui YC, Juan-Rodrigo BO, Hanifa T, Bradley DO, Jens ES (2019) Preparation and characterization of whey protein-based polymers produced from residual dairy streams. Polymers 11:722. https://doi.org/10.3390/polym11040722
93. Prodan D, Miuța F, Mihaela V, Marioara M, Rahela C, Viorica T, Andreea SP (2020) Whey proteins characterization, free amino acids profile and antimicrobial study of some dairy drinks based on milk serum. Preprint 12. https://doi.org/10.20944/preprints202009.0607.v1
94. Morais HA, Silvestre MPC, Amorin LL, Silva VDM, Silva MR, Silva ACS, Silveira JN (2014) Use of different proteases to obtain whey protein concentrate hydrolysates with inhibitory activity toward angiotensin-converting enzyme. J Food Biochem 38:102–109. https://doi.org/10.1111/jfbc.12032
95. Silva MR, Silvestre MPC, Silva VDM, Souza MWS, Junior COL, Afonso WO, Rodrigues DF (2014) Production of ACE-inhibitory whey protein concentrate hydrolysates: use of pancreatin and papain. Int J Food Prop 17:1002–1012. https://doi.org/10.1080/10942912.2012.685821
96. Silvestre MPC, Morais HA, Silva VDM, Silva MR (2013) Whey as source of peptides with high anti-oxidant activity: use of a pancreatin and an Aspergillussojae protease. Publicatio UEPG Ciências Biológicas e da Saúde 19:143–147. https://doi.org/10.5212/Publ.Biologicas.v.19i2.0007
97. Ahaotu NN, Bede NE, Umejesi TJ, Echeta CK, Awuchi CG (2020) Studies on the functional

properties of malted soy-garri as affected by moisture variation during storage. Eur Acad Res 8(8):4616–4625
98. Awuchi CG, Owuamanam IC, Ogueke CC, Hannington T (2020) The assessment of the physical and pasting properties of grains. Eur Acad Res 8(2):1072–1080
99. Ahaotu NN, Echeta CK, Bede NE, Awuchi CG, Anosike CL, Ibeabuchi CJ, Ojukwu M (2020) Study on the nutritional and chemical composition of "*Ogiri*" condiment made from sandbox seed (*Hura crepitans*) as affected by fermentation time. GSC Biol Pharm Sci 11(2):105–113. https://doi.org/10.30574/gscbps.2020.11.2.0115
100. Twinomuhwezi H, Awuchi CG, Kahunde D (2020) Extraction and characterization of pectin from orange (*Citrus sinensis*), Lemon (*Citrus limon*) and tangerine (*Citrus tangerina*). Am J Phys Sci 1(1):17–30

Chapter 13
Corn Byproducts as Source of Nutraceuticals

Santwana Palai and Shyam Sundar Kesh

13.1 Introduction

Nutraceuticals are items that are both nutritional and medicinal in nature. A nutraceutical product is a substance that has physiological benefits or protects the body against enduring disease. Nutraceuticals can be used to recover health, reduce the ageing progression, avert chronic diseases, lengthen lifespan, and keep body's construction, function in good shape and can aid in the prevention of disease and promote health. Natural foods like fortified dairy products, citrus fruits, cereals, vegetables, dietary supplements, antioxidants, vitamins, minerals, herbals, milk are examples of nutraceuticals [1].

Corn, otherwise known as maize (*Zea mays* L.), is an American crop. After being found in 1492 by Christopher Columbus, it was spread to China, Europe, and the rest of the world during the next century. Corn is now an important food source all around the world. Corn has a considerable quantity of bioactive chemicals that have health advantages in addition to its position as a primary food source [2]. A wet milling operation is used to manufacture numerous major products from corn. The gluten (high protein component of corn flour), germ (grain core), bran (seed coat), and many other substances are present among the its byproducts like corn flour, cornmeal, cornflakes, corn oil, cornstarch, corn syrup and other corn products [3]. Table 13.1 lists the food uses of corn byproducts.

Corn phytochemicals have hitherto gotten lesser importance than fruits and vegetables phytochemicals. Corn and its byproducts can decrease chronic illnesses risk

S. Palai (✉)
Department of Veterinary Pharmacology and Toxicology, College of Veterinary Science and Animal Husbandry, Odisha University of Agriculture & Technology, Bhubaneswar, India

S. S. Kesh
Department of Veterinary Clinical Complex (Veterinary Biochemistry), Faculty of Veterinary and Animal Sciences, West Bengal University of Animal and Fishery Sciences, Kolkata, India

© The Author(s), under exclusive license to Springer Nature Switzerland AG 2022

C. Egbuna et al. (eds.), *Food and Agricultural Byproducts as Important Source of Valuable Nutraceuticals*, https://doi.org/10.1007/978-3-030-98760-2_13

Table 13.1 Food uses of corn byproducts

Sl. no.	Name of the by product	Food uses	Aid in food preparation	References
1.	Corn flour	Bread, wafers, pancake breakfast cereals, muffins, mixes, infant food, biscuits, doughnuts, breadings	Thickener for sauces, stews, soups; binder, filler and carrier of meat products	[12]
2.	Corn starch	Fried or breaded foods, marinades, sauces, soups, gravies, casseroles	Added to confectioner's sugar to prevent clumping, thickening agent; forms a crisp crust when used as coating for fried foods	[12]
3.	Cornmeal/ Polenta	Grits, porridge, samp, corn fritters, corn bread, hushpuppies, batter for fried food, spoon bread, Muffins	Coating favorite foods, adds a grainy texture and yellowish color in bread other than cornbread, texture enhancer; prevent pizza and breads from sticking to their pans when baking as a release agent	[24, 13]
4.	Cornflakes	Toasted flakes of corns enjoyed for breakfast with hot or cold milk	buttery, sweet taste, easy to have, super filling, goodness of vitamins and minerals	[25]
5.	Corn oils/Free fatty acids	Salad oil, cooking and frying oil, sautéing and frying, salad dressings and marinades, cakes, bread	Moisture retention, effective cooking in frying, enhances aeration	[26]
6.	High fructose corn syrup	Jams and jellies baked goods, other food products	Better stability, flavor enhancement, pourability, freshness, color, texture and consistency in foods with respect to sucrose	[27]
7.	Dextrins	Food batters, coatings, and glazes	Contributing to flavor, color and crispness crispness enhancer for food processing due to useful thickening properties of bread during the baking process	[28]
8.	Maltodextrins	Baked goods, beer, yogurt, spice mixes, sauces, snack, artificial sweeteners, nutrition bars, weight-training supplements, meal-replacement cereals shake, salad dressings, gelatins, salad dressings. Low-fat and reduced-calorie products, chips, pie fillings, condiments	Used as a thickening or filler to enhance the volume of a processed food, improve consistency, texture, and flavour as a preservative that extends the shelf life of packaged goods	[29]

(continued)

Table 13.1 (continued)

Sl. no.	Name of the by product	Food uses	Aid in food preparation	References
9.	Dextrose	Baking products	An artificial sweetener or preservative that is widely found in packaged foods due to its widespread availability; helps to balance out foods that are otherwise highly spicy or salty	[30, 31]
10.	Fructose or crystalline fructose	Nutritive sweetener in foods and beverages, flavored water, ports and energy drinks, chocolate milk, low-calorie products, breakfast cereals, carbonated beverages, yogurt, baked goods, dry mix beverages, fruit packs, confections	The least cariogenic of the nutritive sugars, sweeten foods and beverages, have a low glycemic index, and improve the sweetness and other qualities of foods and beverages	[32, 33]
11.	Hydrol/ Hydrolyzed corn protein	Soups, spice mixtures, chilis, sauces, meat products, gravies, side dishes and frozen meals.	This flavour enhancer gives processed savoury food products and pre-prepared meals a meat-like flavour; when hydrolyzed protein is chemically broken down, glutamic acid is released, which mixes with salt to generate MSG	[34]
12.	Zein	Ready-to-eat chicken, frozen foods, cheese and liquid eggs	Used as a water-protective barrier for oxygen and moisture in products, such as nuts and candies	[35]
13.	Sorbitol	Beverages, packaged foods, medications	Utilised to keep items moist, provide sweetness, and texture, as well as perhaps supporting digestive and oral health	[36]

of neurodegenerative diseases [4], cardiovascular disease [5], type 2 diabetes [6], obesity [7], and certain malignancies [5]. The health-promoting benefits of corn might be attributed to the phytochemicals like phenolic acids, carotenoids having antioxidant, anti-inflammatory, free radical scavenging, neuroprotective and hepatoprotective and antiproliferative properties [8] and basic elements like dietary fibres, minerals, vitamins and fatty acids improving digestive health [9].

Different corn byproducts contain a diverse range of phytochemicals with superior nutritional value which may extend the health benefits as prevention of chronic diseases. The goal of this chapter is to focus on recent research on corn byproducts, their phytochemicals and likely health benefits when taken as nutraceutical.

13.2 Products of Corn and Byproducts

All the byproducts of corn have its own phytochemical profile as each is made up of endosperm, germ, and bran like key components of corn in different proportions. For e.g., Corn bran contain health-promoting phytochemicals such as phenolics and phenolic acids (ferulic acid, vinnilic acid, courmaric acid, syringic acid, caffeic acid etc.). Corn flour with powdered germ, endosperm, and bran contains more bioactive chemicals than refined cornstarch and corn oil (Fig. 13.1) [10].

13.2.1 Corn Flour

Corn flour is a fine powder derived from finely crushed whole corn kernels. Protein, fibre, carbohydrate, and the vitamins and minerals are present in whole corn. It is usually yellow in colour. Corn flour provides B vitamins, various minerals like iron, potassium, magnesium and extra fibre and protein. Corn flour has a similar earthy, sweet flavour like whole corn [11].

13.2.2 Corn Starch

Corn starch a fine powder made solely from corn's starchy part. Cornstarch is more refined, as it is produced by removing the protein and fibre from the corn kernel, leaving only the starchy endosperm. This is then processed into a white powder. Cornstarch is gluten free and contains mostly carbohydrates, no protein, B vitamins

Fig. 13.1 Corn products

and much smaller amounts of other nutrients, compared with corn flour. Cornstarch is considered a highly processed food. It is flavorless [12].

13.2.3 Cornmeal

The texture is the main distinction between cornmeal and corn flour. Cornmeal is a coarse, gritty powder with a yellowish appearance, whereas corn flour is a fine powder with a white colour. In essence, they're both ground forms of milled, dried corn, with the textural difference owing to the coarseness or fineness with which the corns are crushed. Polenta is a cooked cornmeal used in dishes like bread and desserts. Polenta is a nutritious dish that has various health benefits. It is an adaptable dish that can be served cold or hot [13].

13.2.4 Corn Flakes

Corn flakes are the well-known cereals that is taken by both children and adults. It is one of the world's healthiest snacks. It is consisting of baked toasting corn flakes that are ready to consume. Cornflakes have a crisp feel and a yellowish-orange colour to them [14].

13.2.5 Corn Oil

Corn oil is made from the seeds of corn kernels with barely 3–5 percent oil content. Corn germ is high in oil, and yields all commercial corn oil. Corn oil has a pleasant flavour, low saturated fatty acid levels, low alpha-linolenic acid levels, and high polyunsaturated fatty acid levels. Corn oil is mostly made as a by-product of corn starch industry. It transports soluble vitamins, helps sensory palatability, satiation mechanisms and cholesterol density in cells, such as LDL and HDL. Corn oils and fats include triglyceride phospholipids, sterol, pigments, fatty alcohols, and tocopherol. Tocopherols and tocotrienols are classified as alpha, beta, gamma, and delta. Antioxidant and vitamin E act differently. The antioxidant activity of delta tocopherol is the highest [15].

13.2.6 High Fructose Corn Syrup (HFCS)

Since last 4 decades, high fructose corn syrup has become a popular food ingredient. As it does not crystallize when heated, it is choosed as an ingredient for candies. In comparison to other caloric sweeteners, HFCS raises the risk of obesity and other negative health outcomes. Common HFCS (HFCS-42 and HFCS-55) made up equal proportions of fructose and glucose, and are identical to sucrose. In HFCS, these monosaccharides are free in solution, whereas in sucrose, they are in disaccharide form. As sucrose is easily cleaved in the small intestine, both sucrose and HFCS absorb free fructose and glucose. It helps to keep foods fresh, reduces the freezing point, keeps moisture in bran cereals and breakfast bars, improves fruit and spice aromas, promotes surface browning, and offers fermentability, in addition to its sweetening characteristics [16].

13.2.7 Corn Dextrin

Dextrins are a type of low-molecular-weight carbohydrate that is formed when starch or glycogen is hydrolyzed. Dextrins is made from starch by using amylases in human digestion and malting or by dry heat application in an acidic environment in pyrolysis or roasting. During roasting in acidic conditions, the starch hydrolyzes, and short-chained starch portions partially rebranch with bonds to the damaged starch molecule [17].

13.2.8 Maltodextrin

Maltodextrin is a food additive. It is starch sugar with a short chain made from gelled starch by enzymatic hydrolysis. It is a creamy-white hygroscopic spray-dried powder. It is easily digested, absorbing as quickly as glucose, but lesser sweet or tasteless [18].

13.2.9 Corn Sugar/Dextrose

Dextrose is a basic sugar of corn and other plants. It is used to sweeten foods and extend the shelf life of various products. Dextrose is nearly resembling blood sugar and body quickly use it as a source of energy. Dextrose included in many processed and pre-packaged meals has some health benefits [19].

13.2.10 Fructose or Crystalline Fructose

Crystalline fructose, a nutritive sweetener derived from corn or sugar (sugar cane), is extensively used in foods and beverages as a crystalline form of fructose. Crystalline fructose is often utilised in health food due to its high sweetness among natural sugars, low glycemic index, and low calories. D-fructose is purified and crystallised having fructose and glucose. It has same chemical formula as glucose, with different molecular structure. Pure crystalline fructose and sucrose each have distinct qualities that make them well-suited for most food applications [20].

13.2.11 Hydrol/Hydrolysed Corn Protein

The granular or powdered corn protein has a yellow colour. It has a distinct odour that is similar to that of corn fermentation. It is not easily dissolved in water due to its more hardness and less moisture. Hydrolyzed Corn protein is a hydrolyzed product made from the acidic, alkaline, or enzymatic hydrolysis of corn having proteins, amino acids and peptides. Carbohydrates and lipids are the two main contaminants that can be found in it. Long protein strands' peptide linkages break down into smaller chains or free amino acids which are easier for human bodies to absorb. It is helpful for persons suffering from malabsorption problems [21].

13.2.12 Zein/Corn Protein

Corn contains zein, a type of prolamine protein. It's commonly made from corn gluten meal. Pure zein is clear, odourless, tasteless, hard, insoluble in water, and edible with a wide range of food applications. Corn endosperm cells contain the amino acid zein. Glutamic acid, alanine, leucine and proline are found in larger concentrations in zein proteins [22].

13.2.13 Sorbitol

Sorbitol has about two-thirds the calories of table sugar and about 60% of the sweetness. When compared to items manufactured with typical sweeteners like table sugar, it has relatively minimal influence on blood sugar levels when consumed. Sorbitol is a widely used sugar alcohol frequently substituted for regular sugar in foods and beverages to minimise calorie content [23].

13.3 Phytoconstituents Present in Corn Byproducts with Health Benefits

Blue, purple and red corn possess cyanidin derivatives anthocyanidins. Yellow corn is rich in carotenoids including lutein, zeaxanthin, alfa and beta-carotene [5]. Corn byproducts contain various parts of the corn kernel, such as germ, bran, and endosperm, and each byproduct has its own set of phytochemicals comprising of an unique spectrum of phytochemicals (phenolic acids, syringic acid, courmaric acid, ferulic acid, and caffeic acid) as well as carotenoid (lutein, zeaxanthin, β-cryptoxanthin, α-carotene) and minerals (calcium, magnesium, phosphorus, potassium, sodium, zinc), vitamins (carotenoids, thiamine, riboflavin, niacin, pyridoxine, folate, ascorbic acid, vitamin E, and vitamin K). The most abundant macro- and micronutrients in corn byproducts are carbohydrates and water. Lipids in corn byproducts are mostly mono- and poly-unsaturated forms and lesser saturated form. Table 13.2 lists the majority of phytochemicals found in corn that are helpful to human health.

Corn flour includes more beneficial chemicals than refined cornstarch and refined corn oil because it contains pulverised germ, endosperm, and bran. In terms of flavonoids and carotenoids contents, they have drastically different phytochemical profiles. As a result, consuming various corn byproducts delivers a diverse spectrum of phytochemicals in addition to the nutritious benefits of corn. Bioactive substances found in corn byproducts may be responsible for health benefits such as a reduced risk of diabetes, cancer, and neurological disorders like Alzheimer's and Parkinson's disease [37] (Fig. 13.2). Corn byproducts contains insoluble fibres such as cellulose, amylose, hemicellulose, and lignin, which support digestive health. The appropriate proportion of insoluble and soluble fibres aids digestion. Corn insoluble fibres have the capacity to ferment in the colon and aid a variety of microbiomes.

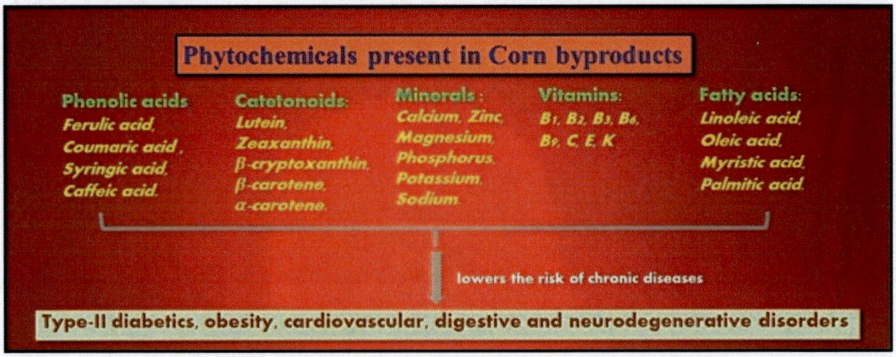

Fig. 13.2 Corn byproducts as source of nutraceuticals

Table 13.2 Phytochemicals present in corn byproducts with health benefits

Sl. no.	Phytoconstituents present	Group	Structure	Pharmacological effects	Disease conditions	References
1.	Ferulic acid (4-hydroxy-3-methoxycinnamic acid)	Phenolic acids	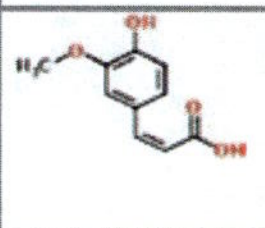	Anticancer, antioxidant, antimicrobial, anti-inflammatory, and antidiabetic effect	Neurodegenerative diseases like Alzheimer and Parkinson	[38]
2.	Coumaric acid (4-hydroxycinnamic acid)	Phenolic acids	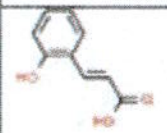	Anti-inflammatory, high free radical scavenging, antimicrobial, antineoplastic, activities	Experimental diabetes and hyperlipidemia	[39, 40]
3.	Syringic acid	Phenolic acids	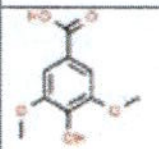	Anti-oxidant, antiendotoxic, antimicrobial, anti-inflammatory, free radical scavenger, neuro and hepatoprotective activities	Diabetes, cerebral ischemia, cancer, neurodegeneration and liver damage	[41]
4.	Vannilic acid (4hydroxy-3-methoxybenzoic acid)	Phenolic acids	–	Antibacterial, antimicrobial and chemopreventive effects; alleviates oxidative stress, cytokine production, and NF-κB activation	Ulcerative colitis, diabetes mellitus, neurodegenerative diseases like Alzheimer's *disease*, rotenone-induced Parkinson's *disease*	[42, 43]
5.	Caffeic acid (quinic acid ester called chlorogenic acid)	Phenolic acids	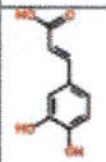	Antioxidant	Inflammation, cancer, diabetes, neurodegenerative *disease*	[44]
6.	*Trans*-cinnamic acid	Phenolic acids	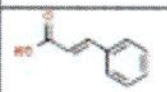	Inhibit intracellular ROS production; ameliorate UVA induced cytotoxicity	UVA irradiation-induced skin damage	[45]

(continued)

Table 13.2 (continued)

Sl. no.	Phytoconstituents present	Group	Structure	Pharmacological effects	Disease conditions	References
7.	Protocatechuic acid	Phenolic acids	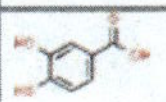	Antioxidant, anti-inflammatory as well as antihyperglycemic and neuroprotective activitie, pro-apoptotic and anti-proliferative effects	Alzheimer's and parkinson's disease, cancer, coronary heart disease	[46]
8.	Lutein	Carotenoids	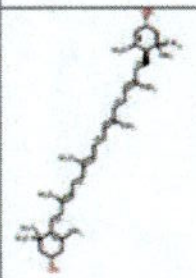	Protective against light-induced oxidative damage, anti-inflammatory properties	Age-related macular degeneration; cataracts atherosclerosis	[47]
9.	Zeaxanthin	Carotenoids	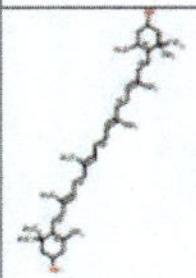	Protection against inflammation, blue light filtration properties and local antioxidant activity	Visual disorders and cognition diseases, retinitis pigmentosa, ischemic/hypoxia induced retinopathy, like age-related macular degeneration, age-related cataract, cognition diseases	[48, 49]
10.	β-Cryptoxanthin	Carotenoids	–	Antioxidant properties, anabolic effects on bone calcification, stimulatory effects on osteoblastic bone growth, and inhibitory effects on osteoclastic bone resorption	Cancers and degenerative disease, osteoporosis	[50]
11.	β-Carotene	Carotenoids	–	Antioxidant by acting against reactive oxygen species	Cardiovascular disease, injuries and accidents, respiratory disease, diabetes mellitus, stroke, cancer	[51]

12.	α-Carotene	Carotenoids		Antioxidant and anti-carcinogenic qualities, inactivating free radical antioxidants, enhance immune function	Cardiovascular diseases and cancer of lung, prostate, breast and liver, erythropoietic protoporphyria	[52]
13.	Calcium	Minerals	Ca	Maintain strong bones and teeth, the ability to move muscles, and the ability for nerves to transport messages from the brain to every area of the body	Paget's disease of bone, osteoporosis hyperparathyroidism	[53]
14.	Magnesium	Minerals	Mg	Lowering blood pressure and increasing serum, lowers chance of developing metabolic syndrome, aids cellular glucose metabolism	Hypokalaemia, hypocalcaemia, hypertension, osteoporosis, cardiac and neurological manifestations, diabetes, hypertension, coronary heart disease, obesity	[54]
15.	Phosphorus	Minerals	P	Mineral metabolism skeletal development,, cell membrane phospholipid content and function, platelet aggregation, cell communication	Rickets in children, osteomalacia in adult	[55]

(continued)

Table 13.2 (continued)

Sl. no.	Phytoconstituents present	Group	Structure	Pharmacological effects	Disease conditions	References
16.	Potassium	Minerals	K	Main cation in intracellular fluid; involved in acid-base balance, muscular contraction,nerve impulse transmission, cell membrane function, osmotic pressure management	Type 2 diabetes, cardiovascular, kidneys and bones diseases	[56]
17.	Sodium	Minerals	Na	Maintain the balance of ions for the proper functioning of the bodily fluids system, nerves, and muscles	High blood pressure, heart disease and stroke, adrenal gland *disorders* like addison's *disease*	[57]
18.	Zinc	Minerals	Zn	Cofactor for a large number of enzymes, zinc metalloenzymes plays a role in every biochemical pathway	Hypogonadism, alopecia, delayed wound healing impaired taste acuity, dermatitis, weight loss	[58]
19.	Thiamine (Vit-B1)	Vitamins	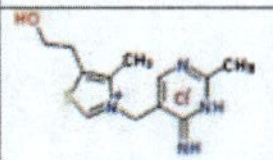	Helps the alteration of carbohydrates into energy, involved in muscular contraction and nerve signal transmission	Beriberi, certain nerve *diseases*, wernicke-korsakoff syndrome	[59]
20.	Riboflavin (Vit-B2)	Vitamins	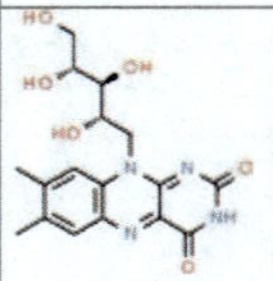	Anti-oxidant, anti-nociceptive anti-inflammatory, anti-aging, anti-cancer properties	Sepsis, ischemia, cancer	[60]

21.	Niacin (Vit-B3)	Vitamins	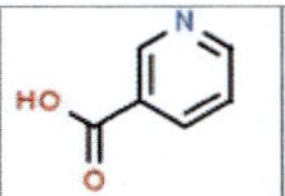	Precursors of the coenzyme nicotinamide adenine dinucleotide and nicotinamide adenine dinucleotide phosphate	Neurological, cardiovascular and skin diseases, diabetes, osteoarthritis	[61]
22.	Pyridoxine (Vit-B6)	Vitamins	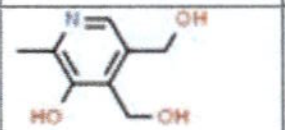	It plays a key function in amino acid metabolism, including important interactions with endogenous redox processes via the glutathione peroxidase system		[62]
23.	Folate (Vit-B9)	Vitamins	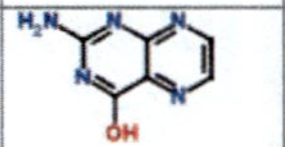	These hydrophilic anionic molecules can accumulate in mammalian cells and tissues through membrane transport systems	Megaloblastic anemia, hemolytic anaemia, chronic kidney disease	[63]
24.	Ascorbic acid (Vit-C)	Vitamins	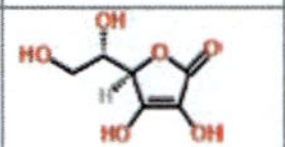	Antioxidant that protects the body from free radicals' effects, required for connective tissue growth and maintenance, aids in bone creation, wound healing, protects the immune system, preserves healthy gums	Scurvy, reduces the severity of allergic reactions and helps fight off infections	[64]
25.	Alpha-tocopherol (vitamin E)	Vitamins		Boost its immune system, act as an antioxidant, scavenging loose electrons as free radicals	Platelet hyperaggregation, cancer, arthritis, ageing, cataracts	[65]

(continued)

Table 13.2 (continued)

Sl. no.	Phytoconstituents present	Group	Structure	Pharmacological effects	Disease conditions	References
26.	Phylloquinone (vitamin K)	Vitamins	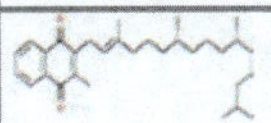	Maintains hemostasis, facilitates ATP generation, cofactor in the activation of vitamin K-dependent coagulation factor and rescues mitochondrial dysfunction as an antioxidant	Cancers, bone metabolism, cardiovascular and chronic kidney disease	[66]
27.	Linoleic acid	Fatty acid	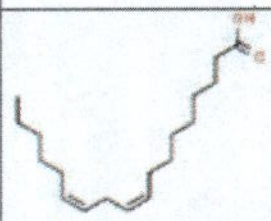	Reduces total and LDL cholesterol, increases insulin sensitivity and blood pressure, and substitutes saturated fats	Heart health, coronary heart diseases, hypertension, diabetes mellitus	[67]
28.	Oleic acid	Fatty acid		Saturated fats are replaced in the diet, which helps to enhance heart health by lowering cholesterol and reducing inflammation	Heart disease, hypercholesteremia	[67]
29.	Myristic acid	Fatty acid		Enhances the levels of long-chain omega-3 fatty acids in plasma phospholipids, which improves cardiovascular health	Cardiometabolic disease, cardiovascular health immunomodulatory functions	[68]
30.	Palmitic acid	Fatty acid		Elevate cholesterol levels and induce fat deposition in coronary arteries and other bodily parts	Cardiovascular diseases, managing body weight	[69]

13.4 Future Perspectives

Polyphenols, vitamins, minerals, carotenoids, and other phytochemicals found in corn have anti-oxidant and anti-inflammatory qualities. Unfortunately, the effects of these natural antioxidants are restricted due to low absorption and degradation along the delivery channel, etc. Nanoformulations of these natural antioxidants can be formed to overcome these difficulties and for effective targeted delivery of antioxidants at the specific targeted site with gradual and sustained release in the treatment of several disease.

13.5 Conclusion

Corn byproduct delivers nutrition in form of vitamins, minerals, fatty acids and phytochemicals like, phenolic acids, carotenoids, flavonoids and dietary fibre. Corn byproducts phytochemicals lower the risk of chronic and neurodegenerative diseases like heart disease, type 2 diabetes, and obesity, and improves digestive health. Addition of corn byproducts as nutraceuticals in daily basis can lower the risk of chronic diseases and paves way towards healthier life.

References

Egbuna C, Tupas GD (eds) (2020) Functional foods and nutraceuticals – bioactive components, formulations and innovations, 1st edn. Springer Nature, Cham, Switzerland, p 646

Li YO, Komarek AR (2017) Dietary fibre basics: health, nutrition, analysis, and applications. Food Qual Saf 1(1):47–59

Watson SA (1988) Corn marketing, processing, and utilization. Corn Corn Improve 18:881–940

Hasanudin K, Hashim P, Mustafa S (2012) Corn silk (*Stigma maydis*) in healthcare: a phytochemical and pharmacological review. Molecules 17(8):9697–9715

Siyuan S, Tong L, Liu R (2018) Corn phytochemicals and their health benefits. Food Sci Human Wellness 7(3):185–195

Wang Y, Ning Y, Yuan C, Cui B, Liu G, Zhang Z (2021) The protective mechanism of a debranched corn starch/konjac glucomannan composite against dyslipidemia and gut microbiota in high-fat-diet induced type 2 diabetes. Food Funct 12(19):9273–9285

Tsuda T, Horio F, Uchida K, Aoki H, Osawa T (2003) Dietary cyanidin 3-O-β-D-glucoside--rich purple corn color prevents obesity and ameliorates hyperglycemia in mice. J Nutr 133(7):2125–2130

Makbul SAA, Jahan N, Ansari AA (2020) Cereals and grains as functional food in unani system of medicine. In: Egbuna C, Dable Tupas G (eds) Functional foods and nutraceuticals. Springer, Cham. https://doi.org/10.1007/978-3-030-42319-3_15

Thomsen MH (2005) Complex media from processing of agricultural crops for microbial fermentation. Appl Microbiol Biotechnol 68(5):598–606

Adom KK, Liu RH (2002) Antioxidant activity of grains. J Agric Food Chem 50(21):6182–6187

Rackis JJ, Sessa DJ, Honig DH (1979) Flavor problems of vegetable food proteins. J Am Oil Chem Soc 56(3Part2):262–271

Streit L (2019) What's the difference between corn flour and cornstarch? Accessed 22 Oct 2021. <https://www.healthline.com/nutrition/cornstarch-vs-corn-flour#bottom-line>

Quratulain, cornmeal vs. corn flour – what's the difference? Accessed 24 Oct 2021. <https://theculinaryprofessor.com/cornmeal-vs-corn-flour/>

Siegel N, Schwarz C (2016) Mushrooms of the Redwood Coast: a comprehensive guide to the fungi of coastal northern California. Ten Speed Press

Huang SW, Frankel EN, German JB (1995) Effects of individual tocopherols and tocopherol mixtures on the oxidative stability of corn oil triglycerides. J Agric Food Chem 43(9):2345–2350

Parker K, Salas M, Nwosu VC (2010) High fructose corn syrup: production, uses and public health concerns. Biotechnol Mol Biol Rev 5(5):71–78

Aiyer PV (2005) Amylases and their applications. Afr J Biotechnol 4(13)

Parekh M, Blaschek HP (1999) Butanol production by hypersolvent-producing mutant *Clostridium beijerinckii* BA101 in corn steep water medium containing maltodextrin. Biotechnol Lett 21(1):45–48

Shahzad MA, Nisa M, Sarwar M (2011) Influence of replacing corn grain by enzose (corn dextrose) on nutrient utilization, thyroid hormones, plasma metabolites, and weight gain in growing lambs. Asian Australas J Anim Sci 24(7):946–951

Crestani CE, Silva AR, Bernardo A, Costa CB, Giulietti M (2021) Crystalline fructose production: a conceptual design with experimental data and operating cost analysis. Chem Eng Commun 27:1–3

Wang XJ, Zheng XQ, Kopparapu NK, Cong WS, Deng YP, Sun XJ, Liu XL (2014) Purification and evaluation of a novel antioxidant peptide from corn protein hydrolysate. Process Biochem 49(9):1562–1569

Shukla R, Cheryan M (2001) Zein: the industrial protein from corn. Ind Crop Prod 13(3):171–192

Hazrol MD, Sapuan SM, Zainudin ES, Zuhri MY, Abdul Wahab NI (2021) Corn starch (*Zea mays*) biopolymer plastic reaction in combination with sorbitol and glycerol. Polymers 13(2):242

Wikipedia, the Free Encyclopedia (2021). Accessed 22 Oct 2021. <https://en.wikipedia.org/wiki/Cornmeal>

Cornflakes with Milk for Breakfast is a Really Healthy Option (2021). Accessed 22 Oct 2021. <https://www.herzindagi.com/diet-nutrition/benefits-cornflakes-for-breakfast-nutrition-weight-loss-healthy-heart-article-170754>

Ogori AF (2020) Source, extraction and constituents of fats and oils. J Food Sci Nutr 6(2):100060

Moeller SM, Fryhofer SA, Osbahr AJ III, Robinowitz CB (2009) Council on science and public health, American medical association. The effects of high fructose syrup. J Am Coll Nutr 28(6):619–626. https://doi.org/10.1080/07315724.2009.10719794

The Surprising Facts About Dietary Fiber Indigestable Dextrin (2021) Accessed 27 Oct 2021. <https://www.otsuka.co.jp/en/health-and-illness/fiber/about/type/dextrin/>

Tiberian J (2021) Read your food labels: watch out for maltodextrin. Accessed 25 Oct 2021. <https://www.mdvip.com/about-mdvip/blog/read-your-food-labels-watch-out-maltodextrin>

Fletcher J (2018) Everything you need to know about dextrose. Accessed 27 Oct 2021 <https://www.medicalnewstoday.com/articles/322243#uses-in-food>

Brennan D (2021) What to know about dextrose. Accessed 27 Oct 2021. <https://www.webmd.com/diet/what-to-know-dextrose#1>

What is Crystalline Fructose in Food: Uses, Health Benefits, Side Effects (2020) Accessed 27 Oct 2021. <https://foodadditives.net/sweeteners/crystalline-fructose/>

High Fructose Corn Syrup and Fructose, the Differences (2019) Accessed 27 Oct 2021. <http://fructosefacts.org/about/high-fructose-corn-syrup-and-fructose-the-differences/>

What is Hydrolyzed Corn Protein and Its Uses in Food? (2019) Accessed 27 Oct 2021. <https://foodadditives.net/flavor-enhancer/hydrolyzed-corn-protein/>

Lorenzo G, Sosa M, Califano A (2018) Alternative proteins and pseudocereals in the development of gluten-free pasta. In: Alternative and replacement foods. Academic Press, Cambridge, MA, pp 433–458

Grembecka M (2015) Sugar alcohols—their role in the modern world of sweeteners: a review. Eur Food Res Technol 241(1):1–4

Choi DJ, Kim SL, Choi JW, Park YI (2014) Neuroprotective effects of corn silk maysin via inhibition of H2O2-induced apoptotic cell death in SK-N-MC cells. Life Sci 109(1):57–64

Fazel Nabavi S, Pandima Devi K, Sheeja Malar D, Sureda A, Daglia M, Mohammad NS (2015) Ferulic acid and Alzheimer's disease: promises and pitfalls. Mini Rev Med Chem 15(9):776–788. https://doi.org/10.2174/1389557515666150522102545

Ferreira PS, Victorelli FD, Fonseca-Santos B, Chorilli M (2019) A review of analytical methods for p-coumaric acid in plant-based products, beverages, and biological matrices. Crit Rev Anal Chem 49(1):21–31. https://doi.org/10.1080/10408347.2018.1459173

Alam MA, Subhan N, Hossain H, Hossain M, Reza HM, Rahman MM, Ullah MO (2016) Hydroxycinnamic acid derivatives: a potential class of natural compounds for the management of lipid metabolism and obesity. Nutr Metabol 13(1):1–3. https://doi.org/10.1186/s12986-016-0080-3

Srinivasulu C, Ramgopal M, Ramanjaneyulu G, Anuradha CM, Kumar CS (2018) Syringic acid (SA)–a review of its occurrence, biosynthesis, pharmacological and industrial importance. Biomed Pharmacother 108:547–557

Calixto-Campos C, Carvalho TT, Hohmann MS, Pinho-Ribeiro FA, Fattori V, Manchope MF, Zarpelon AC, Baracat MM, Georgetti SR, Casagrande R, Verri WA Jr (2015) Vanillic acid inhibits inflammatory pain by inhibiting neutrophil recruitment, oxidative stress, cytokine production, and NFκB activation in mice. J Nat Prod 78(8):1799–1808. https://doi.org/10.1021/acs.jnatprod.5b00246

Ullah R, Ikram M, Park TJ, Ahmad R, Saeed K, Alam SI, Rehman IU, Khan A, Khan I, Jo MG, Kim MO (2021) Vanillic acid, a bioactive phenolic compound, counteracts LPS-induced neurotoxicity by regulating c-Jun N-terminal kinase in mouse brain. Int J Mol Sci 22(1):361. https://doi.org/10.3390/ijms22010361

Birková A, Hubková B, Bolerázska B, Mareková M, Čižmárová B (2020) Caffeic acid: a brief overview of its presence, metabolism, and bioactivity. Bioact Compd Health Dis 3(4):74–81

Hseu YC, Korivi M, Lin FY, Li ML, Lin RW, Wu JJ, Yang HL (2018) Trans-cinnamic acid attenuates UVA-induced photoaging through inhibition of AP-1 activation and induction of Nrf2-mediated antioxidant genes in human skin fibroblasts. J Dermatol Sci 90(2):123–134. https://doi.org/10.1016/j.jdermsci.2018.01.004

Masella R, Santangelo C, D'archivio M, LiVolti G, Giovannini C, Galvano F (2012) Protocatechuic acid and human disease prevention: biological activities and molecular mechanisms. Curr Med Chem 19(18):2901–2917. https://doi.org/10.2174/092986712800672102

Bartlett HE, Eperjesi F (2007) Effect of lutein and antioxidant dietary supplementation on contrast sensitivity in age-related macular disease: a randomized controlled trial. Eur J Clin Nutr 61(9):1121–1127. https://doi.org/10.1038/sj.ejcn.1602626

Mares J (2016) Lutein and zeaxanthin isomers in eye health and disease. Annu Rev Nutr 36:571–602. https://doi.org/10.1146/annurev-nutr-071715-05111

Jia YP, Sun L, Yu HS, Liang LP, Li W, Ding H, Song XB, Zhang LJ (2017) The pharmacological effects of lutein and zeaxanthin on visual disorders and cognition diseases. Molecules 22(4):610. https://doi.org/10.3390/molecules22040610

Yamaguchi M (2008) β-cryptoxanthin and bone metabolism: the preventive role in osteoporosis. J Health Sci 54(4):356–369

Elvira-Torales LI, García-Alonso J, Periago-Castón MJ (2019) Nutritional importance of carotenoids and their effect on liver health: a review. Antioxidants 8(7):229

Galanakis CM (ed) (2016) Nutraceutical and functional food components: effects of innovative processing techniques. Academic Press, Cambridge, MA

Blair HC, Robinson LJ, Huang CL, Sun L, Friedman PA, Schlesinger PH, Zaidi M (2011) Calcium and bone disease. Biofactors 37(3):159–167. https://doi.org/10.1002/biof.143

Geiger H, Wanner C (2012) Magnesium in disease. Clin Kidney J 5(Suppl_1):i25–i38. https://doi.org/10.1093/ndtplus/sfr165

Moe SM (2008) Disorders involving calcium, phosphorus, and magnesium. Prim Care 35(2):215–237. https://doi.org/10.1016/j.pop.2008.01.007

Patience JF (1990) A review of the role of acid-base balance in amino acid nutrition. J Anim Sci 68(2):398–408

Constantin MU, Alexandru I (2011) The role of sodium in the body. Balneo-Res J 2(1):70–74

Evans GW (1986) Zinc and its deficiency diseases. Clin Physiol Biochem 4(1):94–98

Soetan KO, Olaiya CO, Oyewole OE (2010) The importance of mineral elements for humans, domestic animals and plants-a review. Afr J Food Sci 4(5):200–222

Suwannasom N, Kao I, Pruß A, Georgieva R, Bäumler H (2020) Riboflavin: the health benefits of a forgotten natural vitamin. Int J Mol Sci 21(3):950

Gasperi V, Sibilano M, Savini I, Catani MV (2019) Niacin in the central nervous system: an update of biological aspects and clinical applications. Int J Mol Sci 20(4):974

Dalto DB, Matte JJ (2017) Pyridoxine (vitamin B6) and the glutathione peroxidase system; a link between one-carbon metabolism and antioxidation. Nutrients 9(3):189

Arya SS, Pavitra K (2012) Folate: its health benefits. Agro Food Industry Hi-Tech 23(4):XXVIII-XI

Iqbal K, Khan A, Khattak MM (2004) Biological significance of ascorbic acid (vitamin C) in human health-a review. Pak J Nutr 3(1):5–13

Rizvi S, Raza ST, Faizal Ahmed AA, Abbas S, Mahdi F (2014) The role of vitamin E in human health and some diseases. Sultan Qaboos Univ Med J 14(2):e157

Halder M, Petsophonsakul P, Akbulut AC, Pavlic A, Bohan F, Anderson E, Maresz K, Kramann R, Schurgers L (2019) Vitamin K: double bonds beyond coagulation insights into differences between vitamin K1 and K2 in health and disease. Int J Mol Sci 20(4):896

Vilchez C, Touchburn SP, Chavez ER, Chan CW (1990) The influence of supplemental corn oil and free fatty acids on the reproductive performance of Japanese quail (*Coturnix coturnix japonica*). Poult Sci 69(9):1533–1538

Chowdhury R, Steur M, Patel PS, Franco OH (2016) Individual fatty acids in cardiometabolic disease. In: Handbook of lipids in human function. AOCS Press, pp 207–318

Carta G, Murru E, Banni S, Manca C (2017) Palmitic acid: physiological role, metabolism and nutritional implications. Front Physiol 8:902. https://doi.org/10.3389/fphys.2017.00902

Chapter 14
Sorghum Byproducts as Sources of Nutraceuticals

Daniel Okwudili Nnamani, Pascal Chukwuemeka Aleke, and Peculiar Feenna Onyekere

14.1 Introduction

Sorghum is one of the major cereals in the semi-arid regions of the world. It is an important food and feed crop. Sorghum is known by various names: great millet and guinea corn in West Africa, dura in Sudan, ntama in Eastern Africa, jowar in India and kaoliang in China [1]. Sorghum is usually referred to as milo-maize in Northern America. Sorghum species (sorghum valgare and sorghum bicolour) are members of the *poaceae* grass family.

The USA is the major producer of sorghum but it is not consumed as food except for a very small fraction but as an animal feed while in semi-arid tropics of Africa and India the grain forms the staple diet for large populations where most of the produce is used as human food.

Sorghum like other cereals is an excellent source of starch and protein. It is worthy of note that sorghum is gluten-free which bears significance to our present day reality where the occurrence of Celiac Disease, an Immunological response to gluten intolerance is on the rise. Grain sorghum contains phenolic compounds like flavonoids [2], which have been found to inhibit tumor development [3].

D. O. Nnamani
Department of Pharmaceutics, Faculty of Pharmaceutical Sciences, University of Nigeria, Nsukka, Enugu State, Nigeria

P. C. Aleke
Department of Pharmacy, Alex-Ekwueme Federal University Teaching Hospital, Abakaliki, Ebonyi State, Nigeria

P. F. Onyekere (✉)
Department of Pharmacognosy and Environmental Medicine, Faculty of Pharmaceutical Sciences, University of Nigeria, Nsukka, Enugu State, Nigeria
e-mail: peculiar.onyekere@unn.edu.ng; peculiar.onyekere@daad-alumni.de

© The Author(s), under exclusive license to Springer Nature Switzerland AG 2022
C. Egbuna et al. (eds.), *Food and Agricultural Byproducts as Important Source of Valuable Nutraceuticals*, https://doi.org/10.1007/978-3-030-98760-2_14

The flavonoids consist of anthocyains, flavonols and flavonones and the phenolic acids are benzoic and cinnamic derivatives [4]. One type of anthocyanin that is unique to sorghum is 3-deoxyanthocyanin [5]. The starches and sugars in sorghum are released slowly than in other cereals and hence could be beneficial to diabetic patients [6]. Car et al. supplemented the diets of hamsters with sorghum lipids extracted from the grain with hexane. Plasma cholesterol and plasma non-HDL cholesterol fractions consisting mostly of low density lipoprotein (LDL) cholesterol were significantly lowered in a dose dependent manner with grain sorghum lipids at 1% of the diet. This point to the fact that sorghum grain has a role to play in reducing cardiovascular diseases risks.

Because of the high bound of tannins to sorghum starch and proteins, sorghum has the least starch and protein digestibility. Since starch is the principal source of calories from cereal grains and in human feeding in general, starch binding is particularly relevant because it helps lower caloric intake which could contribute to fighting obesity [7, 8]. Sorghum is also an important source of B group of vitamins, fat soluble vitamins like D, E and K and minerals [9, 10].

14.2 Phytoconstituents and Chemical Composition of Sorghum Grains

Phytoconstituents or phytochemicals are bioactive non-nutrient plant compounds in fruits, vegetables, grains and other plant foods that are linked to reducing the risk of many chronic diseases. More than 5000 phytochemicals have been identified but a large percentage still remains undocumented [11].

Phenolic compounds are one of the most starched phytochemicals. Sorghum grains contains large concentrations of phenolic compounds [12].

14.2.1 Phenolic Acid

Phenolic acids are present in sorghum grains [3]. These are divided into hydrobenzoic acids and hyroxycinnamic acids. Hydroxybenzoic acids are derived from benzoic acid (Fig. 14.1a) and include gallic, p-hydroxybenzioc, vanillic, syringic and protocatechuic acids (Table 14.1). Caffeic, ferulic and sinapic acids constitute the hydroxycinnamic acid (Fig. 14.1a; Table 14.2) [13]. Phenolic acids occur in sorghum grain in both free and bound states. Free phenolic acid are located in the pericarp and bound ones are esterified to cell walls in the endosperm [12].

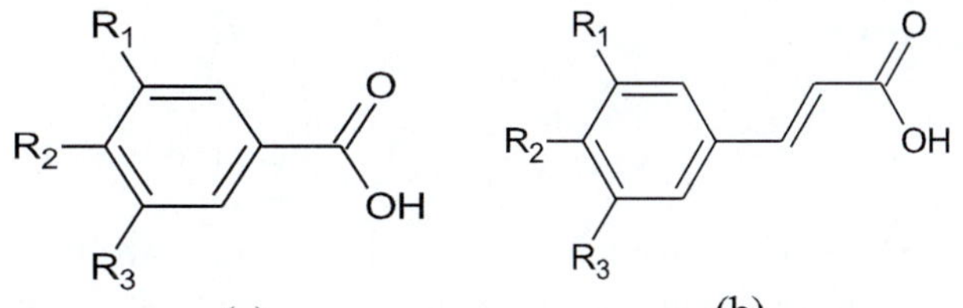

Fig. 14.1 Structure of (**a**) benzoic acid and its derivatives and (**b**) cinnamic acid derivaties adapted from LIU, R.H.J, Nutri. 2004: 34795–34,855

Table 14.1 Benzoic acid derivatives

Benzoic acid Derivatives	Substitutions		
	R1	R2	R3
P-hydroxybenzoic	H	OH	H
Protocatechuic	H	OH	OH
Vannilic	CH_3O	OH	H
Syringic	CH_3O	OH	CH_3O
Gallic	OH	OH	OH

Table 14.2 Benzoic acid derivatives

Cinnamic acid Derivatives	Substitutions		
	R1	R2	R3
p-coumaric	H	OH	H
Caffeic	H	OH	OH
Ferulic	CH_3O	OH	H
Sinapie	CH_3O	OH	CH_3O

14.2.2 Flavonoids

These are compounds with a C6–C3–C6 skeleton that consist of two aromatic rings joined by a three carbon link. There are subdivided by their generic structure of the heterocycle C ring as flavonoid, flavones, flavonones, anthocyanidins and isoflavonoids.

Flavonoids possess various biological activities including anti-cancer, antimicrobial, anti-inflammatory, immunomodulatory and anti-thromobotic effect [14].

It is known, however that different varieties of sorghum have significant difference in both their type and contest of flavonoid [15]. Black sorghum grain was reported to contain highest amount of flavan-4-ols, apiforol predominantly [16]. Two common flavones identified in sorghum include apigenin and luteolin [13].

14.2.3 Tannins

These convert animal skin to leather during the tannin process, hence their name. There are two types of tannins which includes condensed tannins and hyrolysable tannin. The latter has never been found in sorghum. Condensed tannins or proanthocyanidins were named because Anthocyanidins were released when tannin polymers are treated with mineral acids [17]. The condensed tannin in sorghum are mainly polymerized flavan-3-ol and or flavan-3,4-diol [18].

Due to the butter taste of tannin in foods, many types of high-tannin-sorghum may prove useful in astringent taste foods like dark chocolate [17].

14.2.4 Phytosterols

These are cholesterol like compounds that are natural compound of plant cell membrane. These compounds mimic and compete with cholesterol absorption in the GI tract and can produce a cholesterol lowering effect. Sitosterol, campesterol and stigmasterol have been found in sorghum [12].

14.2.5 Chemical Composition of Sorghum

Sorghum grains exhibit a wide range of variation in their chemical composition (Table 14.3).

Submitted as C.P No 108 by the International Crops Research Institute for the Semi-Arid Tropics.

From the table above, it could be deduced that starch is the major constituent of sorghum followed by amylose and protein.

Table 14.3 Proximate composition of sorghum grain

Constituent (%)	Range
Protein	4.4–21.1
Water soluble protein	0.3–0.9
Starch	55.6–75.2
Soluble sugars	0.7–4.2
Crude fibre	10–3.4
Fat	2.1–7.6
Ash	1.3–3.3
Amylose	21.2–30.2

14.3 Nutritional Benefit of Sorghum Grains

Sorghum is an important cereal crop in Africa and Asia and it is consumed in different forms like fortillas, porridges, couscous and baked goods [19]. Preparation of extruded products from Sorghum has been reported [20]. Sorghum has been used in pasta processing [21].

Sorghum products including expanded snacks, cookies, and ethnic foods are gaining serious popularity in areas like Japan [22]. In India, 70% of the total sorghum produced is consumed in the form of *roti* which is an unleavened flat bread [22].

The deep fried sorghum products such as tortillar chips and jowar crunch made from sorghum grains are popular in Asia and Africa [23]. In addition sorghum has been traditionally used in home made alcoholic beverages. Especially in Nigeria where it has been used to produce beer such as lager and stout on a large industrial scale. These beers are gluten-free and have a pleasant fruity flavor and have achieve a huge commercial success [24].

14.4 Sorghum Bran and Its Health-Promoting Benefits

Sorghum bran is a mixture of grain pericarp (bran) and variable amounts of grain fragments (endosperm, germ). It is a byproduct of dry milling in sorghum flour manufacturing [25]. Therefore, it is an excellent source of health-promoting nutrients. This is because it contains relatively high levels of phytochemicals, including polyphenolic compounds, bioactive lipids, flavonoids, phytosterols, policosanols, and starch/carbohydrate fractions [12, 26, 27]. However, the predominant phytochemicals found in sorghum are phenolic compounds, including phenolic acids and flavonoids [28]. These predominant phenolic acids found in sorghum are mainly benzoic and cinnamic acid derivatives [12, 26], while the flavonoids, categorized into several subclasses, are typified with two phenyl rings and a heterocyclic ring differentiated by the position and type of functional group [29]. The bright colors of sorghum grains indicate the presence of these polyphenols, especially flavonoids, and rightly so, according to Awika [30] and Girard and Awika [26], sorghum is the most diverse cereal regarding the amounts and types of polyphenols present in them. Therefore, it has one of the widest ranges of health-beneficial components compared to other cereals [26]. These phytochemicals are vital because they play essential roles in sorghum's antioxidant and anti-inflammatory activity and have other potential health-promoting effects.

Studies have also reported that phenolic extracts (due to their high redox potential) exert a protective effect to help prevent the onset of related neurodegenerative diseases, lower low-density lipoprotein (LDL), reduce swelling, tackle rheumatism and arthritis, and lower the incidence of oesophageal cancer [31–37]. Other properties of sorghum that indicate its potential as a healthy food include the absence of gluten, a protein that provides no essential nutrient found in most grains. In addition, the rich and diverse bioactive polyphenols in sorghum are responsible for its reduced risk of nutrition-linked chronic diseases, including type 2 diabetes, cardiovascular disease, and some types of cancer [38]. Aside from bioactive compounds, sorghum endosperm also generally has a slower starch digesting profile than other cereal grains, a property demonstrated to modulate postprandial blood glucose response in humans [8].

14.5 Sorghum Brewers Grain and Its Uses

Sorghum brewer's grain makes up about 85% of the brewing waste [39]. It is the solid residue left after processing germinated and dried sorghum for the production of beer. They are collected at the end of the mashing process once all sugars have been removed from the grain; thus, they contain a high level of insoluble material. This includes 15–26% protein and 35–60% fiber on dry basis [40]. In addition, it was shown that brewer's grain is a good source of phenolic antioxidants, including ferulic acid and *p*-coumaric acid; since most of the phenolic compounds are trapped in the cell walls of cereals, those are not solubilized in brewing, thus get concentrated in the brewer's grain [39]. Furthermore, brewer's grain is used for feed or as a substrate in the production of bioethanol, lactic acid, xylitol, and enzymes [41]. Sorghum brewers spent grain has also been used as a potential boiler fuel source for bioelectricity and energy conversion [42–46].

14.6 Sorghum Distillers Dried Grains and Solubles (DDGS) and Its Uses

Distiller grain is the undigested portion of the sorghum grain after undergoing ethanol/starch production processing. This process involves grinding whole grain and mixing with water and enzymes for starch degradation and ethanol fermentation. The resulting unfermented components of the kernel (germ, fiber, and proteins) are concentrated into the distiller's dried grains with solubles (DDGS), the principal byproduct of the process [47]. DDGS is a heterogeneous material rich in carbohydrates, fiber, lipids, crude protein, and other nutrients such as vitamins. The heterogeneous nature of DDGS makes it an essential source of nutrients, particularly proteins, for livestock [48]. DDGS are good sources of carbon and nitrogen hence used as a valuable feedstock for microbial fermentation [49, 50]. Furthermore, DDGS has been reported to be a potential fertilizer [51]. Other uses of DDGS include the production of organic acids (such as succinic acid, fumaric acid, and lactic acid) [52–54], biofuels (such as methane, biodiesel, and biohydrogen) [55–58], hydrolytic enzymes (such as proteases, cellulases, peptones, and hemicellulases) [59, 60].

14.7 Sorghum Wine Residue and Its Uses

Sorghum wine residue is also known as sorghum distillery residue or kaoliang residue in Taiwan [61]. It is the byproduct of the manufacture of sorghum liquor. Studies of its use have shown good application in agriculture [62, 63], such as maintaining the effectiveness of fertilizers, promoting vegetable growth, and improving the quality of eggs. In addition, some studies focused on using its biochar for environmental treatment, water purification [64, 65], and bioethanol production, which can be utilized in transportation, thereby serving as an alternative and reducing environmental pollution, and global warming [61]. Furthermore, aqueous and alcoholic extracts of sorghum wine residue have shown antibacterial and antioxidant properties [66].

14.8 Sorghum Gluten Feed and Its Application

Sorghum gluten feed is a byproduct of the manufacture of sorghum starch or syrup by wet milling. The byproducts are usually recombined and dried to produce sorghum bran, gluten feed, and germ meal [67]. It is the sorghum equivalent of the corn gluten feed and consists of a mixture of bran, steep liquor, and other residues. Because sorghum starch is not widely used in the starch industry, sorghum gluten feed and gluten meal production is circumscribed [68]. However, it finds application as food and non-food products [69]. One such use is the potential use of sorghum gluten feed as a replacement for soybean meal in poultry birds [70]. It also has good feed value in ruminants, pigs, and rabbits [71, 72].

14.9 Sorghum Germ Meal and Its Uses

Sorghum germ meal, also known as sorghum germ cake, is a byproduct of sorghum starch extraction, consisting of the germ of sorghum grains from which part of the oil has been pressed. It is the product obtained in the wet milling manufacturing process of starch, syrup, and other sorghum products. The chemical composition of sorghum germ meal varies to a large extent showing low protein and high oil content, intermediate content, or high protein and low oil content [73]. Like other sorghum byproducts, sorghum germ meal is used in poultry feeds due to its chemical composition; however, it decreases growth performance in broilers [74].

14.10 Conclusion

Sorghum and its bye products are important source of valuable nutraceuticals. Their health promoting benefits have found use in reduction of risks of nutrition-linked chronic diseases including type 2 diabetes, cardiovascular disease, and some types of cancer. Their pharmacological effects include anti-oxidant, anti-inflammatory, anti-rheumatoid and immune-modulatory activities. Their phytochemical constituents are abundantly rich in phenols and phenolic compounds including flavonoids. The nutritional potentials and numerous wide applications of sorghum and its bye products are in great demand in the agricultural sectors, pharmaceutical and brewery industries.

References

1. Purseglove JW (1972) Tropical crops: monocotyledons. Longman Group Limited, Longres
2. Shahidi F, Naczk M (1995) Phenolics in cereals and legumes. In: Shahidi F (ed) Food phenolics. Technomic Publishing, Lancaster, pp 9–52
3. Makbul SAA, Jahan N, Ansari AA (2020) Cereals and grains as functional food in Unani system of medicine. In: Egbuna C, Dable TG (eds) Functional foods and nutraceuticals. Springer, Cham. https://doi.org/10.1007/978-3-030-42319-3_15
4. Egbuna C, Ifemeje JC, Udedi SC, Kumar S (eds) (2018) Phytochemistry, volume 1: fundamentals, methods, and applications, 1st edn. Apple Academic Press/CRC Taylor & Francis, New York, 684 p
5. Awika JM, Rooney LW, Waniska RD (2004) Properties of 3-deoxyanthocyanins from sorghum. J Agric Food Chem 52(14):4388–4394
6. Carr TP et al (2005) Grain sorghum lipid extract reduces cholesterol absorption and plasma non-HDL cholesterol concentration in hamsters. J Nutr 135(9):2236–2240
7. Stefoska-Needham A, Beck EJ, Johnson SK, Tapsell LC (2015) Sorghum: an underutilized cereal whole grain with the potential to assist in the prevention of chronic disease. Food Rev Int 31:401–437
8. Simnadis TG, Tapsell LC, Beck EJ (2016) Effect of sorghum consumption on health outcomes: a systematic review. Nutr Rev 74:690–707
9. Ochanda SO, Akoth OC, Mwasaru MA, Kagwiria OJ, Mutiso MF (2010) Effects of Maltingan fermentation treatments on group B-vitamins of red sorghum, white sorghum and pearl millets in Kenya. J Appl Biosci 34(2128–2134):8
10. Martino HSD, Tomaz PA, Moraes EA, Conceiçao LL, Oliveira DS, Queiroz VAV, Rodrigues JAS, Pirozi MR, Pinheiro-Sant'Ana HM, Ribeiro MR (2012) Chemical characterization and size distribution of sorghum genotypes for human consumption. Rev Inst Adolfo Lutz 71:337–344
11. Liu RH (2004) Potential synergy of phytochemicals in cancer prevention: mechanism of action. J Nutr 134(12):3479S–3485S
12. Awika JM, Rooney LW (2004) Sorghum phytochemicals and their potential impact on human health. Phytochemistry 65(9):1199–1221
13. Dykes L, Rooney LW (2006) Sorghum and millet phenols and antioxidants. J Cereal Sci 44(3):236–251
14. Kim HP et al (2004) Anti-inflammatory plant flavonoids and cellular action mechanisms. J Pharmacol Sci 96(3):229–245

15. Lo SCC, De Verdier K, Nicholson RL (1999) Accumulation of 3-deoxyanthocyanidin phytoalexins and resistance to Colletotrichum sublineolum in sorghum. Physiol Mol Plant Pathol 55(5):263–273
16. Dykes L et al (2005) Phenolic compounds and antioxidant activity of sorghum grains of varying genotypes. J Agric Food Chem 53(17):6813–6818
17. Hahn DH, Rooney LW, Earp CF (1984) Tannins and phenols of sorghum. Cereal Foods World 29(12):776–779
18. Awika JM et al (2003) Processing of sorghum (Sorghum bicolor) and sorghum products alters procyanidin oligomer and polymer distribution and content. J Agric Food Chem 51(18):5516–5521
19. Anglani C (1998) Sorghum for human food – a review. Plant Foods Hum Nutr 52(1):85–95
20. Youssef MMA, Moharram YG, Moustaffa H, Bolling H, El-Baya A, Harmuth AE (1990) New extruded products from sorghum. Food Chem 37(3):189–199
21. Miche JC, Alary R, Jeanjean MF, Abecassis J (1977) Potential use of sorghum grains in pasta processing. Dendy, D.A.V., Ed.; Trop. Prod. Inst., London
22. Murty DS, Subramanian V (1981) Sorghum Roti: I. Traditional methods of consumption and standard procedures for evaluation. ICRISAT proceedings of the international symposium on sorghum grain quality, Hyderabad, 28–31 October 1981, pp 73–78
23. Serna Saldivar SO, Canett R, Vargas J, Gonz'alez M, Bedolla S, Medina C (1988) Effect of soybean and sesame addition on the nutritional value of maize and decorticated sorghum tortillas produced by extrusion cooking. Cereal Chem 65(1):44–48
24. Ilori MO, Makinwa EO, Irefin IA (1996) Indigenous technological capability development in the brewing industry in Nigeria: an engineering economic assessment and policy implications. Food Rev Intl 12(4):511–523
25. Lazaro E, Favier J (2000) Alkali Debranning of Sorghum and Millet. Cereal Chem 77:717–720
26. Girard AL, Awika JM (2018) Sorghum polyphenols and other bioactive components as functional and health promoting food ingredients. J Cereal Sci 84:112–124
27. Leguizamon C, Weller CL, Schlegel VL, Carr TP, Leguizamón C, Weller CL, Schlegel VL, Carr TP (2009) Plant sterol and policosanol characterization of hexane extracts from grain sorghum, corn and their DDGS. J Am Oil Chem Soc 86:707–716
28. Teferra T, Awika J (2019) Sorghum as a healthy global food security crop: opportunities and challenges. Cereal Foods World 64
29. Althwab S, Carr TP, Weller CL, Dweikat IM, Schlegel V (2015) Advances in grain sorghum and its co-products as a human health promoting dietary system. Food Res Int 77:349–359
30. Awika JM (2017) Sorghum: its unique nutritional and health-promoting attributes. In: Gluten-free ancient grains [Internet]. Elsevier, p 21–54. Available from: https://linkinghub.elsevier.com/retrieve/pii/B9780081008669000030
31. Oboh G, Akomolafe T, Adetuyi A (2010) Inhibition of cyclophosphamide-induced oxidative stress in brain by dietary inclusion of red dye extracts from sorghum (Sorghum bicolor) stem. J Med Food 13:1075–1080
32. Shih C, Siu S, Ng R, Wong E, Chiu LC, Chu I et al (2007) Quantitative analysis of anticancer 3-deoxyanthocyanidins in infected sorghum seedlings. J Agric Food Chem 55:254–259
33. Taylor JR, Belton P, Beta T, Duodu KG (2014) Increasing the utilization of sorghum, millets and pseudocereals: developments in the science of their phenolic phytochemicals, biofortification and protein functionality. J Cereal Sci 59:257–275
34. Burdette A, Garner P, Mayer E, Hargrove J, Hartle D, Greenspan P (2010) Anti-inflammatory activity of select sorghum (Sorghum bicolor) brans. J Med Food 13:879–887
35. Rao B, Bharti N, Srinivas K (2017) Reinventing the commercialization of sorghum as health and convenient foods: issues and challenges. Indian J Econ Dev 13:1–10
36. Kaur K, Jha A, Sabikhi L, Singh A (2014) Significance of coarse cereals in health and nutrition: a review. J Food Sci Technol 51:1429–1441

37. Aruna C, Visarada KBR (2019) Other industrial uses of sorghum. In: Aruna C, Visarada KBR, Bhat B, Tonapi V (eds) Breeding sorghum for diverse end uses. Elsevier, Amsterdam, pp 271–292
38. Adebo OA (2020) African sorghum-based fermented foods: past, current and future prospects. Nutrients 12(4):1111
39. Mussatto S (2014) Brewer's spent grain: a valuable feedstock for industrial applications. J Agric Sci 94:1264–1275
40. Ikram S, Huang L, Zhang H, Wang J, Yin M (2017) Composition and nutrient value proposition of brewers spent grain. J Food Sci 82:2232–2242
41. Nigam P (2017) An overview: recycling of solid barley waste generated as a by-product in distillery and brewery. Waste Manag 62:255–261
42. Manyuchi M, Frank R, Mbohwa C, Muzenda E (2017) Potential to use sorghum brewers spent grains as a boiler fuel. Bioresources 12(4):7228–7240
43. Dido M, Rieker C, Bursche J, Lopez V, Hamhaber J (2014) Anaerobic digestion of spent grains: potential use in small-scale biogas digesters in Jos, Nigeria. Biogas:19–25
44. Buffington J (2014) The economic potential of brewers spent grain (BSG) as a biomass feedstock. Adv Chem Eng Sci 4(3):308–318
45. Mussatto S, Dragone G, Roberto I (2006) Brewers' spent grain: generation, characteristics and potential applications. J Cereal Sci 43:1–14
46. Weger A, Binder S, Franke M, Hornung A, Rub W, Mayer W (2014) Solid biofuel production by mechanical pre-treatment of brewers spent grain. Chem Eng Trans 37:661–666
47. Chatzifragkou A, Charalampopoulos D (2018) Distiller's dried grains with solubles (DDGS) and intermediate products as starting materials in biorefinery strategies. Sustain Recovery Reutil Cereal Process - Prod:63–86
48. Schingoethe D, Kalscheur K, Hippen A, Garcia A (2009) Invited review: the use of distillers products in dairy cattle diets. J Dairy Sci 92:5802–5813
49. Chatzifragkou A, Kosik O, Prabhakumari P, Lovegrove A, Fraziera R, Shewry P et al (2015) Biorefinery strategies for upgrading distillers' dried grains with solubles (DDGS). Process Biochem 50:2194–2207
50. Iram A, Cekmecelioglu D, Demirci A (2020) Distillers' dried grains with solubles (DDGS) and its potential as fermentation feedstock. Appl Microbiol Biotechnol 104(14):6115–6128
51. Nelson K, Motavalli P, Smoot R (2009) Utility of dried distillers grain as a fertilizer source for corn. J Agric Sci 1(1):3–12
52. Liu H, Yue X, Jin Y, Wang M, Deng L, Wang F et al (2017) Preparation of hydrolytic liquid from dried distiller's grains with solubles and fumaric acid fermentation by Rhizopus arrhizus RH 7–13. J Environ Manag 201:172–176
53. Brock S, Kuenz A, Prüße U (2019) Impact of hydrolysis methods on the utilization of agricultural residues as nutrient source for D-lactic acid production by Sporolactobacillus inulinus. Fermentation 5(1):12
54. Nghiem N, Montanti J, Kim T (2016) Pre-treatment of dried distiller grains with solubles by soaking in aqueous ammonia and subsequent enzymatic/dilute acid hydrolysis to produce fermentable sugars. Appl Biochem Biotechnol 179:237–250
55. Lei H, Ren S, Wang L, Bu Q, Julson J, Holladay J et al (2011) Microwave pyrolysis of distillers dried grain with solubles (DDGS) for biofuel production. Bioresour Technol 102:6208–6213
56. Cheng C, Lo Y, Lee K, Lee D, Lin C, Chang J (2011) Biohydrogen production from lignocellulosic feedstock. Bioresour Technol 102:8514–8523
57. Gu J, Xin Z, Meng X, Sun S, Qiao Q, Deng H (2015) Studies on biodiesel production from DDGS-extracted corn oil at the catalysis of Novozym 435/super absorbent polymer. Fuel 146:33–40
58. Gyenge L, Crognale S, Lanyi S, Abraham B, Raduly B (2014) Anaerobic digestion of corn-DDGS: effect of pH-control, agitation and batch repetition. UPB Sci Bull Ser B Chem Mater Sci 76:163–172

59. Romero E, Bautista J, Garcia-Martinez A, Cremades O, Parrado J (2007) Bioconversion of corn distiller's dried grains with solubles (CDDGS) to extracellular proteases and peptones. Process Biochem 42:1492–1497
60. Hoskins B, Lyons M (2009) Improving bioethanol yield: the use of solidstate fermentation products grown on DDGS. J Inst Brew 115:64–70
61. Chen W-H, Lo H-J, Yu K-L, Ong H-C, Sheen H-K (2021) Valorization of sorghum distillery residue to produce bioethanol for pollution mitigation and circular economy. Environ Pollut 285:117196
62. Kuo W, Yeh N, Fujimoto N, Lin H (2019) Physicochemical properties and fertility index of culture media containing distillery residue biochar and their applications to plug seedling. J Fac Agric Kyushu Univ 64:127–135
63. Lin H, Jhao R, Fujimoto N, Lin HC, Jhao RY, Fujimoto N (2019) Mixing sorghum distillery residue biochar with old corrugated cardboard and cattle manure to develop degradable biochar plug. J Fac Agric Kyushu Univ 64:335–344
64. Lin H, Yeh J, Tsai K, Lian J, Chen M, Huang W et al (2018) Blending agricultural and forestry wastes in wine sludge to develop degradable sludge biochar medium pellet. J Fac Agric Kyushu Univ 63:371–378
65. Lin H, Hu J, Lee W, Lai Y, Wu S, Fujimoto N (2015) A study of sorghum distillery residue activated carbon for water purification. J Fac Agric Kyushu Univ 60:435–443
66. Lee M-Y, Tsai S-H, Lai Y-J (2020) Antibacterial and antioxidant properties of sorghum distillery residue. J Food Drug Anal [Internet] 19(4). Available from: https://www.jfda-online.com/journal/vol19/iss4/5
67. Souilah R, Belhadi B, Khoudja E, Ben El Aakri S, Djabali D, Nadjemi B (2019) In vitro starch digestibility of byproducts of sorghum starch extraction. Anim Res Int 16(1):3162–3173
68. Taylor J, Schober T, Bean S (2006) Novel food and non-food uses for sorghum and millets. Cereal Sci 44:252–271
69. Wronkowska M (2016) Wet-milling of cereals. J Food Process Preserv 40(3):572–580
70. Khalifa N, Elzubeir E, Mustafa E (1994) Use of sorghum gluten feed as a substitute for soyabean meal in layer diets. Anim Feed Sci Technol 48(1–2):165–168
71. Elzubeir E, Mustafa E (1992) The replacement value of sorghum gluten meal for soya-bean meal in broiler chick diets. Anim Feed Sci Technol 36(3–4):339–342
72. Hamid I, Elzubeir E (1990) Effect of graded levels of sorghum gluten meal on the performance of broiler chicks. Anim Feed Sci Technol 29(3–4):289–294
73. Elzubeir E, Jubarah S (1993) Nutritional evaluation of sorghum germ meal as a substitute for sorghum in broiler diets. Anim Feed Sci Technol 44(1–2):93–100
74. Jubarah S, Elzubeir E (1992) Effect of dietary sorghum germ meal on performance and meat quality of broiler chicks. J Sci Food Agric 58(3):301–305

Chapter 15
Rice Husk as a Source of Nutraceuticals

Kirankumar Shivasharanappa, Jayashree V. Hanchinalmath, Sooraj Shivakumar, Sonal Kudva, Sathwik C. Jain, Manoj Girish, D. G. W. M. H. M. M. Wijekoon, Rhishika Dutta, T. Pramod, and Sharangouda J. Patil

Abbreviations

BET	Brunauer–Emmett–Teller theory
DNA	Deoxyribo nucleic acid
FIM	Foundation for Innovation in Medicine
GABA	γ–aminobutyric acid
PZC	Point of zero charge

15.1 Introduction

The concept of "nutraceutical" arose first in the survey from the UK, Germany and France, where the diet was rated higher by consumers, than fitness training sessions or genetic heritage to achieve good health [1]. The term 'nutraceutical' is an amalgamation of the word 'nutrient' (food or component of food that provides necessary nutrients) and 'pharmaceutical' (a medical concoction). Stephen De Felice, who was the prime architect and president of the Foundation for Innovation in Medicine (FIM) (an American organization that encourages medical health), invented the

K. Shivasharanappa (✉) · R. Dutta
Department of Life Science, School of Sciences, Garden City University, Bangalore, Karnataka, India

J. V. Hanchinalmath · S. Shivakumar · S. Kudva · S. C. Jain · M. Girish · D. G. W. M. H. M. M. Wijekoon
Department of Life Science, School of Sciences, Jain University, Bangalore, Karnataka, India

T. Pramod
Department of Microbiology, The Oxford College of Science, Bangalore, Karnataka, India

S. J. Patil
Department of Zoology, NMKRV College for Women (Autonomous), Bangalore, Karnataka, India

© The Author(s), under exclusive license to Springer Nature Switzerland AG 2022
C. Egbuna et al. (eds.), *Food and Agricultural Byproducts as Important Source of Valuable Nutraceuticals*, https://doi.org/10.1007/978-3-030-98760-2_15

term in 1980. De Felice states that a nutraceutical could be described as "food or a part of food that provides aesculapian or other benefits to maintain a good physical condition, which includes the deterrence and treatment of disease." Another definition for nutraceuticals given by Health Canada states "development of a product from food but introduced and marketed in a variety of medicinal conformations such as pills, or powder (potions) not generally linked to foods". Such products may range from nutrients that extracted, supplements for the diet, and user-personalized diets to foods and phyto-products that are genetically modified (termed as "designer" food) [2].

In Britain, the Ministry of Agriculture, Fisheries and Food has developed a definition of a function; good as "a food that has a component incorporated into it to give it a specific medical or physiological benefit, other than purely nutritional benefit. Functional foods and nutraceuticals can be differentiated by a fine line. When food is cooked using scientific intelligence with or without the understanding the method or the reason it is being used for, that food is termed as "functional food". Functional food provides the body with the required amount of essential food classes for living a healthy lifestyle. In contrast, a nutraceutical is a functional food that has helped in preventing and/or treating a disease(s) and/or disorder(s) besides anemia. Example of nutraceutical includes fortified dairy products (milk) and citrus fruits (orange juice) [3]. Nutraceuticals are important for the following reasons:

- Appearance of side effects seldom occurs.
- They have a relatively long half-life.
- Absorption into the intestine is fairly easy for them [4].
- They don't necessarily have to be described by a medical practitioner for its consumption.
- Many people believe this approach is more natural than using a prescribed drug.
- They feel dietary supplements will help them feel stronger and healthier, give them more energy and prevent illness.
- They feel dietary supplements will help them feel stronger and healthier, give them more energy and prevent illness [5].
- Some people resort to these products when they feel regular medical treatments have turned futile to treat their illness.

Some people resort to these products when they feel regular medical treatments have turned futile to treat their illness [6].

15.2 Nutraceutical Categorization

Nutraceuticals are biological therapies that are not specific, and are used for the purpose of maintaining wellness, preventing invasive processes and controlling the manifestations of a medical condition. They can be categorized as follows:

Nutrient: A component of food in a conformation and at an extent that will aid in supporting the life of an animal. The main classes of nutrients comprise of proteins, fats, carbohydrates, minerals and vitamins.

Dietary Supplement: It is a food product that holds either one or a combination of the given healthy ingredients: minerals, vitamins, herbs or other phyto parts, amino acids (protein) and also includes some concentrations, parts, isolates or metabolites produced by these compounds.

***Nutraceutical*:** The component of food that does not display any toxicity and has substantiating scientific evidence regarding benefits to health, including treatment and prevention of various medical conditions.

Herbals: Herbs or botanical products as concentrates and extracts. Herbs and their concoctions have existed since the times of the ancient civilizations and can be called as a treasury of cure-alls for several long-term and short-terms diseases. India has the oldest written tradition of nature's remedies called 'Ayurveda' which possesses a number of effective ways of carrying out proper health care. Numerous nutraceuticals are present in the main constituents of medicinal herbs.

15.3 Rice

Rice, a plant that is monocotyledonous, is scientifically known as Oryza. Twenty-one different wild species and two frequently grown species, *Oryza sativa* and *Oryza glaberrima* constitute the entire genus of *Oryza*. The two cultivated species are native to Asia and Africa, respectively, although they can be grown in different geographical conditions [7]. *Oryza sativa* produces a higher yield and quality after being milled, and is grown for commercial processes in 112 countries globally while *Oryza glaberrima* is endemic to the Western Africa part of the African continent.

Oryza sativa is one of the crops that produces majority of the cereals and one of the most important staple foods for a lot of the people in the world, particularly in Asian countries [8]. The amount of rice that is harvested globally can be estimated to about 600 million tons and the worldwide production for rice, with respect to area, is about 150 million hectares. 90% of the global rice produce is generated by the Asian continent, out of which approximately 28.7% and 19.5% are produced by China and India respectively. Rice is also emerging as an important crop in North and South America, Africa and Europe [9].

15.4 Rice Husk

Rice Husk or rice hull is the hard-shell that covers the seed of rice, which is the provider of nourishment and accumulates the appropriate metabolites as the grain matures, and provides protection against physical trauma, and infections caused by

(a)

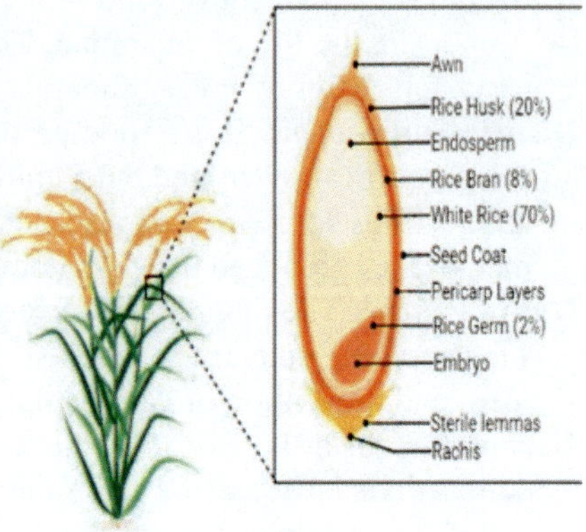

(b)

Fig. 15.1 (**a**) Rice husk, (**b**) Structure of a Rice Grain

microbes, insects and pests as shown in Fig. 15.1 [10]. It has found wide application in different sectors as depicted in Fig. 15.2. Rice husk comprises two major modified structures of the leaf known as the lemma and palea which forms a shell around the caryopsis [11]. The structures layers of rice husks are segregated into four parts, namely (1) a part that is very rich in silica, the coarse superficial epidermis that has surface hairs; (2) sclerenchyma; (3) cells of the parenchyma that are elastic and porous; and (4) inner layer of the epidermis, that has a very smooth surface and is hairless [12].

Rice husk is made up of 24.3% hemicellulose, which is utilized as an alternate resource of xylose, activated carbon, and silicon oxide, 19.2% lignin, 18.85% ash, and 3.25% comprises of trace elements [13]. The elements that encompass rice husk contains are Carbon 37.05%, Hydrogen 8.80%, Nitrogen 11.06%, Silicone 9.01% and Oxygen 35.03%. The density of rice husk in bulk is 96–100 kg/m^3, hardness (Mohr's scale) 5–6, ash 22.29%, Oxygen 31–37%, Nitrogen 0.23–0.32%, Sulphur 0.04–0.08%, Hydrogen 4–5% [14]. The mature rice husk has a varying composition that is linked to several factors such as the variety of rice, the kind of fertilizer used, chemistry of the soil and even the geographical location of the production.

Rice kernel mainly consists of endosperm, husk, bran, and germ. Endosperm accounts for 70% of the entire weight of the seed. Rice husk accounts for 20–21% of it while rice bran and rice germ account for 6–8% and 1% of the entire weight of the seed respectively.

A rice husk is about 8–10 mm in length, 2–3 mm in width and 0.2 mm in thickness. Rice husk is the tough outer shell of paddy rice seed. It is the provider of nourishment and accumulates the appropriate metabolites as the grain matures, and provides protection against physical trauma, and infections caused by microbes, insects and pests. The leaf gets modified to develop two structures called lemma and palea in the rice husk. These structures form a protective shell encasing the caryopsis completely [15]. Outer surface of the rice husk is relatively globular. It is a corrugated structure that is well organized.

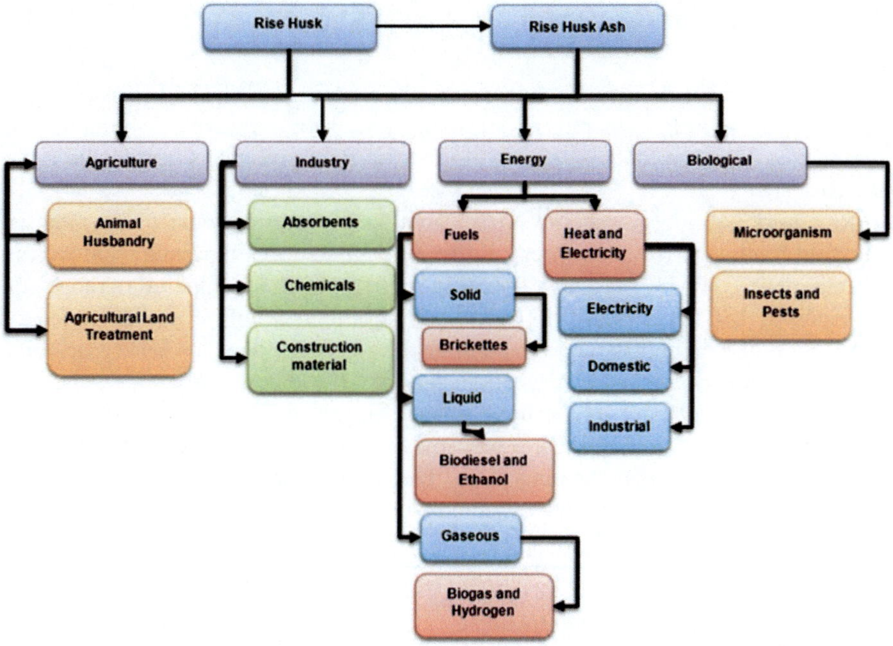

Fig. 15.2 Application of rice husk

The arrangement of the epidermal cells in the lemma is in the form of linear ridges and furrows. The ridges have globular extensions at regular intervals. A carcass made of SiO that is stable relatively, and biomass aggregated in such a way that it surrounds the carcass is responsible for the tight structure of the husk [16].

Rice husk has a set of four different structural layers, outer epidermis, sclerenchyma, spongy parenchyma cells and inner epidermis. Outer is rough with surface hairs. And it is highly concentrated with silica. Silica is mostly distributed in the outer most layer of the husk. But mid region and inner epidermis contain less silica. In contrast to the outer surface the surface of the inner epidermis is smoother and does not have any hair. The rice husk surface is relatively nonporous. Though there are silicon atoms in the surface, they are contained in protuberances and hairs. Silica exists in amorphous form. Silica with high purity can be obtained by pretreating the husk of rice with mineral acids followed by controlled ashing. There is an interlayer between the inner and outer surface. It is composed of interlaced plates and sheets which are loose and honeycombed. It contains many holes about 10 micrometers long [17].

15.4.1 Physical Properties

The bulk density of rice husks ranges from 100 to 160 kg/m^3 and true density ranges from 670 to 740 kg/m^3 whereas rice husk can only be compressed to 400 kg/m^3. In comparison to other fuels rice husk produces a high amount of ash in the range 10–20% [18].

Rice husk has a large specific surface area, good wear resistance, high intensity and void fraction. Rice husks present thermal degradation temperatures in two different ranges (270–330 °C and 330–650 °C). First transition temperature range corresponds to the temperature at which cellulose/hemicellulose get degraded on application of heat and second temperature range corresponds to thermal degradation of lignin. Rice husks are resistant to decomposition in the ground.

The moisture content of the husk of rice ranges in 5–16%. Moisture content affects the quality of the husk as a fuel source. An increase in moisture content in the rice husk decreases its heating value. The particle size of a rice husk ranges in length from 6.2 to 10.0 mm and in width from 1.7 to 2.4 mm [19].

15.4.2 Chemical Properties

Rice husk contains lignin, cellulose, hemicellulose and silica (Fig. 15.3). It consists of 35% cellulose, 35% hemicellulose, 20% lignin, 10% ash by dry weight basis.

Rice husk comprises 80% organic substance and 20% inorganic materials. Crude protein and fat amounts in rice husk are very low. Crude protein ranges from 2.0% to 2.8% and fat ranges from 0.3% to 0.8% whereas crude fiber ranges from 34.5% to 45.9%. Moisture content of a rice husk varies from 4.55% to 10.57% while volatile matter ranges from 58.22% to 71.24% and ash ranges from 9.29% to 30.18%. A rice husk contains elements such as H, N, O, C, and S. But C, O and H are

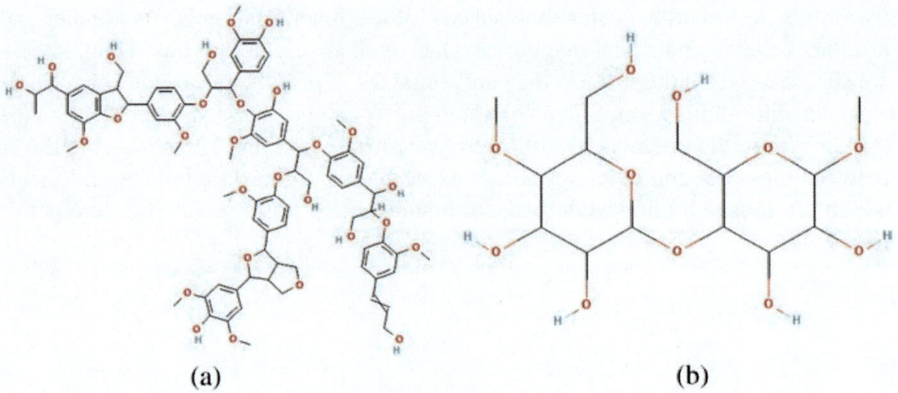

Fig. 15.3 Major components of rice husk and straw (**a**) Lignin; (**b**) Cellulose

concerned as major elemental constituents. Rice husk contains 29.98%–50.455% of C, 4.40%–6.58% of H, 35.20%–59.46% of O, 0.05%–4.26% of N and 0%–0.64% of S.

Rice husk can absorb water significantly because of its hydrophilic characteristics. Also, rice husk can effectively remove metals including Cd, Pb, Zn, Cu, Ni, Hg and Co. Rice husks are difficult to decompose with bacteria and digest. Its nutritional value even for animals is low.

The acid value (pH) of rice husk is 6.91 ± 0.23. This value indicates that the outer layer of the husk has balancing solutions that are neutral. Change in pH could affect the chemistry of the outer layer of the rice husk. Measured value of PZC rice husk is 6.18. Indicates that in an aqueous solution rice husk has a negative charge leading to a pH above 6.18 and that of the positively charged ones is lower than a pH of 6.18. BET surface area value of the husk of rice is 0.1403 m^2g^{-1} [20].

15.4.3 Thermochemical Properties

A proximate analysis is performed to determine the high volatile matter content, high ash content and low fixed carbon. These contents are relatively high in rice husks. Due to the high volatile content, rice husk is more readily devolatilized.

The major elements in rice husk (C, H and O) are important in assessing the heat value and the suitability of rice hull as a fuel. Information on producing undesirable products during the thermochemical conversion process is given by other elements (N and S).

Ash composition is important in thermochemical conversions as it indicates the possible disposal problems of ash and the potential for the formation of undesirable bonded deposits. Rice husk ash is very light and displays high levels of porosity, comprises 87–97% of silica and has a very high external surface area. K_2O, Al_2O_3, CaO, MgO, Na_2O, Fe_2O_3 are other constituents in rice husk ash and they make up less than 1%.

Ash fusibility is important because inorganic compounds in rice husks melt and high temperatures and on cooling form clinker and slag, resulting in serious operational problems. Thus, melting behavior of the fuel ash must be determined in both oxidizing and reducing modes for gasification purposes. Deformation temperature, softening temperature, hemispherical temperature and fluid temperature are determined. In 1982, Commercial Testing and Engineering Co. measured the initial deformation and fluid temperatures of the ash of rice hull in oxidizing and reducing atmospheres. According to the report of this group, in an aerobic surrounding the initial deformation temperature is 144 °C and the fluid temperature is 165 °C. In reducing atmosphere initial deformation initial deformation temperature is 1400 °C and fluid temperature is 1600 °C.

15.4.4 Biochemical Properties

Biochemicals in rice husk such as momilactones A and B act as plant growth inhibitors [21]. They are also involved in drought and salinity tolerance in rice. As rice husk contains 4-Hydroxybenzoic acid and trans-4-hydroxycinnamic acid, it shows an antimicrobial potential.

Glycosyl flavonoid and phytic acid, 2, 3, 6-trimethylanisole (anisole), m-hydroxybenzaldehyde, 4-hydroxy-3-methoxybenzaldehyde (vanillin), and 4-hydroxy-3,5-dimethoxybenzaldehyde (syringaldehyde) are responsible for the anti-oxidizing property of husk [22]. The polysaccharide fraction in rice husk shows the qualities of adhesives, films, and biofuel. Rice husks also exhibit weed-suppressing abilities.

15.5 Rice Hull as Source of Nutraceuticals

Rice bran, which is derived from the outer layer of the rice grain, is composed of an aleurone layer of the rice kernel, with some proportion of the endosperm and germ accounting for approximately 10% of the rice grain's mass. Lipophilic anti-oxidizing compounds (tocopherols, γ-oryzanol, and tocotrienols) and phenolics constitute this part [23]. These substances provide protection against long-lasting diseases that compromise the health of the heart and aid in deactivating the free radicals and anticancer effects.

The content of vitamin E in the rice bran is 5 times lower than that of the rice germ. The crucial component of vitamin E in rice germ is α-tocopherol (the most functional conformation of vitamin E), whereas in rice bran, γ-tocopherol makes up the vital constituent of vitamin E. Additionally, the rice germ also has a considerable amount of vitamins (thiamine—B1, pyridoxine—B6 and riboflavin—B2), fiber and neurotransmitter γ–aminobutyric acid (GABA), which has a large number of positive effects on the health, such as decreasing the pressure of blood, increasing the efficiency of cognitive behavior, and lowering the levels of glucose in the blood serum [24].

The byproducts of rice include cellulose, hemicellulose, pectin, hydrocolloids and lignin which are elements of the dietary fibers. These can be segmented into two classes, on the basis of their ability to dissolve in water. The structural or matrix fibers, such as lignin, cellulose and some hemicellulose, are insoluble, while the natural gel-forming fibers (pectin, gum, mucilage and the remainder of the hemicelluloses) are soluble. Generally, the fiber that is soluble forms a gelatin-like compound within the intestine and leads to formation of stool that has a higher liquid content than normal. Experiments have revealed that it also has the ability to decrease the blood cholesterol and sugar after meals in diabetics. Insoluble fiber plays a key role in adding bulk or softening stool, which helps to reduce constipation and hemorrhoids and is also effective in creating a feeling of fullness.

The rice hull is mainly rich in the fibers lignin, hemicellulose, cellulose and moisturized silica [25]. The rice hull that is generated from defatted concentrated rice bran has a similar composition of fibers. They are not hydrolyzed by the brush border enzymes or human pancreatic enzymes and therefore do not undergo compete absorption. However, the serum blood glucose and concentrations of lipid stay in control upon consumption by making the contents of the gastrointestinal tract more viscous.

Antioxidants play an important role in preventing damage to the cellular components caused by the chemically reactive free radicals. The rind of plants, that consists their husks, shells and peels, can prevent the seeds from getting damaged caused by oxidation due to a high concentration of strong antioxidants, such as flavonoids, hydrocinnamic acid derivatives, isovitexin, phytic acid, anisole, vanillin and syringaldehyde. They are good sources of natural antioxidants. However, these by-products, such as rice husk, get wasted [26] and are utilized as a feedstock since they are not easily hydrolyzed, have a unique size, a lower density of the bulk, a large content of ash/silica, and rough characteristics [27].

Rice hull contains defense agents with anti-oxidizing properties [28], which also includes polyphenolic compounds that provide protection to the inner components from stress caused by oxidation [29]. Phenolic compounds that are isolated from methanol extracts of rice hull display high antioxidant reactions towards compounds that scavenge singlet oxygen and inhibit DNA damage caused by hydrogen peroxide human lymphocytes. Numerous primary compounds with nutraceutical properties isolated from rice bran also display high concentrations of plant chemicals, which exert antioxidant activities. These phytochemicals are α, β, γ, δ-tocopherol, tocotrienols (vitamin E) and γ-oryzanol [30]. Vitamin E protects the cell membrane by preventing the oxidation of unsaturated fatty acids and by scavenging free radicals. Gamma-oryzanol, however, has been shown to have higher antioxidant activity than tocopherols or tocotrienols.

15.6 Conclusion

Nutraceuticals are compounds that provide corporeal benefits and protect the body against several long-lasting medical conditions, besides being a source for the required nourishment. They also improve and maintain health, slow down senescence, stop the onset of persisting diseases, increasing the longevity, and even provide functional and structural aid to the body. At present, nutraceuticals have gained a lot of attention due to their non-toxic and nutritional characteristics. The beneficial prospect of rice husk makes it suitable as a source of nutraceuticals. The anti-oxidizing property and presence of a large quantity of vitamin E in rice hull, provides protection against damage caused by oxidation reactions and reactivity with free radicals. Several studies have given evidences of the possible potential rice has to be used as nutraceutical. The mechanisms with which nutraceuticals work is not clearly understood, however the possibility that they are involved in a lot of processes in the body is the driving motivation behind all research being done on nutraceuticals.

References

1. Pandey M, Verma R, Saraf S (2010) Nutraceuticals: new era of medicine and health. Asian J Pharmaceut Clin Res
2. Egbuna C, Tupas GD (eds) (2020) Functional foods and nutraceuticals - bioactive components, formulations and innovations, 1st edn. Springer Nature, Cham, 646 pp
3. Chellappan K, Dinesh, Prabu L, Sakthivel, Timmakondu, Suriyaprakash, Sureshkumar S, Ragavendran T (2012) Nutraceuticals: a reviews. Elixir Pharmacy 46:8372–8377
4. Muntohar AS (2001) Swelling characteristics of expansive clay stabilized with LRHA. In: Proceeding National Seminar on the Engineering Science, Yogyakarta, pp 62–68
5. Muntohar AS, Hashim R (2002) Silica waste utilization in ground improvement: a study of expansive soil treated with LRHA. Fourth International Conference on Environmental Geotechnics, 11–15 August 2002, Rio de Janeiro
6. Kumar S, Sangwan P, Dhankhar R, Mor V, Bidra S (2013) Utilization of rice husk and their ash: a review. Res J Chem Env Sci 1(5):126–129
7. Esa NM, Ling TB, Peng LS (2013) By-products of rice processing: an overview of health benefits and applications. J Rice Res 1:107. https://doi.org/10.4172/jrr.1000107
8. Babaso PN, Sharanagouda H (2017) Rice husk and its applications: review. Int J Curr Microbiol App Sci 6(10):1144–1156
9. Bao J (2012) Nutraceutical properties and health benefits of rice. In: Shahidi F, Yu L, Tsao R, Shahidi F (eds) Cereals and pulses
10. Vuong QV (ed) (2017) Utilisation of bioactive compounds from agricultural and food waste, 2nd edn. CRC Press
11. Abraham G, Ebenezer I, Murugesan A, Ponsamuel J, Dayanadan P (1990) Role of Palea and lemma in the development of rice caryopsis. J Indian Bot Society 60:245–250
12. Matuez F, Kinga A, Ales P, Dora K (2016) Rice husks - structure, composition and possibility of use them at surface treatment. Mater Sci Forum 844:153–156
13. Muthadhi A, Anitha R, Kothandaraman S (2007) Rice husk ash – properties and its uses: a review. J Institut Engineers (India): Civil Engineering Division 88:50–56
14. Bining AS, Jenkins BM (1992) Thermo chemical reaction kinetics for rice straw from an approximate integral technique. Am. Soc. Agric. Eng. ASAE paper No. 92–6029. St. Joseph, MI
15. Juliano BO (1972) The rice caryopsis and its composition. Rice Chem Technol 2:16–74
16. Zou Y, Yang T (2019) Rice husk, rice husk ash and their applications, rice bran and rice bran oil. AOCS Press, pp 207–246
17. Rama Rao G, Sastry ARK, Rohatgi PK. Nature and reactivity of silica available in rice husk and its ashes. Bull Mater Sci 12:469–479
18. Prachayawarakorn J, Yaembunying N (2005) Effect of recycling on properties of rice husk-filled-polypropylene. J Sci Technol 27(2)
19. Mansaray KG, Ghaly AE (1997) Physical and thermochemical properties of Rice husk. Energy Sources 19(9):989–1004
20. Minh TN, Xuan TD, Ahmad A, Elzaawely AA, Teschke R, Van TM (2018) Efficacy from different extractions for chemical profile and biological activities of rice husk. Sustainability 10(5):1–15
21. Kato-Noguchi H, Ot K (2013) Biological activities of rice allelochemicals momilactone A and B. Rice Res 1(108)
22. Gao Y, Guo X, Yu L, Fang Z, Zhang M, Zhang R, You L, Li T, Liu RH (2018) A full utilization of rice husk to evaluate phytochemical bioactivities and prepare cellulose nanocrystals. Nat Sci Rep 8
23. Hong S-H, Ng L-T. An improved high-performance liquid chromatographic method for simultaneous determination of tocopherols, tocotrienols and γ-oryzanol in rice, vol 1218(29), pp 4701–4713

24. Moongngram A, Daomukda N, Khumpika S (2012) Chemical compositions, phytochemicals, and antioxidant capacity of rice bran, rice bran layer, and rice germ. APCBEE Procedia, vol 2, pp 73–79
25. Gupta NK, Prakash P, Kalaichelvi P, Sheeba KN (2016) The effect of temperature and hemicellulose-lignin, cellulose-lignin, and cellulose-hemicellulose on char yield from the slow pyrolysis of rice husk, vol 38(10), pp 1428–1434
26. Goodman BA (2020) Utilization of waste straw and husks from rice production: a review. J Bioresour Bioprod 5(3):145–169
27. Van Soest PJ (2006) Rice straw, the role of silica and treatments to improve quality. Anim Feed Sci Technol 130:137–171
28. Tan BL, Norhaizan ME (2020) Phytonutrients and antioxidant properties of rice by-products, rice by-products: phytochemicals and food products application, pp 41–68
29. Wanyo P, Meeso N, Siriamornpun S (2014) Effects of different treatments on the antioxidant properties and phenolic compounds of rice bran and rice husk. Food Chem 457–463
30. Shao-Hua H, Lean TN. Quantification of tocopherols, tocotrienols, and γ-oryzanol contents and their distribution in some commercial rice varieties in Taiwan. J Agric Food Chem 59(20):11150–11159

Chapter 16
Byproducts of Groundnut as Source of Nutraceuticals

Neelma Munir, Maria Hasnain, Maria Hanif, Surajudeen Abiola Abdulrahman, and Chukwuebuka Egbuna

16.1 Introduction

The term "nuts" indicates the hard seeds that grow on trees, while some nuts grow in soil and these are known as groundnuts. These nuts include almonds (*Prunus amygdalus*), pecan (*Carya illinoinensis*), pistachios (*Pistacia vera*), hazelnuts (*Corylus avellana*), walnuts (*Juglans regia*) and peanuts (*Arachis hypogaea*), also known as groundnut. According to the botanical rules, peanuts are not considered as a nut, but their chemical composition and nutrient profile is similar to that of tree nuts [1]. Peanuts (groundnuts) are the edible seeds of legumes. According to the annual production rate, India is the largest peanut producer in the world. Around 7.131 million metric tons per year of peanut is produced in India. Peanut (*Arachis hypogaea*) belongs to the fabaceae family of legume. Due to the presence of high oil content in groundnuts, it is generally classified among oilseeds. An abundant amount of protein, fibers and oil are found in groundnuts. Groundnuts have also been used in the manufacturing of peanut butter, roasted groundnuts, snacks, soups, confectionaries and desserts [2].

N. Munir (✉) · M. Hasnain · M. Hanif
Department of Biotechnology, Lahore College for Women University, Lahore, Pakistan

S. A. Abdulrahman
Health Education England, East Midlands Deanery, Leicester, UK

C. Egbuna
Chukwuemeka Odumegwu Ojukwu University, Awka, Nigeria

© The Author(s), under exclusive license to Springer Nature Switzerland AG 2022
C. Egbuna et al. (eds.), *Food and Agricultural Byproducts as Important Source of Valuable Nutraceuticals*, https://doi.org/10.1007/978-3-030-98760-2_16

Many kinds of peanut cultivars are found around the globe. Among them, few cultivars are used for many purposes because of the differences in flavor, nutritional content and disease resistance. The most popular groundnut cultivars are Spanish, Runner, Virginia and Valencia. Valencia cultivar is preferred due to its large shell size and good taste. Spanish groundnuts have been used to produce peanut candy and salted nuts [3].

16.1.1 Nutrition Profile of Groundnut

The major components of peanuts are proteins, fats and fiber. These nutritional contents play a very beneficial role in diet. Plant-based protein is found in groundnut while fiber is present in the form of complex carbohydrate. These are best for human nutrition.

16.1.1.1 Proteins

Groundnuts (peanuts) are legumes, and they contain more protein in comparison to other nuts. After the extraction of peanut oil, protein content can reach up to 50% [4]. All 20 amino acids are found in groundnuts in variable proportion of which arginine is most abundant [5]. According to Protein Digestibility Corrected Amino Acid Score (PDCAAS), the nutritional content of groundnut protein and other proteins in beans such as soy proteins are considered equal to the nutritional content of meat and egg protein. The amino acid content of peanut meals indicates that it can be an ingredient for protein fortification [6]. The plant-based protein in peanuts contains additional components like fiber, rare bioactive components and animal protein. These components show positive health benefits. Groundnut meals offer significant emulsifying activity and stability as well as water retention and foaming capacity [7]. In the food industry, groundnut is a key ingredient in the manufacturing of such foods that have high protein content.

16.1.1.2 Fats

According to the data obtained from American peanut council (APC), peanut contains 50% monounsaturated fatty acids (MUFAs), 33% of Paraformaldehyde (PFAs) and 14% saturated fatty acids [3]. Products made from groundnut are considered to be more favorable for cardiac health in comparison to diets that contain low fat. The MUFA content of peanut diets has been documented to lower total body cholesterol level by up to 11% while (good) HDL cholesterol level is maintained by the decrease in triglycerides [8]. The peanut diet is as much beneficial for lowering the body cholesterol level as olive oil diet. The various types of fat can affect health in numerous ways. The peanut fats and peanut butter have been shown to healthy and beneficial for malnourished infants and children [9].

16.2 Groundnut Digestibility

Groundnut (peanut) contains highly digestible components. The digestibility of groundnut is comparable to animal protein. While amino acids content of peanuts such as lysine, methionine or threonine [10] may vary, the overall quality of protein is generally based on its digestibility and the sequence of amino acids. It was reported that the PDCAAS for peanuts is estimated to be 0.70 out of 1, while the PDCAAS for wheat is reported to be 0.46. Table 16.1 shows the comparative digestibility of peanut-based protein and other grains. The structure of different fatty acids determines the digestibility of fat. 50% of the monounsaturated fats present in groundnuts are easily digestible. These monounsaturated fats contain single unsaturated hydrogen bond which can be broken easily. Peanuts are also known as legumes, and they contain phytic acid which can decrease the availability of many other nutrients because of their binding properties. Phytic acids are found in smaller quantity in groundnuts as compared to the other legumes like soybean. Groundnut contains lower amount of soluble fiber, hence its fiber content is mainly insoluble [13].

16.2.1 Fiber

According to the data obtained from United States Food and Drug Administration (FDA), groundnuts are also an important source of fiber. Sucrose and starch are found in an abundant quantity in groundnuts while reducing sugars are present in a trace amount. This indicates that groundnuts have low glycemic index (GI) and low glycemic load (GL). The glycemic index of groundnut is 14 [14] while its glycemic load is 1 (one). It has been reported that adding groundnuts to the meals containing high glycemic load can keep the blood sugar stabilized. Carbohydrates-containing foods such as groundnuts can increase the blood-glucose level, while some simple carbohydrates like sugar may have a rapid effect on blood sugar. Groundnut is considered to be healthy for human diet as it contains magnesium, oils and fiber which do not significantly affect human blood glucose level [3].

Table 16.1 The comparative digestibility and average PDCAAS for groundnut and other legumes

Food item	Digestibility percent	PDCAAS	References
Peanuts	94	0.70	[11]
Soy	86	0.91	[12]
Whole wheat	86	0.46	[12]
Maize	85	0.43	[3]

16.2.2 Vitamins

Table 16.2 shows the amount of vitamins present in groundnut. Groundnut has been shown to be a great source of niacin. Niacin is vital for optimal performance of the digestive system and skin. It helps in the digestion of food and its conversion to energy, and has also been documented to offer protection against Alzheimer's disease and cognitive decline. Groundnut also contains significant amount of Vitamin E, the deficiency of which may lead to coronary heart disease. Peanut contains an abundant amount of folate which plays a very important role in pregnancy, and for maintenance of cells [14].

Table 16.2 Nutritional profile of groundnut per 100 g

Nutritional profile	Nutrient value	Percentage of RDA	References
Energy	567 kcal	29	[15]
Carbohydrates	16.13 g	12	
Protein	25.80 g	46	
Total fat	49.24 g	165	[16]
Cholesterol	0 mg	0	
Dietary fiber	8.5 g	22	
Vitamins			
Folates	240 µg	60	[17]
Niacin	12.066 mg	75	[18]
Pantothenic acid	1.767 mg	35	[19]
Pyridoxine	0.348 mg	27	
Riboflavin	0.135 mg	10	[20]
Thiamin	0.640 mg	53	
Vitamin A	0 IU	0	[21]
Vitamin C	0	0	
Vitamin E	8.33 mg	55	
Electrolytes			
Sodium	18 mg	1	[22]
Potassium	705 mg	15	
Minerals			
Calcium	92 mg	9	[20]
Copper	1.144 mg	127	
Iron	4.58 mg	57	
Magnesium	168 mg	42	[23]
Manganese	1.934 mg	84	[24]
Phosphorus	76 mg	54	
Selenium	7.2 µg	13	
Zinc	3.27 mg	30	[19]

16.2.3 Minerals

Table 16.2 summarizes the mineral composition of groundnut. This required amount can compete for many minerals mentioned in RDA that are vital for human health. It was observed that 100 g of groundnut contains an RDA level of 129% copper, 82% manganese, 56% iron, 53% phosphorous, 43% magnesium. The intake of groundnut can help reduce inflammation in the body as well as to decrease the risk of metabolic syndrome and type II diabetes [25].

16.3 Groundnuts as Functional Food

Various compounds are found in groundnuts as well as in its skin. These compounds are beneficial for human health beyond basic nutrition. Groundnut is considered as a functional food as it contains Coenzymes Q10. This compound helps protect the heart in periods of reduced oxygen circulation in human body. Groundnuts are also a good source of fiber, and it contains various essential nutrients and vitamins such as vitamin E. Numerous minerals are found in groundnuts such as iron, magnesium, zinc and potassium, as well as antioxidant minerals like selenium, copper and manganese. Many other antioxidant components such as flavonoids and resveratrol are also present in groundnut. These above-mentioned bioactive compounds possess significant disease preventive properties. Groundnuts exhibit antioxidant capacity due to the totality of biological agents present in its seeds such as vitamin E, coumaric acid, ferulic acid, caffeic acid and flavonoids. Extensive studies of its antioxidant activity has been conducted using the fermented groundnut meal [26].

16.4 Bioactive Compounds in Groundnuts

16.4.1 Arginine

The most important amino acid in the body is arginine which keeps many body parts healthy such as skin, liver and joints, while also helping to fortify the immune system, improve circulation, normalize hormones and blood sugar regulation, and preserves male potency [27]. Arginine triggers the T-cells from the thymus gland to improve the immune system [28]. It has been observed in many studies that arginine has potential to treat AIDS [29], cancer [30] and other immunosuppressive diseases. It also functions to enhance the absorption of nutrients in the gastro intestinal tract [31]. Arginine neutralizes toxic substances such as ammonia, and detoxifies the liver. Groundnut has the highest fraction of arginine which is a precursor to nitrous oxide (NO) which helps to preserve the arteries and improve flow of blood [32].

16.4.2 Resveratrol

In response to any kind of stress, damage or infections, plants produce resveratrol which is a polyphenolic compound of class stilbenes [33]. Groundnuts are brilliant source of resveratrol which plays a defensive role against cancers [34], heart disease, arteriosclerosis, degenerative nerve disease and Alzheimer's disease [35]. Bioflavonoids reduce 30% chance of brain strock by improving brain blood flow [36]. Resveratrol also acts as chemo-preventive mediator by controlling tumor initiation, promotion and progression phases, as well as prolonging the lifespan of *Saccharomyces cerevisiae* [37], *Drosophila melanogaster* [38] and mouse [39]. All parts of groundnut, including its shell, comprise resveratrol. Scientists have claim that exposing groundnuts to different types of stress could increase its resveratrol content [40].

16.4.3 Phytosterols

Phytosterols are mainly present in cell membranes of plant cells and structurally analogous to cholesterol [41]. When phytosterols are consumed by human they compete with body's cholesterol for absorption causing a reduction in body's cholesterol metabolism and ultimately lower body cholesterol levels [42]. Hence, for patients with heart disease, phytosterols-containing diet is highly recommended to achieve as much as 10% and 14% reduction in total cholesterol and LDL cholesterol levels, respectively [43].

Groundnuts are full of phytosterols mainly beta sitosterol, campesterols and stigmasterol [44] which inhibit cholesterol absorption, improve serum lipid profile and reduce cardiovascular disease risk [45]. Phytosterols suppress inflammation and slow down the growth of numerous cancers mainly lung [46], stomach, ovarian, prostrate [47], colon and breast cancer [48].

16.4.4 Phenolic Acids and Flavonoids

Peanut skin contains huge fraction of polyphenolic antioxidants mainly p-coumaric acid which accounts for up to 22% of its overall antioxidant content [49]. Roasted peanuts skin has more antioxidant capacity than unroasted peanuts. Flavonoids are present in all groundnut plants, and provide protection against cancer and heart disorders [50]. Table 16.3 summarizes phytochemicals and their quantities in byproduct of groundnuts while Table 16.4 highlights their antioxidant activities.

Table 16.3 Phenolic compounds in groundnuts by-products

Source	Phytochemical	Quantity	References
Almond skin	Catechin	90 μg/g	[51]
	Epicatechin	36 μg/g	[52]
	Condensed tannins	75 μg/g	[53]
	Insoluble dietary fiber	40%	
	Insoluble dietary fiber	8%	[54]
Hazelnut skin	Catechin	250 mg/g	[55]
	Condensed tannins	2.9 mg/g	
	Insoluble dietary fiber	52%	[56]
	Quercetin	85 mg/g	[57]
	Kaempherol	35 mg/g	
Hazelnut shell	Total phenols	211 mg/g	[58]
Hazelnut green cover	Total phenols	134 mg/g	[59]
Peanut skin	Total phenols	169 mg/g	[60]
Peanut Shell	Condensed tannins	17%	[61]
	Insoluble dietary fiber	55%	[62]
	Insoluble dietary fiber	70%	
Pecan Shell	Condensed tannins	340 mg/g	[63]
	Insoluble dietary fiber	49%	
Pistachio skin	Flavonoids	70 mg/g	[64]
	Condensed tannins	460 mg/g	

16.5 Health Benefits of Groundnuts

Groundnuts has gained enormous popularity due to significant lipid profile, higher USFAs and cholesterol-free oil with high fraction of oleic acid. Peanut oil reduces risk of cardiovascular disease (40%) and LDL oxidation improves serum lipid profiles [68] as shown in Fig. 16.1. Regular use of groundnuts reduces colorectal cancer risk due to its high antioxidant properties as its skin contains large fraction of antioxidant [69]. Boiling of peanuts increased its antioxidants biochanin A content by two-fold and genistein content by fourfold [70].

16.5.1 Diabetes and Inflammation

Daily consumption of peanut reduces the risk of diabetes by a quarter. Magnesium and dietary fibers in peanuts improve human health by reducing inflammation [71]. Inflammatory markers such as C-reactive proteins acts as a predictor of cardiovascular disease [72]. Phytosterols present in groundnut regulate inflammation [73].

Table 16.4 Antioxidant activity of phenolic compounds in groundnuts by-products

Source	Phytochemical	Antioxidant activity	References
Hazelnut skin	Total phenolic compounds and total tannins	DPPH EC_{50} = 143 μg/mL	[58]
		DPPH 99% radical inhibition	[59]
Hazelnut shell	Total phenolic compounds and total tannins	DPPH 74% radical inhibition	[55]
		ABTS 157.6 μmol/g	[65]
		DPPH EC_{50} = 2.2 mg/mL	
Hazelnut green cover	Total phenolic compounds and total tannins	DPPH 99% radical inhibition	[66]
Peanut shell	Poliphenolic rich extract	DPPH 95% radical inhibition	[60]
		80% carotene bleaching	
Pecan nut shell	Phenolic compounds and condensed tannins	ABTS 644 μmol/g	[64]
		DPPH 720 μmol/g	
		ORAC 227 μmol/g	
		ABTS 2228 μmol/g	[63]
		DPPH 529 μmol/g	
Pistachio shell	Flavonoids	DPPH EC_{50} = 2.5 μg/mL	[64]
		Carotene EC_{50} = 7.8 μg/mL	
Walnut husk	Phenolic compounds	DPPH 60% radical inhibition	[67]

16.5.2 Cancer

Peanut kernel consists of vitamins, minerals and bioactive components which have cancer preventive properties, of which phytosterols have been studied widely [74]. Phytosterols has been shown to reduce growth of prostrate tumor by up to 40% and also reduces the chance of metastasis to other parts of the body by up to 50% [75]. Resveratrol prevents blood supply to cancerous cell thereby inhibiting their growth. Free and transformed phenolic compounds also play an important role in cancer prevention in comparison to complex and bound phenolic compounds [76], as illustrated in Fig. 16.2.

16.5.3 Alzheimer Disease

Groundnuts contain large quantities of niacin which has protective effects against Alzheimer's disease [77]. One recent study showed that niacin slowed cognitive decline rate in people aged 65 years old and above [16]. It has been observed that vitamin E consumption from supplements rather than food had no positive effect on Alzheimer's disease. People who had peanuts in the top fifth of their diet have been

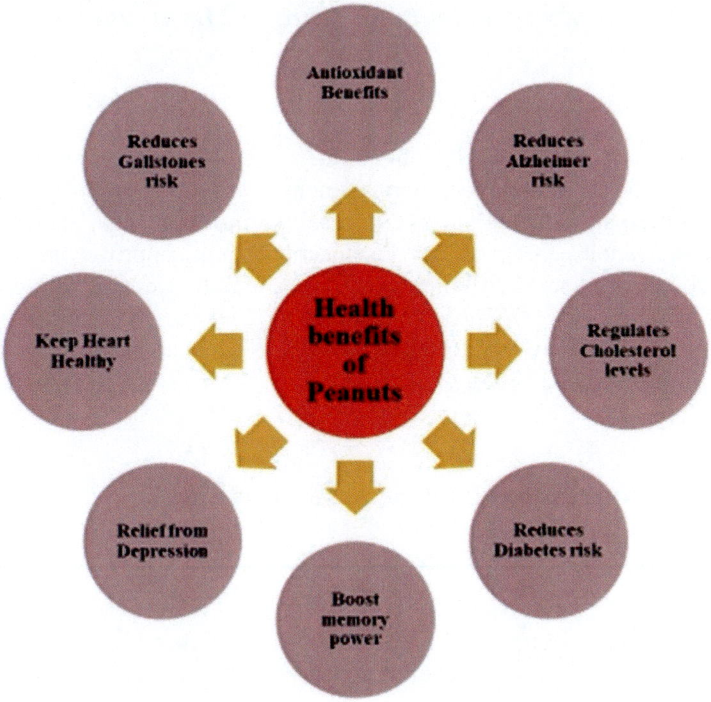

Fig. 16.1 Health benefits of peanuts

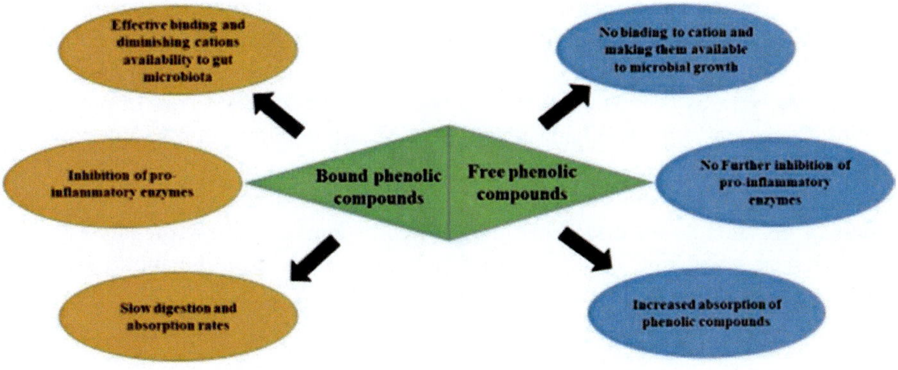

Fig. 16.2 Role of free and transformed phenolic compounds in cancer prevention

shown to have up to 70% lower risk of Alzheimer's disease [23]. Resveratrol content in nuts has been shown to be useful in prevention of Alzheimer's disease. People who consumed groundnut 5 days a week have been shown to have up to 25% reduced risk of gallbladder disease [26].

16.5.4 Weight Management and Hunger Maintenance

Numerous studies have shown that peanut in daily diet does not cause body weight gain [17]. In a recent study it has been shown that groundnut-containing diet can help people of all age groups to maintain their body weight. Comparatively lower weight over a 2-year period was reported among intervention group served groundnut as compared to control subjects [18]. Peanuts consumption has also been shown to reduce LDL cholesterol level [45]. Peanut consumption improves fullness feeling than carbohydrates snacks in same quantity [19] due to presence of monounsaturated fat that might trigger a hormone that aids to feel satisfaction after groundnut intake. In the same vein, groundnut was shown to curb the appetites of consumer because of fullness effect [20].

16.6 Anti-nutritional Factors in Groundnuts

16.6.1 Allergens

Groundnuts contain water-soluble and saline-soluble proteins i.e. albumins and globulins. Mostly, 87% of the globulins make up the total protein content of groundnut. The two major proteins that are found in globulins are arachin and conarachin. Groundnut allergy is caused by the action of immunoglobulin E (IgE) and other anaphylatoxins. Due to degranulation, histamine and other mediator substances are released from mast cells. Histamine can induce vasodilation and constriction of bronchioles in the lungs. Symptoms of bronchospasm include vomiting, urticarial, diarrhea, swelling of face and throat as well as asthma. In order to treat the groundnut allergy, various new techniques have been discovered such as use of probiotics, anti IgE therapy, soy-based immunotherapy, plasmid DNA immunotherapy and oligodeoxynucleotide-based immunotherapy. Still, greater focus is needed to create good techniques that increase the overall function of extracted nutritional components of groundnuts for the preparation of neutraceuticals. These prepared neutraceuticals are beneficial for those individuals that suffer from metabolic disorders and cannot consume groundnuts directly [22].

16.6.2 Food Poisoning

Mostly, groundnuts are contaminated by the *Aspergillus flavus*, which produces aflatoxin. Groundnuts may get infected during transportation and storage of groundnut meals. In food safety, aflatoxins are considered to be highly toxic secondary metabolites. Extensive research has been conducted to analyze the mycotoxin contamination in groundnuts. Results from a recent study in which random samples of groundnuts were analyzed showed that 82% of samples contained aflatoxins which indicate that a greater proportion of the samples were contaminated with fungal growth while only a small proportion were contaminated with mycotoxins [17].

16.7 Potential Uses of Groundnuts by-Products

Groundnut oil is one of the beneficial oils available in the market. Various groundnut by-products are produced from crushed groundnut, including groundnut meal, skin, groundnut hull and vine. Some of the groundnut by-products are used in food processing industry. The by-products of groundnut contain some nutritional compounds such as protein, polyphenolics and fiber. These compounds can be used in processed foods as functional ingredients. Mostly, groundnuts are used for oil production, peanut butter, roasted peanuts, as a snack product also used in soups and desserts. Groundnut byproducts can also be used as animal feed and treated as a fertilizer. Most often, groundnuts are used to produce edible oil. After the extraction of oil, groundnut kernels are used for the separation of groundnut meals. The groundnut skin is used as a functional ingredient in foods and dietary supplements as it contains polyphenols. Groundnut vines are rich source in dietary fibre and flavonoid compounds [3].

16.8 Conclusion

Groundnuts are a huge source of nutrition and provide a variety of health benefits especially in prevention and control of non-communicable diseases such as cancers and cardiovascular diseases that are currently ravaging many nations of the world. Groundnut byproducts have high fraction of active compounds with significant benefits in prevention and control numerous fatal diseases. Hence, the use of groundnut in functional food production has gained enormous popularity among industry practitioners, researchers and the wider public. If all by-products of groundnuts are carefully managed and applied, it will not only boost economic profit in agro industry, but also provide huge cost-savings on health care budgets spent on treatment and care of chronic health conditions like cancers and cardiovascular disease which can be prevented and controlled using nutraceuticals such as by-products of groundnuts.

References

1. De La Rosa LA (2018) Plant food by-products, 3rd edn. Apple Academic Press, pp 201–220
2. Suchoszek-Łukaniuk K, Jaromin A, Korycińska M, Kozubek A (2011) Nuts and seeds in health and disease prevention, 2nd edn. Elsevier, pp 873–880
3. Arya SS, Salve AR, Chauhan S (2016) Peanuts as functional food: a review. J Food Sci Tech 53(1):31–41
4. Zhao G, Liu Y, Zhao M, Ren J, Yang B (2011) Enzymatic hydrolysis and their effects on conformational and functional properties of peanut protein isolate. Food Chem 127(4):1438–1443
5. Shapouri H, Duffield J, Mcaloon A (2001) Secondary United states department of agriculture, pp 1–18
6. Yu J, Ahmedna M, Goktepe I, Dai J (2006) Peanut skin procyanidins: composition and antioxidant activities as affected by processing. J Food Compos Ana 19(4):364–371
7. Wu H, Wang Q, Ma T, Ren J (2009) Comparative studies on the functional properties of various protein concentrate preparations of peanut protein. Food Res Int 42(3):343–348
8. Pelkman CL, Fishell VK, Maddox DH, Pearson TA, Mauger DT, Kris-Etherton PM (2004) Effects of moderate-fat (from monounsaturated fat) and low-fat weight-loss diets on the serum lipid profile in overweight and obese men and women. Am J Clini Nutri 79(2):204–212
9. Matilsky DK, Maleta K, Castleman T, Manary MJ (2009) Supplementary feeding with fortified spreads results in higher recovery rates than with a corn/soy blend in moderately wasted children. J Nutri 139(4):773–778
10. Venkatachalam M, Sathe SK (2006) Chemical composition of selected edible nut seeds. J Agric Food Chem 54(13):4705–4714
11. Suárez L, Kizlansky A, López L (2006) Assessment of protein quality in foods by calculating the amino acids score corrected by digestibility. Nutricion Hospitalaria 21(1):47–51
12. Schaafsma G (2000) The protein digestibility–corrected amino acid score. J Nutri 130(7):1865–1867
13. Schlemmer U, Frølich W, Prieto RM, Grases F (2009) Phytate in foods and significance for humans: food sources, intake, processing, bioavailability, protective role and analysis. Molec Nutri Food Res 53(S2):330–375
14. Gao X, Wilde PE, Lichtenstein AH, Bermudez OI, Tucker KL (2006) The maximal amount of dietary α-tocopherol intake in US adults (nhanes 2001–2002). J Nutri 136(4):1021–1026
15. Zhang Y, Zhang H, Wang L, Guo X, Qi X, Qian H (2011) Influence of the degree of hydrolysis (dh) on antioxidant properties and radical-scavenging activities of peanut peptides prepared from fermented peanut meal. Europ Food Res Technol 232(6):941–950
16. Rusu ME, Gheldiu A-M, Mocan A, Vlase L, Popa D-S (2018) Anti-aging potential of tree nuts with a focus on the phytochemical composition, molecular mechanisms and thermal stability of major bioactive compounds. Food Funct 9(5):2554–2575
17. Vázquez-Flores AA (2018) Byproducts of the nut and peanut agro-industry as sources of phytochemicals suitable for the nutraceutical and food industries. Inst Ciencias Biomédicas
18. Flores-Cordova M, Muñoz-Márquez E, Muñoz-Márquez E, Ojeda-Barrios DL, Soto-Parra JM, Preciado-Rangel P (2017) Phytochemical composition and antioxidant capacity in Mexican pecan nut. Emirates J Food Agric:346–350
19. Elhadef K, Akermi S, Ben Hlima H, Ennouri K, Fourati M, Ben Braïek O et al (2021) Tunisian pistachio Hull extracts: phytochemical content, antioxidant activity, and foodborne pathogen inhibition. J Food Qual 2021
20. Laxmi S, Sahu RM, Kumar T (2019) Groundnut benefits: trend and growth situation in Madhya Pradesh. INDIA Plant Arch 19(2):2753–2758
21. Ojiewo CO, Janila P, Bhatnagar-Mathur P, Pandey MK, Desmae H, Okori P et al (2020) Advances in crop improvement and delivery research for nutritional quality and health benefits of groundnut (Arachis hypogaea L.). front. Plant Sci 11:29
22. Udeh EL, Nyila MA, Kanu SA (2020) Nutraceutical and antimicrobial potentials of Bambara groundnut (Vigna subterranean): a review. Heliyon 6(10):e05205

23. Rehan M, Elshemy NS, Haggag K, Montaser AS, Ibrahim GE (2020) Phytochemicals and volatile compounds of peanut red skin extract: simultaneous coloration and in situ synthesis of silver nanoparticles for multifunctional viscose fibers. Cellulose 27(17):9893–9912
24. Rusu ME, Fizeşan I, Pop A, Gheldiu A-M, Mocan A, Crişan G et al (2019) Enhanced recovery of antioxidant compounds from hazelnut (Corylus avellana L.) involucre based on extraction optimization: phytochemical profile and biological activities. Antioxidants 8(10):460
25. Larsson S, Wolk A (2007) Magnesium intake and risk of type 2 diabetes: a meta-analysis. J Intern Med 262(2):208–214
26. Arya SS, Salve AR, Chauhan S (2016) Peanuts as functional food: a review. J Food Sci Technol 53(1):31–41
27. Mirmiran P, Bahadoran Z, Gaeini Z, Azizi F (2021) Habitual intake of dietary L-arginine in relation to risk of type 2 diabetes: a prospective study. BMC Endocr Disord 21(1):1–7
28. Gao Q, Hong J, Xiao X, Cao H, Yuan R, Liu Z et al (2020) T cell epitope of arginine kinase with CpG co-encapsulated nanoparticles attenuates a shrimp allergen-induced Th2-bias food allergy. Biosci Biotechnol Biochem 84(4):804–814
29. De LC, De T (2019) Healthy food for healthy life. J Glob Biosci 8:6453–6468
30. Chen C-L, Hsu S-C, Ann DK, Yen Y, Kung H-J (2021) Arginine signaling and cancer metabolism. Cancers (Basel) 13(14):3541
31. Unita S, Hirashima N, Shimada M, Tsunekawa T, Tanaka D, Kondo T et al (2020) Successful treatment of adult-onset type II citrullinemia with a low-carbohydrate diet and L-arginine after DNA analysis produced a definitive diagnosis. Clin J Gastroenterol 13(5):823–833
32. Mulwa PM, Njue W (2020) Assessment of L-Citrulline, L-arginine and L-glutamic acid content in selected fruits, vegetables, seeds, and nuts sold in Markets in Nairobi City County, Kenya. Eur J Agric Food Sci 2(5)
33. Dybkowska E, Sadowska A, Swiderski F, Rakowska R, Wysocka K (2018) The occurrence of resveratrol in foodstuffs and its potential for supporting cancer prevention and treatment. A review. Rocz Państwowego Zakładu Hig 69(1)
34. Chen Y-A, Lien H-M, Kao M-C, Lo U-G, Lin L-C, Lin C-J et al (2017) Sensitization of radioresistant prostate cancer cells by resveratrol isolated from arachis hypogaea stems. PLoS One 12(1):e0169204
35. Kumar S, Chang Y-C, Lai K-H, Hwang T-L (2021) Resveratrol, a molecule with anti-inflammatory and anti-cancer activities: natural product to chemical synthesis. Curr Med Chem
36. Leon-Galicia I, Diaz-Chavez J, Albino-Sanchez ME, Garcia-Villa E, Bermudez-Cruz R, Garcia-Mena J et al (2018) Resveratrol decreases Rad51 expression and sensitizes cisplatin-resistant MCF-7 breast cancer cells. Oncol Rep 39(6):3025–3033
37. Lu Y, Shao D, Shi J, Huang Q, Yang H, Jin M (2016) Strategies for enhancing resveratrol production and the expression of pathway enzymes. Appl Microbiol Biotechnol 100(17):7407–7421
38. Bonilla E, Contreras R, Medina-Leendertz S, Mora M, Villalobos V, Paz M et al (2012) Manganese toxicity in Drosophila melanogaster: extension of the life span by resveratrol. Toxicol Environ Chem 94(4):742–747
39. Almeida TC, da Silva GN (2021) Resveratrol effects in bladder cancer: a mini review. Genet Mol Biol 44
40. Guthrie AR, Chow HS, Martinez JA (2017) Effects of resveratrol on drug-and carcinogen-metabolizing enzymes, implications for cancer prevention. Pharmacol Res Perspect 5(1):e00294
41. Islam MA, Jeong B-G, Jung J, Shin E-C, Choi S-G, Chun J (2017) Phytosterol determination and method validation for selected nuts and seeds. Food Anal Methods 10(10):3225–3234
42. Huang J, Xu M, Fang Y-J, Lu M-S, Pan Z-Z, Huang W-Q et al (2017) Association between phytosterol intake and colorectal cancer risk: a case–control study. Br J Nutr 117(6):839–850
43. Rajput H, Goswami D, Nayik GA (2020) Peanut 26. Antioxidants Veg Nuts-Properties Heal Benefits 509

44. Toomer OT (2018) Nutritional chemistry of the peanut (Arachis hypogaea). Crit Rev Food Sci Nutr 58(17):3042–3053
45. Bonku R, Yu J (2020) Health aspects of peanuts as an outcome of its chemical composition. Food Sci Hum Wellness 9(1):21–30
46. Bhatt A, Gairola S, Carón MM, Santo A, Murru V, El-Keblawy A et al (2020) Effects of light, temperature, salinity, and maternal habitat on seed germination of Aeluropus lagopoides (Poaceae): an economically important halophyte of arid Arabian deserts. Botany 98(2):117–125
47. Singh A, Raina SN, Sharma M, Chaudhary M, Sharma S, Rajpal VR (2021) Functional uses of peanut (Arachis hypogaea L.) seed storage. Proteins
48. Lee J, Shin A, Oh JH, Kim J (2018) The relationship between nut intake and risk of colorectal cancer: a case control study. Nutr J 17(1):1–10
49. de Camargo AC, Regitano-d'Arce MAB, Rasera GB, Canniatti-Brazaca SG, do Prado-Silva L, Alvarenga VO et al (2017) Phenolic acids and flavonoids of peanut by-products: antioxidant capacity and antimicrobial effects. Food Chem 237:538–544
50. Juliano FF, Alvarenga JFR, Lamuela-Raventos RM, Massarioli AP, Lima LM, Santos RC et al (2020) Polyphenol analysis using high-resolution mass spectrometry allows differentiation of drought tolerant peanut genotypes. J Sci Food Agric 100(2):721–731
51. Akhihiero ET, Ebhodaghe SO (2020) Effect of blending ratio on the fuel properties of almond biodiesel. Eur J Sustain Dev Res 4(3):em0119
52. Prgomet I, Gonçalves B, Domínguez-Perles R, Pascual-Seva N, Barros AI (2017) Valorization challenges to almond residues: phytochemical composition and functional application. Molecules 22(10):1774
53. Moradi B, Heidari-Soureshjani S, Asadi-Samani M, Yang Q (2017) A systematic review of phytochemical and phytotherapeutic characteristics of bitter almond. Int J Pharm Phytopharm Res 7:1–9
54. Roncero JM, Álvarez-Ortí M, Pardo-Giménez A, Rabadán A, Pardo JE (2020) Review about non-lipid components and minor fat-soluble bioactive compounds of almond kernel. Foods 9(11):1646
55. Bottone A, Cerulli A, D'Urso G, Masullo M, Montoro P, Napolitano A et al (2019) Plant specialized metabolites in hazelnut (Corylus avellana) kernel and byproducts: an update on chemistry, biological activity, and analytical aspects. Planta Med 85(11/12):840–855
56. Susilastuti D, Meri A, Marhaeni LS (2021) Functional food support government policies to prevent stunting. In: Proceeding International Conference of Innovation Science, Technology, Education, Children and Health
57. Gyurova D (2021) Dietary fiber in nuts and seeds–analytical data and comparisons. J Biosci Biotechnol 10(1):31–34
58. Otunola GA, Martirosyan D (2021) Choosing suitable food vehicles for functional food products. Funct Foods Heal Dis 11(2):44–55
59. Xu D, Wei M, Peng S, Mo H, Huang L, Yao L et al (2021) Cuminaldehyde in cumin essential oils prevents the growth and aflatoxin B1 biosynthesis of aspergillus flavus in peanuts. Food Control 125:107985
60. Hajikazemi H, Mirzaei M, Mirdamadi S (2020) An investigation on the effects of hydrolysis conditions on the extraction of antioxidant peptides from peanuts. J Food Technol Nutr 17:5–12
61. Li W, Huang L, Liu N, Pandey MK, Chen Y, Cheng L et al (2021) Key regulators of sucrose metabolism identified through comprehensive comparative transcriptome analysis in peanuts. Int J Mol Sci 22(14):7266
62. Salve AR, LeBlanc JG, Arya SS (2020) Effect of processing on polyphenol profile, aflatoxin concentration and allergenicity of peanuts. J Food Sci Technol 1–11
63. Syed F, Arif S, Ahmed I, Khalid N (2021) Groundnut (Peanut) (Arachis hypogaea). In: Oilseeds: health attributes and food applications. Springer, pp 93–122
64. Katzman EW, Nielsen SJ (2021) The association between peanut and peanut butter consumption and cognitive function among community-dwelling older adults. J Prev Alzheimers Dis:1–6

65. Liu Z, Li C (2021) A predictive model for the risk of cognitive impairment in patients with gallstones. Biomed Res Int 2021
66. Salve AR, Arya SS (2020) Bioactive constituents, microstructural and nutritional quality characterisation of peanut flat bread. J Food Meas Charact 14(3):1582–1594
67. Shibli S, Siddique F, Raza S, Ahsan Z, Raza I (2019) Chemical composition and sensory analysis of peanut butter from indigenous peanut cultivars of Pakistan. Pak J Agric Res 32(1):159
68. Rohimah A, Setiawan B, Palupi E, Sulaeman A, Handharyani E (2021) Comparison of peanut and black oncom biscuit: nutritional characteristics and aflatoxin evaluation with the potential health benefits. Ann Agric Sci
69. Balasubramanian P, Mariappan VE, Lourdusamy DK, Chinnamuthu CR, Swetha S (2020) Peanut as a smart food and their nutrients aspects in planet: a review. Agric Rev 41(4)
70. Oduma OV, Inyang UE, Okongoh ON (2020) Impact of partial replacement of Peanut paste with sesame seed paste on the nutritional and anti-nutritional components of butter made from the blends. Eur J Nutr Food Saf:53–66
71. Eshun AAA (2019) Development and evaluation of the functional, nutritional quality and consumer perception and acceptability of a sesame seed and Peanut based complementary food for infants 6–24 months. University of Ghana
72. Tutunchi H, Ostadrahimi A, Saghafi-Asl M (2020) The effects of diets enriched in monounsaturated oleic acid on the management and prevention of obesity: a systematic review of human intervention studies. Adv Nutr 11(4):864–877
73. Wang J, Wang S, Henning SM, Qin T, Pan Y, Yang J et al (2021) Mixed tree nut snacks compared to refined carbohydrate snacks resulted in weight loss and increased satiety during both weight loss and weight maintenance: a 24-week randomized controlled trial. Nutrients 13(5):1512
74. Kamal S, Akhter N, Yaqoob N, Qaisar U, Bibi I, Shahzadi S et al (2020) Quantification and characterization of Myo-inositol in locally available peanuts and pine nuts in Pakistan. Pak J Pharm Sci 33(3):1289–1295
75. Barnett D, Baldo BA, Howden ME (1983) Multiplicity of allergens in peanuts. J Aller Clin Immunol 72(1):61–68
76. Nowak-Węgrzyn A, Sampson HA (2011) Future therapies for food allergies. J Aller Clin Immunol. 127(3):558–573
77. Abbas M, Asi M, Anwar F, Mahmood T, Khan A, Yaqub T (2019) Retracted article: assessment of aflatoxins in peanuts grown in the pothohar area of Pakistan. Food Additiv Contam: Part B 12(3):1–6

Chapter 17
Banana Peel as a Source of Nutraceuticals

Babatunde Oluwafemi Adetuyi, Adebanke E. Ogundipe, Olubanke Olujoke Ogunlana, Chukwuebuka Egbuna, Odoh Uchenna Estella, Abhay Prakash Mishra, Muhammad Akram, and Raghu Ram Achar

17.1 Introduction

Banana serve as a source of food, but people utilize it more than just for food purposes. Banana by-products have been used for wrappings foods, clothes and in various ceremonial occasions with its usage expanding through cultural diversification [1]. Banana (*Musa* spp) which belongs to the Musaceae family is one of the most important subsistence crops [2, 3]. The production of banana all over the world yield 30 million per year and banana peels make up to 30% of the ripe fruit. The

B. O. Adetuyi (✉)
Department of Natural Sciences, Biochemistry Unit, Precious Cornerstone University, Ibadan, Oyo State, Nigeria

A. E. Ogundipe · O. O. Ogunlana
Department of Biochemistry, Covenant University, Ota, Ogun State, Nigeria

C. Egbuna
Chukwuemeka Odumegwu Ojukwu University, Awka, Nigeria

O. U. Estella
Department of Pharmacognosy and Environmental Medicines, Faculty of Pharmaceutical Sciences, University of Nigeria, Nsukka, Nigeria

A. P. Mishra
Adarsh Vijendra Institute of Pharmaceutical Sciences, Shobhit University, Saharanpur, UP, India

Department of Pharmacology, Faculty of Health Science, University of Free State, Bloemfontein, South Africa

M. Akram
Department of Eastern Medicines, Government College University, Faisalabad, Pakistan

R. R. Achar
Division of Biochemistry, School of Life Sciences, JSS Academy of Higher Education and Research, Mysuru, Karnataka, India

© The Author(s), under exclusive license to Springer Nature Switzerland AG 2022

C. Egbuna et al. (eds.), *Food and Agricultural Byproducts as Important Source of Valuable Nutraceuticals*, https://doi.org/10.1007/978-3-030-98760-2_17

uses of banana waste include protein production by biotechnological process, ethanol, alpha-amylase and cellulose [4]. There are lots of phytochemicals and phytonutrient compounds which are mainly antioxidant in banana peel. Various types of compounds such as anthocyanins, delphinidin, cyaniding, and catecholamines are found in ripen banana peel. Carotenoids, beta- carotene, alpha- carotene have also been identified by different methods [5–7]. Various parts of the banana plants such as roots, leaves, flowers, peels have been used for medicinal purposes [4].

It has been reported that dessert banana pulp and peel contains some secondary metabolites e.g. catecholamines [8], phenolics [9], and carotenoid compounds [10], as well as pyridoxine (vitamin B6). Many of banana's volatile compounds such as esters [11] and alcohols [12] play important role in the aromatic properties of dessert bananas. According to Kumar et al. [13], banana fruits from India are traditionally used to help overcome or prevent a substantial number of illnesses and health conditions such as depression (related to the banana tryptophan content), anemia (high iron content which stimulates the production of hemoglobin), and blood pressure control (high potassium content and low in salt). Bananas can be a source of vitamin B6. The vitamins B6 and B12, as well as the potassium and magnesium content help the body recover from the effects of nicotine withdrawal. Bananas can help people trying to give up smoking. The fruits are also used against constipation (high in fiber, including bananas in the diet can help restore normal bowel action), hangover (to build up depleted blood sugar levels), heartburn and ulcers (bananas have a natural antacid effect in the body), and against stress condition, because potassium content is a vital mineral [13, 14].

The sap of banana plant is used to treat diarrhea, epilepsy, dysentery and hysteria. The root of banana plant is used in the treatment of general diseases and anemia. Also, the leaves acts as an abortifacient with the leaf juice used to treat fresh wound cuts and insect bites. Research has also shown that it can be used as antiscorbutic, aphrodisiac, and diuretic [15]. Banana stem extracts has been suggested to be beneficial for the treatment of hypeoxaluricurolithiasis, high blood pressure and kidney diseases.

17.2 Nutraceuticals

In 1989, Dr. Stephen DeFelice coined the term "nutraceutical" from the words nutrition and pharmaceutical and defined it as "a food or part of a food that provides medical or health benefits, including the prevention and/or treatment of a disease." [16]. Nutraceuticals help enhance overall health and wellness, stimulate immune health and boost energy, bone and heart health [16]. The Foundation for Innovation in Medicine (New York, US) coined the word "Nutraceutical" (defined as any substance that may be considered a food or part of a food and provides medical or health benefits including the prevention and treatment of disease), to describe the growing area of biomedical research centered around the isolation of nutrients/compounds, natural products (from fruits and plants, dietary supplements and diets to genetically engineered "designer" foods), herbal products and processed products (such as cereals, soups and beverages) [17].

Nutraceuticals, by having antioxidant, anti-inflammatory, immunomodulatory, sedative, adaptogenic, anticancer, and other such properties, are used against many ailments, including arthritis, diabetes, dermatitis, gastrointestinal, cardiovascular, respiratory, neurodegenerative, reproductive, and metabolic diseases [18]. Nutraceuticals have many beneficial effects; hence, they have been used for treating and prevent many health problems, such as cancer, inflammation, hypertension, cardiovascular diseases, atherosclerosis, obesity, diabetes and others [19]. Some nutraceuticals, such as silymarin, curcumin, vitamin E, docosahexaenoic acid, choline and phosphatidylcholine are used in treating and preventing steatosis [20]. Additionally, many nutraceutical products, such as gallic acid, caffeine, curcumin and others act as anti-aging and antioxidant agents [21, 22]. PUFA-rich fish oils reduce the risk of coronary cardiovascular diseases and enhance the brain functions [23]. Nutraceuticals are famous for their anticancer efficacy; hence, many nutraceutical ingredients, such as epigallocatechin gallate, curcumin, pomegranate and others, treat different types of cancer, such as breast cancer, prostate cancer and other types of cancer [24–26].

The global spread of nutraceutical products has dramatically increased recently. The main factor that lead to inflating the market share of these products, is that nutraceuticals have no strict regulations to control them [27]. On the other hand, pharmaceutical products are controlled by strict regulations and are closely monitored. Pharmaceutical products are also strictly regulated and have a governmental sanction [19]. Moreover, nutraceuticals have been advertised under the claim of being safe, effective and being a drug substitute. Additionally, it has been claimed that these products can be used in preventing and treating many health problems without any side effects [28]. The patients also have been concerned about the use of pharmaceutical products because of their high price and several side effects [27]. Therefore, the market share of nutraceuticals has been tremendously expanded [30]. Approximately 80% of global population preferred using dietary supplements and nutraceuticals [28]. Nutraceutical products now are the fastest growing market with an estimated worth of USD 117 billion globally in 2017 [29]. DSHEA (Dietary Supplement Health and Education Act of 1994) and Drug Administration Modernization Act are responsible for confirming the safety of nutraceuticals before commercializing them [31, 32].

17.3 Banana Peels as Source of Nutraceuticals

Research in ancient literatures have shown that banana possess many medicinal properties and have also been found effective in cure of many diseases [33]. Banana is a staple fruit because it is available all through the year and it also serves as source of income and source of food to a great number of people [34]. Banana peel is the major by-product of banana processing industry, which account for 30% of the banana fruit and also constitute to environmental hazard. Banana peels are promising by-product for different applications in nutraceuticals and medicinal due to the high dietary fiber and phenolic content present in them. According to the criteria established by the National Cancer Standard Institute, banana peel extract is non-toxic to normal human cells; thus, it can be safely used as a natural source of antioxidants for value addition. Banana peels are also used for the production of biogas [35], production of ethanol by hydrolysis and fermentation [36], antibacterial and antioxidant activities and biomass production. There are some specific enzymes extracted from banana peels for their nutraceuticals application. One of these enzymes include oxalate oxidase. The enzyme oxalate oxidase degrades stone from kidney and plasma present in banana peel (*Musa paradisica*) [37]. Banana peels are also rich sources of potassium and contain more soluble and insoluble fiber than their flesh [38]. Banana peels are also a good source of carbohydrates and fibers due to the high value of organic content such as lipids, carbohydrate and proteins present in them. High fiber content also indicates that the peels could help treat constipation and improve general health and wellbeing [37]. Taken a piece of banana peel and place it on the wart, with the yellow side out, can be a natural alternative to kill off a wart and to reduce swelling and irritation after a mosquito bite by rubbing the affected area with the inside of a banana skin [13].

Indeed, among the numerous sources of bioactive compounds, banana peel could be considered one of the complex plant matrix rich in high-value compounds [39]. Banana peel is rich in phytochemical compounds, mainly antioxidants. Ripe banana peel contains anthocyanins delphinidin and cyaniding [40], and catecholamines [8].

Carotenoids Carotenoids are also natural antioxidants that contribute to the stability of foods. Such pigments are not evenly distributed in the food itself as various investigators have found that carotenoids are usually more concentrated in the peel than in the pulp of fruits and vegetables [41]. Banana peel has substantially higher carotenoid values in some varieties than the underlying fruit pulp with pro-vitamin A presence as trans-betacarotene.

Phenolic compounds Babbar et al. [42] investigated the antioxidant potential in terms of ABTS and DPPH scavenging abilities, ferric reducing antioxidant power, and phenolic contents of residue extracts of four important fruits grown in Ludhiana, India, between them the banana peel. The antioxidant activities in terms of ABTS determined for banana peel were found to be 5.67 ± 0.32 mg trolox equivalent/g dry weight, respectively. The methanolic extracts considerably differed in the antioxidant activity as measured by the DPPH method. Such findings reveal an important trait of banana peel as source of compounds able in scavenging free radicals.

Besides, banana peels might be an important source of antioxidant compounds considering a growing interest of the food and pharmaceutical industries on medicinal plant biomasses for the development of new therapeutic and prophylactic products. In order to pursue a further technological usage of that residual biomass, in vivo studies in both pre-clinical and clinical levels on the toxicology, bioavailability, distribution, metabolization, and excretion of phenolic compounds from banana peels extracts are needed to encourage an eventual industrial application.

The antioxidant compounds from commercial banana peel *M. cavendishii* Lamb were studied by Someya et al. [43] and the antioxidant gallocatechin was identified. In fact, gallocatechin was more abundant in peel (158 mg/100 g dry weight.) than in pulp (29.6 mg/100 g dry weight.) in *M. cavendishii* Lamb genotypes as the antioxidant activity of the banana peel extract against lipid auto-oxidation was stronger than that of the banana pulp extract. This result was consistent with the gallocatechin analysis and its higher content may account for the better antioxidant effects. Thus, banana peels might be considered as a good source of natural antioxidants for foods, as well as among others possible applications. The antioxidant activity of banana peel extracts (Musa x paradisiaca L.) was studied using an experimental model of rats subjected to a normal diet compared to rats with a diet rich in fatty acids. Animals treated orally with banana peel extract showed significantly decreased concentrations of peroxidation products (MDA), hydroperoxides, and conjugated dienes. At the same time, the enzymatic activities of catalase and superoxide dismutase increased significantly in treated animals, as well as the concentration of reduced glutathione [44]. Unripe banana peel contains leucocyanidin, a flavonoid that induces cell proliferation by increased incorporation of thymidine into cellular DNA [45], accelerating the healing of skin wounds [46].

Antioxidant activity Research has shown that the total amount of phenolic compound in banana (*Musa accuminata*, colla AAA) peel ranges from 0.90 to 3 g/100 g of DW [47]. Gallocatechin was also identified in banana peel at a concentration of 160 mg/100 g DW. The antioxidant activity in banana peel was evaluated with the measure as the effect of lipid auto-oxidation, in relation to its gallocatechin content. Various studies on the antioxidant activity of banana peels are further described. Unripe banana peels exhibit a high antioxidant activity as increasing the polarity, the extracts displayed a stronger antioxidant activity. Polyphenols or flavanones and flavonoids are partly responsible for these activities with glycoside and monosaccharide components in water soluble extracts also displaying significant antioxidant and antimicrobial activities. Another report showed that antioxidant activity and bioactive compounds in banana peels vary [48]. Bioactive compounds, such as carotenoids, antioxidative enzymes and carbohydrate contents are found richly in red banana and Karpooravalli. All these results established the fact that banana may protect itself from the oxidative stress caused by strong sunshine and high temperature by producing large amounts of antioxidants [37].

17.4 Conclusion and New Insight into the World of Nutraceuticals

Banana peel is an underutilized source of phenolic compounds. It can be safely utilized as a natural source of antioxidants and enzyme to cure illness. It is rich in pectin 99–22% which makes it useful in the production of non-conventional product without the addition of any gel additive. Standardized and developed jellies from banana peel are deemed to be nutritive, health-beneficial, and more favorable than tablets or pills. This implies that banana peels can be a source of essential nutrients, which include minerals and amino acids.

However, nutraceuticals formulations are hampered by many obstacles that negatively affect their efficacy [49]. Many natural active ingredients suffer from low solubility, poor permeability, fast metabolism, short half-life and others [49]. For instance, curcumin has low solubility, poor oral bioavailability, limited tissue distribution, short half-life and rapid metabolism.

References

1. Kennedy J (2009) Bananas and people in the homeland of genus Musa: not just pretty fruit. Ethnobot Res Appl 7:179–197
2. Novak FJ (1992) *Musa* (bananas and plantains). In: Hammerschlag FA, Litz RE (eds) Biotechnology of perennial fruit crops. CAB International, University Press, Cambridge, pp 449–488
3. Husain MD, William R (2010) Status of banana cultivation and disease incidences in Malaysia. Crop Protection and Plant Quarantine Division, Department of Agriculture, Malaysia, p 60
4. Kibria AA, Kamrunnessa M, Rahman M, Kar A (2019) Extraction and evaluation of phytochemicals from banana peels (Musa sapientum) and banana plants (*Musa paradisiaca*). Malaysian J Halal Res 2(1)
5. Krishna C (1999) Production of bacterial cellulose by solid state bioprocessing of banana waste. Bio Resour Technol 69:231–239
6. Wolfe K, Xianzhong WU, Liu RH (2003) Antioxidant activity of apple peels. J Agric Food Chem 51:609–614
7. Randy CP, Angela KK, Jeff D, Scot CN (2007) Banana and plantain—an overview with emphasis on Pacific island cultivars. Species Profiles for Pacific Island Agroforestry, 1, pp 2–27
8. Kanazawa K, Sakakibara H (2000) High content of dopamine, a strong antioxidant, in Cavendish banana. J Agric Food Chem 48:844–848
9. Verde-Mendez CDM, Forster MP, Rodríguez-Delgado MA, Rodríguez-Rodríguez EM, Romer CD (2003) Contento f free phenolic compounds in bananas from Tenerife (Canery Island) and Ecuador. Eur Food Res Technol 217:287–290
10. Van den Berg H, Faulks R, Granado HF, Hirschberg J, Olmedilla B, Sandmann G, Southon S, Stahl W (2000) The potential for the improvement of carotenoid levels in foods and the likely systemic effects. J Sci Food Agric 80:880–912
11. Pérez AG, Cert A, Rios JJ, Olias JM (1997) Free and glycosidically bound volatile compounds from two banana cultivars: Valery and Pequena Enana. J Agric Food Chem 45:4393–4397
12. Nogueira JM, Fernandes PJ, Nascimento AM (2003) Composition of volatiles of banana cultivars from Madeira Island. Phytochem Anal 14:87–90

13. Kumar KPS, Bhowmik D, Duraivel S, Umadevi M (2012) Traditional and medicinal uses of banana. J Pharmacognosy Phytochem 1:51–63
14. Bhutani MKK, Atal CK (1984) Chemical constituents of banana. J Sci Ind Res 43:316–323
15. Salawu L, Bolarinwa RA, Adegunloye AB, Moraine HA (2010) HBs Ag, anti-HCV, anti-HIV and VDRL in blood donors: prevalence and trends in the last three and a half years in a tertiary health care facility in Ile-Ife. Nigeria Int J Med Sci 2(11):335–341
16. Egbuna C, Tupas GD (eds) (2020) Functional foods and nutraceuticals - bioactive components, formulations and innovations, 1st edn. Springer Nature, Cham, p 646
17. Egbuna C, Mishra AP, Goyal MR (eds) (2021) Preparation of phytopharmaceuticals for the management of disorders. Academic. https://doi.org/10.1016/C2019-0-01420-3
18. Pandey M, Verma RK, Saraf SA (2010) Nutraceuticals: new era of medicine and health. Asian J Pharmaceut Clin Res 3:11–15
19. Nasri H, Baradaran A, Shirzad H, Rafieian-Kopaei M (2014) New concepts in nutraceuticals as alternative for pharmaceuticals. Int J Prev Med 5(12):1487–1499
20. Stellavato A, Pirozzi AVA, de Novellis F et al (2018) In vitro assessment of nutraceutical compounds and novel nutraceutical formulations in a liver-steatosis-based model. Lipids Health Dis 17:24
21. Bourbon AI, Pinheiro AC, Cerqueira MA, Vicente AA (2018) In vitro digestion of lactoferrin-glycomacropeptide nanohydrogels incorporating bioactive compounds: effect of a chitosan coating. Food Hydrocoll 84:267–275
22. Zhang X, Liu J, Qian C, Kan J, Jin CH (2019) Effect of grafting method on the physical property and antioxidant potential of chitosan film functionalized with gallic acid. Food Hydrocoll 89:1–10
23. Liu F, Zhu Z, Ma C et al (2016) Fabrication of concentrated fish oil emulsions using dual-channel microfluidization: impact of droplet concentration on physical properties and lipid oxidation. J Agric Food Chem 64(50):9532–9541
24. Egbuna C, Patrick-Iwuanyanwu KC, Onyeike EN, Khan J, Alshehri B (2021) FMS-like tyrosine kinase-3 (FLT3) inhibitors with better binding affinity and ADMET properties than sorafenib and gilteritinib against acute myeloid leukemia: in silico studies. J Biomol Struct Dyn 6:1–12. https://doi.org/10.1080/07391102.2021.1969286
25. Khan N, Bharali DJ, Adhami VM et al (2014) Oral administration of naturally occurring chitosan-based nanoformulated green tea polyphenol EGCG effectively inhibits prostate cancer cell growth in a xenograft model. Carcinogenesis 35(2):415–423
26. Jia Q, Yimam M, Ping J, Hong M, Moore B (2017) Compositions, methods, and medical compositions for treatment of and maintaining the health of the liver. US20170035829A1
27. Nounou MI, Ko Y, Helal NA, Boltz JF (2018) Adulteration and counterfeiting of online nutraceutical formulations in the United States: time for intervention? J Diet Suppl 15(5):789–804
28. Ekor M (2014) The growing use of herbal medicines: issues relating to adverse reactions and challenges in monitoring safety. Front Pharmacol 4:177
29. Swaroopa G, Srinath D (2017) Nutraceuticals and their health benefits. Int J Pure App Biosci 5(4):1151–1155
30. Santini A, Cammarata SM, Capone G et al (2018) Nutraceuticals: opening the debate for a regulatory framework. Br J Clin Pharmacol 84(4):659–672
31. Santini A, Novellino E (2018) Nutraceuticals - shedding light on the grey area between pharmaceuticals and food. Expert Rev Clin Pharmacol 11(6):545–547
32. U.S. National Institutes of Health (NIH) (2007) Dietary supplement health and education act of 1994. In: Department of Health and Human Services (ed) United States: U.S. National Institutes of Health (NIH), p. Public law 103–417
33. Loizzo MR, Di LG, Boselli E, Menichini F, Frega NG (2011) Inhibitory activity of phenolic compounds from extract virgin olive oils on the enzyme involved in diabetes, obesity and hypertension. J Food Biochem 35:381–399
34. Chadha KL (2002) Directorate of Information & Publications of Agriculture, ICAR. Banana Handbook of Horticulture, New Delhi, pp 143–153

35. Ilori MO, Adebusoye SA, Lawal AK, Awotiwon OA (2007) Production of biogas from banana and plantain peels. Adv Environ Biol 1:33–38
36. Sirkar A, Das R, Chowdhury S, Sahu SJ (2008) An experimental study and mathematical modeling of ethanol production from banana peels by hydrolysis and fermentation. IE J 88:4–10
37. Anjum S, Sundaram S, Rai GK (2014) Nutraceuticals application and value addition of banana (*Mus paradica L. Variety Bhusawal keli)* peel: a review. Int J Pharm Pharm Sci 6(10)
38. Anhwange BA, Ugye JI, Nyiaatagher TD (2009) Chemical composition of Musa sapientum (banana) peels. Electr J Environ Agri Food Chem 8(6):437–442
39. Laufenberg G, Kunz B, Nystroem M (2003) Transformation of vegetable waste into value added products: (A) the upgrading concept; (B) practical implementations. Bioresour Technol 87:167–198
40. Seymour GB (1993) Banana. In: Seymour JE, Tucker GA (eds) Biochemistry of fruit ripening. Chapman and Hall, New York, pp 83–106
41. Rodriguez-Amaya DB 2001 A guide to carotenoid analysis in foods. OMNI Research, 71 p
42. Babbar N, Oberoi HS, Uppal DS, Patil RT (2011) Total phenolic content and antioxidant capacity of extracts obtained from six important fruit residues. Food Res Int 44:391–396
43. Someya S, Yoshiki Y, Okubo K (2002) Antioxidant compounds from bananas (Musa Cavendish). Food Chem 79(3):351–354. https://doi.org/10.1016/s0308-8146(02)00186-3
44. Vijayakumar S, Presannakumar G, Vijayalakshmi NR (2008) Antioxidant activity of banana flavonoids. Fitoterapia 79:279–282
45. Novak FR, Almeida JAG, Silva RS (2003) Casca da banana: uma possível fonte de infecção no tratamento de fissuras mamilares. J Pediatr 79:221–226
46. Lewis DA, Fields WN, Shaw GP (1999) A natural flavonoid present in unripe plantain banana pulp (Musa sapientum L. var. paradisiaca) protects the gastric mucosa from aspirin-induced erosions. J Ethnopharmacol 65:283–288
47. Nguyen TBT, Saichol K, Doorn WGV (2003) Relationship between browning and the activities of polyphenoloxidase and phenylalanine ammonia lyase in banana peel during low temperature storage. Postharvest Biol Technol 30(2):187–193. https://doi.org/10.1016/s0925-5214(03)00103-0
48. Arora A, Choudhary D, Agarwal R, Singh VP (2008) Compositional variation in β-carotene content, carbohydrate and antioxidant enzymes in selected banana cultivars. J Food Sci Technol 43:1913–1912
49. Flora SJS, Agrawal S (2017) Arsenic, cadmium, and lead. In: Gupta RC (ed) Reproductive and developmental toxicology. Academic Press/Elsevier, Amsterdam, pp 537–566

Index

© The Editor(s) (if applicable) and The Author(s), under exclusive license to Springer Nature Switzerland AG 2022

C. Egbuna et al. (eds.), *Food and Agricultural Byproducts as Important Source of Valuable Nutraceuticals*, https://doi.org/10.1007/978-3-030-98760-2